MOLECULAR ORGANIC MATERIALS
From Molecules to Crystalline Solids

The interest in molecular organic materials is tremendous, driven by the need to find new materials with desirable properties. In recent years, the surface science community has brought fresh insights into the field. This book aims to bring the materials and surface science communities together, integrating physics and chemistry in a non-technical manner, ensuring this fascinating field can be understood by a multidisciplinary audience.

Starting with basic physical concepts and synthetic techniques, the book describes how molecules assemble into highly ordered structures as single crystals and thin films, with examples of characterization, morphology and properties. Special emphasis is placed on the importance of surfaces and interfaces. The final chapter gives a personal view on future possibilities in the field.

Written for beginners and experienced chemists, physicists and material scientists, this will be a useful introduction to the field of molecular organic materials.

JORDI FRAXEDAS is a tenured scientist at the Institut de Ciència de Materials de Barcelona (ICMAB) of the Consejo Superior de Investigaciones Científicas (CSIC).

T0215432

MOLECULAR ORGANIC MATERIALS

From Molecules to Crystalline Solids

JORDI FRAXEDAS

CAMBRIDGE UNIVERSITY PRESS
Cambridge, New York, Melbourne, Madrid, Cape Town, Singapore, São Paulo

Cambridge University Press
The Edinburgh Building, Cambridge CB2 8RU, UK

Published in the United States of America by Cambridge University Press, New York

www.cambridge.org
Information on this title: www.cambridge.org/9780521834469

© Jordi Fraxedas 2006

This publication is in copyright. Subject to statutory exception
and to the provisions of relevant collective licensing agreements,
no reproduction of any part may take place without the written
permission of Cambridge University Press.

First published 2006
This digitally printed version 2008

A catalogue record for this publication is available from the British Library

ISBN 978-0-521-83446-9 hardback
ISBN 978-0-521-06744-7 paperback

Cambridge University Press has no responsibility for the persistence or accuracy of URLs
for external or third-party Internet websites referred to in this publication, and does not
guarantee that any content on such websites is, or will remain, accurate or appropriate.

Dedicated to my wife Montse
and to my sons Roger and Marc

Contents

Preface

Válame Dios, y con cuánta gana debes de estar esperando ahora, lector
ilustre, o quier plebeyo, este prólogo.

Now God defend, reader, noble or plebeian, whate'er thou art! How
earnestly must thou needs by this time expect this prologue.

Miguel de Cervantes, Don Quijote de la Mancha
(Translation by Thomas Shelton)

Winston Smith, the main character in George Orwell's apocalyptic novel *1984*,
perceived while reading the secret Brotherhood's book that the best books are
those that tell you what you know already. If you are encountering the field of
molecular organic materials for the first time and do not want to contradict Orwell,
you are invited to read the book you are holding in your hands more than once. If
you are an expert in the field I hope you will discover new and perhaps unexpected
issues.

This book, my *opera prima*, was conceived for both beginner and experienced
chemists, physicists and material scientists interested in the amazing field of molecular organic materials. Some basic notions of solid-state physics and chemistry and
of quantum mechanics are required, but the book is written trying to reach a broad
multidisciplinary audience.

When entering this field for the first time one is faced with myriad names and
materials and it becomes an arduous task to find one's way. This volume is based
on my own experience, aware of the difficulty that non-specialists encounter.

The underlying philosophy is the following. Chapter 1 is devoted to a general
introduction intended for those wishing to discover what molecular organic materials are about or learn more about this subject. The most relevant molecules used
are shown and the materials they build are also discussed, classified according to
their physical properties. This chapter concludes with some fundamental quantum

mechanical concepts to settle, in a simplified way, the basic concepts related to electronic structure inspired by R. Hoffmann's ideas, bringing together molecular orbital theory with solid-state theory concepts in the spirit of reaching both chemists' and physicists' communities. This is one of my objectives, and I would be glad if my attempt succeeds. Chapter 2 describes the basic strategies used by chemists in order to synthesize the organic and organo-metallic molecules that will form solids and Chapter 3 reviews the preparation methods that lead to materials in practical forms, that is, high quality single crystals and thin films. We thus start with molecules and end up with solids.

As will be evident in Chapter 1, most of the important molecules exhibit quasi-planar geometries, self-assembling in anisotropic solids. This anisotropy translates into 1D and 2D physical properties, highlighting the relevance of interfaces. Interfaces play a key role in the majority of these materials and Chapter 4 is devoted to this subject. Chapter 5 deals with some trends in the characterization of the growth of thin films and Chapter 6 gives some selected examples of physical properties of well-known molecular organic materials, with the intention of discussing different aspects of electronic structure, magnetism, structure–property relationships, polymorphism, etc. Finally, a personal view on future perspectives in the field of molecular organic materials is given in the Afterword.

This field is extremely active and the principal driving force is the permanent quest for new materials with improved or even exotic properties, such as superconductivity with higher transition temperatures, ferromagnetism, multifunctionality, etc. In recent years the surface science community has become increasingly interested in molecular organic materials, bringing new insights to the field. The experimental determination of the band structure or the characterization of the *in situ* formation of sharp organic/inorganic interfaces are two examples of the immense potential of surface science techniques. Another objective is to bring the molecular organic materials and surface science communities together.

I would like to express my sincere gratitude to many collegues who have encouraged me throughout the process of writing, and among them I am specially indebted to Drs E. Canadell and P. Batail for having introduced and guided me in this marvellous field. Their experience and criticism have helped me a lot in the preparation of the text. I also thank Drs D. de Caro, R. Fasel, A. Figueras, M. Fourmigué and E. Hernández for their comments on the manuscript. I would also like to express my gratitude to my mentor, Professor M. Cardona, for having initiated me in solid-state physics and in scientific research. A very special reference goes to Dr F. Sanz and his atomic force microscopy team in Barcelona for some of the images shown here and for our long scientific and non-scientific discussions. Thanks are also due to Drs M. Grioni and J. M. Fabre for the XPS data on $(TMTTF)_2PF_6$ single crystals shown in Fig. 1.31 and to Dr. W. Kang for recuperating and digitizing

Fig. 1.36. I also acknowledge the Cambridge Crystallographic Data Centre for kindly supplying many crystallographic data used in this book.

My wife Montse has also helped me very much with her (almost infinite) patience and permanent support. Hence, I wish to dedicate this book to her and to my beloved sons Roger and Marc. I would like to extend my dedication to my mother and to my father, who passed away nearly 30 years ago, and to the Fraxedas colony spread in the New World, in particular to my cousin Joaquín.

Let me conclude this preface with a very special dedication to my friend and colleague André Deluzet. During the process of proofreading I received the sad news of his death after a long illness. Chapter 3 describes the work on confined electrocrystallization that we performed together. André, all those who have had the privilege of knowing you will miss you.

Abbreviations and symbols

E_{corr}	Correlation gap energy
E_{ex}	Exciton binding energy
E_F	Fermi energy
E_g	Fundamental band gap
E_{gsp}	Single particle energy gap
E_K	Kinetic energy
E_{opt}	Optical band gap
E_{P+}	Hole polaron binding energy
E_{P-}	Electron polaron binding energy
E_{trans}	Transport band gap
E_{vac}	Vacuum level
E_{zb}	Bang gap at zone boundaries
E_α	Atomic/molecular orbital energy
E_β	Nearest-neighbour transfer integrals
EC	Electrocrystallization
EELS	Electron Energy Loss Spectroscopy
EPR	Electron Paramagnetic Resonance
ESA	European Space Agency
ESQC	Electron Scattering Quantum Chemistry
ETL	Electron Transporting Layer
EXAFS	Extended X-ray Absorption Fine Structure
FET	Field-Effect Transistor
FS	Fermi Surface
FTIR	Fourier Transform Infrared
FWHM	Full Width at Half Maximum
g	g- or Landé factor
G	Gibbs free energy
$\mathbf{G}$	Reciprocal lattice vector
G_c	Energy barrier of nucleation
GGA	Generalized Gradient Approximation
GIC	Graphite Intercalation Compound
GIXRD	Grazing Incidence X-Ray Diffraction
$\hbar$	Planck constant
$\hbar\omega$	Photon energy
h_c	Critical thickness
H, H_0	Hamilton operator
HFW	Hooks–Fritz–Ward
HMBD	Hyperthermal Molecular Beam Deposition
HOMO	Highest Occupied Molecular Orbital
HOPG	Highly Oriented Pyrolytic Graphite

HPLC	High-Performance Liquid Chromatography
HREELS	High-Resolution Electron Energy Loss Spectroscopy
HTL	Hole Transporting Layer
HV	High Vacuum
I_D	Drain current
I_E	Ionization energy
I_t	Tunnelling current
IDIS	Induced Density of Interface States
ISO	Infrared Space Observatory
ISTS	Inelastic Scanning Tunnelling Spectroscopy
IUPAC	International Union of Pure and Applied Chemistry
J	Magnetic exchange coupling
k	Wave vector
k_B	Boltzmann factor
k_F	Fermi wave vector
K_ρ	Luttinger-liquid exponent
L_c	Critical length of nucleation
L_S	Observation size
LAPW	Linearized Augmented Plane Wave
LB	Langmuir–Blodgett
LCAO	Linear Combination of Atomic Orbitals
LCW	Lithographically Controlled Wetting
LDA	Local Density Approximation
LDOS	Local Density Of States
LEAD	Low-Energy Atom Diffraction
LEED	Low-Energy Electron Diffraction
LEEM	Low-Energy Electron Microscopy
LMTO	Linear Muffin-Tin Orbital
LUMO	Lowest Unoccupied Molecular Orbital
m_e	Electron mass
m_e^*	Effective electron mass
M	Ion mass
$\mathcal{M}$	Total magnetization
MALDI	Matrix-Assisted Laser Desorption/Ionization
MAPLE	Matrix-Assisted Pulsed Laser Evaporation
MBE	Molecular Beam Epitaxy
MH	Mott–Hubbard
ML	Monolayer
MO	Molecular Orbital
MOM	Molecular Organic Material
MRDT	Modified Reticulate Doping Technique

MWNT	Multiwalled Nanotube
N_{at}	Band filling
$N(E_F)$	DOS at the Fermi level
NASA	National Aeronautics and Space Administration
NEXAFS	Near-Edge X-ray Absorption Fine Structure
N–I	Neutral-to-Ionic
NIL	Nanoimprint Lithography
NMR	Nuclear Magnetic Resonance
OFET	Organic Field-Effect Transistor
OLED	Organic Light Emitting Diode
OMBD	Organic Molecular Beam Deposition
P	Applied hydrostatic pressure
PAH	Polycyclic Aromatic Hydrocarbon
PDOS	Partial Density Of States
PEEM	Photoelectron Emission Microscopy
PLD	Pulsed Laser Deposition
PSD	Power Spectral Density
q	Nesting vector
q_r	Reduced wave vector
R	Lattice vector
R_c	Critical radius of nucleation
R_F	Frank radius
R_H	Hooke radius
R_{hc}	Radius of a hollow core
RCS	Radical Cation Salt
RGA	Residual Gas Analyzer
RHEED	Reflection High-Energy Electron Diffraction
RIS	Radical Ion Salt
RT	Room Temperature
S	Spin
SAED	Selected Area Electron Diffraction
SAM	Self-Assembled Monolayer
SEM	Scanning Electron Microscopy
SDW	Spin-Density Wave
SP	Spin-Peierls
SPM	Scanning Probe Microscope
SQUID	Superconducting Quantum Interference Device
STM	Scanning Tunnelling Microscope
SWNT	Single-Walled Nanotube
T_{ann}	Annealing temperature
T_{ao}	Anion-ordering transition temperature

T_c	Superconductivity transition temperature
T_C	Curie transition temperature
T_N	Néel transition temperature
T_{N-I}	Neutral-to-ionic transition temperature
T_{R_n}	Translation operator
T_{sub}	Substrate temperature
TB	Tight Binding
TDS	Thermal Desorption Spectroscopy
TMAFM	Tapping Mode Atomic Force Microscope
TOF	Time Of Flight
$\tilde{u}$	Strain energy function
U	Coulombic repulsion energy
U_{e-ph}	Electron–phonon interaction energy
UHV	Ultra High Vacuum
UPS	Ultraviolet Photoemission Spectroscopy
UV	Ultraviolet
v	Group velocity
v_{ip}	In-plane velocity
v_R	Radial growth velocity
V_D	Source–drain voltage
V_G	Source–gate voltage
V_t	Tunnelling bias voltage
VBM	Valence Band Maximum
VUV	Vacuum Ultraviolet
W	Bandwidth
W_c	Work of cohesion
XANES	X-ray Absorption Near-Edge Spectroscopy
XPD	X-ray Photoelectron Diffraction
XPS	X-ray Photoemission Spectroscopy
XRD	X-Ray Diffraction
0D	Zero-dimensional
1D	One-dimensional
2D	Two-dimensional
3D	Three-dimensional
α, β, γ	Lattice parameters (angles)
γ_l	Line tension
γ_s	Surface or interface tension
γ_0	Water surface tension
$\gamma_\parallel$	Interface tension parallel to substrate's surface
$\gamma_{\parallel ad}$	Specific free energy of adhesion
$\gamma_{\parallel ns}$	Substrate–nucleus interface tension

$\gamma_{\parallel nv}$	Nucleus–vapour interface tension
$\gamma_{\parallel sv}$	Substrate–vapour interface tension
$\gamma_{\perp}$	Interface tension perpendicular to substrate's surface
$\gamma_{\perp nv}$	Nucleus edge–vapour interface tension
$\gamma_{\perp n\sigma}$	Nucleus edge–step interface tension
$\gamma_{\perp\sigma v}$	Step–vapour interface tension
δ_{nm}	Kronecker delta-function
δ_Y	Deformation at plastic yield
Δ	Dipole barrier
ϵ	Dielectric constant
θ_e	Emission angle
θ_E	Angle of incidence of synchrotron light
Θ_W	Weiss temperature
Θ_D	Debye temperature
ϑ	Coverage
Λ	Burger's vector
μ	Chemical potential
μ_B	Bohr magneton
μ_e	Electron mobility
μ_h	Hole mobility
μ_m	Magnetic moment
$\xi(L,t)$	Correlation length
Π	Surface pressure
ρ	Electrical resistivity
ρ_c	Charge density
ϱ	Degree of charge transfer
σ	Electrical conductivity
$\sigma(\omega)$	Optical conductivity
ς	Surface roughness
$|\phi\rangle$	Bloch function
ϕ_M	Metal work function
$|\Phi\rangle$	Crystal orbital
Φ_m	Molecular flux
χ	Magnetic susceptibility
$|\psi\rangle$	Atomic orbital
$|\Psi^{1e}\rangle$	One-electron wave function
$|\Psi^{Ne}\rangle$	N_e-electron wave function
ω_s	Interface plasmon frequency
ω_D	Debye frequency
Ω	Molecular volume

Chemical abbreviations

BaAA	barium arachidate
BCP	dimethyl-diphenyl-phenanthroline (bathocuproine)
BCTBPP	bis(cyanophenyl)-bis(ditertiarybutylphenyl)porphyrin
bdc	benzenedicarboxylate
BETS	BEDT-TSF
Bu	butyl
CBP	dicarbazolylbiphenyl
$CHCl_3$	chloroform
CH_2Cl_2	dichloroethylene
CH_3CN	acetonitrile
CH_3NH_2	methylamine
CH_4	methane
C_2Cl_4	tetrachloroethylene
C_2H_2	acetylene (ethyne)
C_2H_4	ethylene (ethene)
$C_4H_8O_2$	dioxane
C_5H_5N	pyridine
$C_6H_3Cl_3$	trichlorobenzene
$C_6H_4Br_2$	dibromobenzene
C_6H_5CN	benzonitrile
C_6H_5Cl	chlorobenzene
C_6H_5I	iodobenzene
C_6H_6	benzene
CTBPP	cyanophenyl-tris-ditertiarybutylphenyl-porphyrin
DBTTF	dibenzotetrathiafulvalene
dmit	dithiole-thione-dithiolate
Et	ethyl
ET	BEDT-TTF

Et$_3$N	triethylamine
hfac	hexafluoroacetylacetonate
H$_2$-TBPP	tetrakis-ditertiarybutylphenyl-porphyrin
HtBDC	hexatertbutyldecacyclene (C$_{60}$H$_{66}$)
ITO	indium tin oxide
KAP	potassium acid phthalate
LDA	lithium diisopropylamide
Me	methyl
NPB	diphenyl-bisnaphthyl-biphenyl-diamine
OTS	octadecyltrichlorosilane
pao	pyridinealdoximate
Pc	phthalocyanine
PDMS	polydimethylsiloxane
PPh$_3$	triphenyl phosphine
PPV	polyphenylene-vinylene
PTFE	polytetrafluoroethylene
PVBA	pyridylvinyl benzoic acid
saltmen	tetramethylethylene-bis(salicylideneiminate)
TBA	tetrabutylammonium
TBE	tribromoethane
TCE	trichloroethane
THF	tetrahydrofuran
tmdt	trimethylene-tetrathiafulvalenedithiolate
TMSA	trimethylsilylacetylene
TTF-LB1	bis(phosphonic acid-butylthio)-bis(tetradecylthio)tetrathiafulvalene
TTTA	trithiatriazapentalenyl

1

An introduction to molecular organic materials

Welch Schauspiel! aber ach! ein Schauspiel nur!
  Wo fass ich dich, unendliche Natur?

What pageantry! Yet, ah, mere pageantry!
  Where shall I, endless Nature, seize on thee?

Johann Wolfgang von Goethe,
Faust (Translation by G. M. Priest)

Broadly speaking, crystalline molecular organic materials (MOMs) are soft solids with a 3D periodic distribution of organic molecules exhibiting weak intermolecular forces, their cohesion being essentially mediated by dipolar (permanent or fluctuating charges), hydrogen bonding and $\pi-\pi$ interactions. The molecules involved in the formation of such materials may be of a purely organic nature (e.g., metal-free) or based on hybrid organic–inorganic combinations (e.g., organo-metallic with transition metals). Solids can be built from molecules of a single species or binary or ternary combinations, and inorganic molecules can also be introduced forming hybrid organic–inorganic materials. Such regular solids are often seen by chemists as supramolecular entities, the solid as a macromolecule, in spite of their discrete character, while physicists tend to think in terms of a weakly interacting ordered gas, with cohesion energies larger than about 0.2 eV per molecule ($\simeq$20 kJ mol^{-1}),[1] the typical energies of noble gas crystals. Although describing the same objects, the terminology used by chemists and physicists is usually distinct and scientific discussions are not always fully synthesized, hindering the desired effective flow of ideas. One of the objectives of this book is to bring both scientific communities to a common neutral playground to better understand the inherently interdisciplinary, rich, complex and exciting field of crystalline MOMs, which involves both

---

[1] 1 eV $= 1.6 \times 10^{-19}$ J $= 8065$ cm$^{-1} = 1.16 \times 10^4$ K
  1 eV molec$^{-1} \simeq 100$ kJ mol^{-1}.

experimental and theoretical chemists and physicists and material scientists. When performing research in crystalline MOMs one should go beyond the conventional physics and chemistry frameworks and try to think as an interdisciplinary scientist. In this book we shall restrict our studies to solids made out of low molecular weight organic molecules with dimensions typically smaller than 2 nm, because they have the wonderful tendency to self-assemble in highly ordered structures, so that high-quality single crystals can be obtained. In this sense they are model systems. The structure–property relationships can thus be systematically explored in the ideal situation where disorder can be ignored, an essential issue for the crystal engineering quest to design materials with predefined properties. However, disorder might be an essential parameter when e.g., trying to stabilize metastable polymorphs, a point that will be discussed in Section 5.5.

The paramount advantage of molecular solids over their more classical inorganic counterparts is that their constituents, the building blocks, are molecules or clusters that can be designed and synthesized; in other words they can be intentionally modified. Therefore, we can talk about molecular and crystal engineering and the goal is to be able to produce materials with predetermined physical properties. We are not yet at this desired level but the scientific and technical bases are certainly at hand.

Quantum dots are the engineered counterparts to inorganic materials such as groups IV, III-V and II-VI semiconductors. These structures are prepared by complex techniques such as molecular beam epitaxy (MBE), lithography or self-assembly, much more complex than the conventional chemical synthesis. Quantum dots are usually termed artificial atoms (0D) with dimensions larger than 20–30 nm, limited by the preparation techniques. Quantum confinement, single electron transport, Coulomb blockade and related quantum effects are revealed with these 0D structures (Smith, 1996). 2D arrays of such 0D artificial atoms can be achieved leading to artificial periodic structures.

Neither polymers nor liquid crystals will be studied in this book, but at some stages will be recalled for the sake of comparison. Many books and review articles devoted to these subjects can be found in the literature and interested readers are referred to them (e.g., Nalwa, 1997: Vols. 2-4; Pope & Swenberg, 1999; Heeger, 2001).

1.1 Complex simplicity

Let us start this section with the example of the formation of a molecular solid from the archetypal inert homonuclear linear diatomic nitrogen molecule, N_2, the most abundant gas in air. N_2 exhibits very strong intramolecular bonding, with a binding energy of c. 10 eV ($\simeq$1000 kJ mol^{-1}), a large value due to the triple bond. Solid-phase N_2 will help us to introduce some concepts throughout the book, profiting

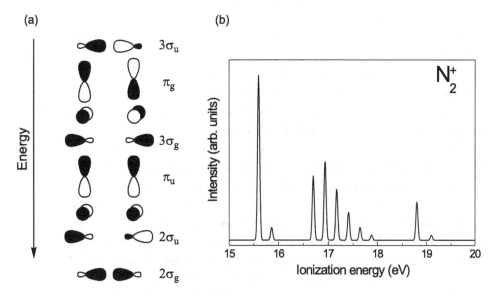

Figure 1.1. (a) Energy levels of N_2. The low-lying $1\sigma_g$ and $1\sigma_u$ orbitals are not shown. (b) Spectrum of N_2^+. Adapted from Eland, 1984; p. 7.

from its conceptual simplicity but at the same time warning us of the complexity in the formation of 2D and 3D networks when even simple molecules are involved. Many of the phenomena to be discussed in the book will be introduced in this section with simple molecules.

Figure 1.1(a) shows the energy levels of N_2. The core N1s orbitals are combined to produce $1\sigma_g$ and $1\sigma_u$ molecular orbitals (MOs) (the number 1 indicates that these are the lowest energy orbitals of their respective symmetries). Subscript g (from German *gerade*, meaning even) refers to symmetry with respect to inversion (the centre of the molecule acting as the inversion centre), while subscript u (from German *ungerade*, meaning uneven) is used for antisymmetry with respect to inversion.

The N2s and N2p orbitals of the two nitrogen atoms combine into pairs of sp hybrids and of p_π atomic orbitals on each nitrogen atom. When the larger lobes of the sp hybrids are directed towards each other along the molecular axis they give rise to bonding σ ($2\sigma_g$) and antibonding σ^* ($3\sigma_u$) orbitals, while the face-to-face combination of the smaller lobes gives rise to non-bonding $2\sigma_u$ and $3\sigma_g$ (lone-pair combinations). The p_π orbitals, which consist of N2p orbitals directed perpendicularly to the molecular axis, produce bonding π (π_u) and antibonding π^* (π_g) orbitals.

The orbital energies can be experimentally determined e.g., through ionization mediated by light excitation (photoelectron emission). According to Koopmans' theorem, each ionization energy I_E corresponds to an orbital energy, hence the photoelectron spectrum of a molecule should be a direct representation of the MO energy diagram. However, the photoelectron spectra of molecules are rather complex, as can be appreciated from Fig. 1.1(b). Ionized N_2 exhibits photoemission lines at 15.6, 16.7–18.0 and at 18.8 eV. The 15.6 eV line corresponds to the $3\sigma_g^{-1}$ MO, where the -1 superscript represents the removal of one electron. The group of peaks between 16.7 and 18.0 eV correspond to π_u^{-1} and finally the line at 18.8 eV corresponds to $2\sigma_u^{-1}$. In fact, each MO gives a group of peaks in the spectrum because of the possibility of vibrational as well as electronic excitations. Each resolved peak stands for a single vibrational line and represents a definite number of quanta of vibrational energy of the molecular ion (Eland, 1984). Ionization is a fast process, about 10^{-15} s being required for the ejected electron to leave the immediate neighbourhood of the molecular ion. The time is so short that motions of the atomic nuclei that make up vibrations and proceed on a much longer time scale of 10^{-13} s can be considered as frozen during ionization. This results in many accessible vibrational states and the effect is known as the Franck–Condon principle.

The fact that the number of peaks largely exceeds the number of MOs shows the limitation of Koopmans' theorem. In the case of the $2\sigma_g$ MOs, at least five features are associated to them in the photoemission spectra (Krummacher *et al.*, 1980), which is a consequence of the strong electron correlation in the ionic state. Bonding orbitals should exhibit distinct vibrational structure as compared to non-bonding orbitals because removal of one electron should strongly perturb the orbitals involved. This is clearly observed for the 16.7–18.0 eV features, corresponding to bonding π_u orbitals, as well as for the features around 15.6 and 18.8 eV, associated with the non-bonding $3\sigma_g$ and σ_u MOs.

Since we are here interested in solid nitrogen, we can ask ourselves which are the aggregation states of a simple molecule such as N_2. The answer comes in the form of a phase diagram, shown in Fig. 1.2. This pressure vs. temperature phase diagram reveals a certain degree of complexity, including several solid crystallographic phases or polymorphs (Bini *et al.*, 2000; Gregoryanz *et al.*, 2002). From the fluid phase, which exists up to *c.* 2 GPa (1 GPa = 10 kbar) at RT, nitrogen solidifies in a disordered hexagonal β-phase, where the N_2 molecular axes are randomly distributed. This is a common feature of diatomic molecular crystals. Low-pressure and low-temperature cubic α- and γ-phases represent two ways of ordered quadrupolar packing. At higher pressures, other classes of structures with non-quadrupolar-type ordering exist (δ, δ_{loc}, ε, ζ).

Because of the weak intermolecular interactions involved, the energy bands associated with solid N_2 can be generated, as a first approximation, by gently broadening the N_2 MOs, resulting in small finite bandwidths W of a few tenths of eV. The solid bands would originate from the perturbation of the molecular levels

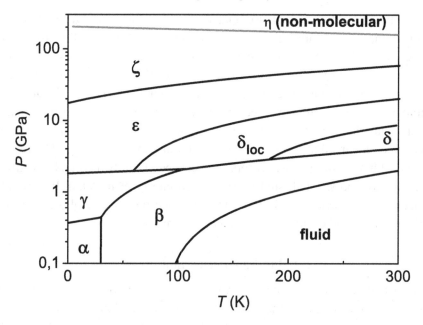

Figure 1.2. Phase diagram of nitrogen. Adapted from Bini *et al.*, 2000; Gregoryanz *et al.*, 2002.

following the physicists approach, an approximation that remains valid as long as the bands do not overlap, which implies that the free molecule electronic structure still largely determines the electronic structure of the crystal and that electrons are essentially localized within the molecules. This localization prevents electron mobility and the resulting solids are electrical insulators. However, when pressure is further raised, molecule–molecule distances are reduced and intermolecular interactions increase and the N_2 molecules eventually dissociate forming a non-molecular (atomic) framework or polymeric structure, called the η-phase. This non-molecular phase of nitrogen is semiconducting up to at least 240 GPa, with an energy gap of ~ 0.4 eV at 240 GPa as determined by transport measurements (Eremets *et al.*, 2001). This band gap compares to those of inorganic semiconductors such as InSb (0.16 eV), InAs (0.35 eV), Ge (0.67 eV), Si (1.1 eV), GaAs (1.4), GaP (2.2 eV), etc., measured at RT. At 300 K the insulator–semiconductor transition starts at *c.* 140 GPa. The perturbation picture no longer applies because of strong orbital overlap, and a conventional covalent solid approximation, which enables the delocalization of electrons through the solid, (e.g., the tight binding (TB) approximation) has to be considered. In this limit physicists would approximate the problem by considering quasi-free electrons.[2]

---

[2] Diatomic hydrogen, H_2, also exhibits a rich and remarkable complex solid phase diagram. At low temperatures, when a H_2-based solid is formed, electrons are confined in the H_2 molecules and the solid is thus a large-gap insulator. E. Wigner and H. B. Huntington predicted in 1935 the possibility of preparing

As mentioned before, we shall use small molecules to introduce the fundamentals for more complex molecules, the real core of this book, which will be listed in the next section. Such molecules form solids with remarkable properties (metallicity, superconductivity, ferromagnetism, etc.), some of them at ambient conditions or at much lower hydrostatic pressures than those found for H_2 and N_2, and some technological applications have been already developed, deserving the name of functional materials. Most of the molecules studied in this book are planar, or nearly planar, which means that the synthesized materials reveal a strong 2D structural character, although the physical properties can be strongly 1D, and because of this 2D distribution we shall study surfaces and interfaces in detail. In particular, interfaces play a crucial role in the intrinsic properties of crystalline molecular organic materials and Chapter 4 is devoted to them.

1.2 Building blocks: molecules

Tables 1.1 and 1.2 show the most relevant organic and hybrid organic–inorganic molecules used throughout the body of the book in alphabetical order. The schemes of the molecules go hand in hand with their most commonly accepted acronyms as well as reduced chemical names. Instead of using the well-established International Union of Pure and Applied Chemistry (IUPAC) convention for unequivocally identifying the molecules, we follow a reduced formulation, thinking more of the non-chemist communities. For instance, 7,7,8,8-tetracyano-*p*-quinodimethane will be simplified to tetracyano-quinodimethane and 2,3,6,7-tetramethyl-1,4,5,8-tetraselenafulvalene to tetramethyl-tetraselenaful-valene. In general the reduction will consist in removing the numeric locators. This modification should not affect the comprehension of the issues described in the book. In case of doubt it is recommended to consult the schemes in the mentioned tables. The molecules will be classified in the next sections

---

metallic hydrogen under high pressures (Wigner & Huntington, 1935). However, the experimental realization of metallic hydrogen has remained elusive despite studies performed at extremely elevated pressures (~ 350 GPa). N. W. Ashcroft (Aschcroft, 1968) even predicted that the hypothetical metallic phase of hydrogen might become superconducting at high temperatures (even at RT) based on the Bardeen–Cooper–Schrieffer (BCS) theory. The BCS theory is based on the coherent motion of paired electrons with a phonon-mediated attractive interaction. In the weak electron–phonon coupling limit, the transition temperature T_c is given by the expression $T_c \simeq 1.14\Theta_D e^{-1/[UN(E_F)]}$. For metallic hydrogen Θ_D is very high, ~ 3500 K, because $\Theta_D = (\hbar/k_B)\omega_D \propto m^{-1/2}$, where m represents the mass and $UN(E_F) > 0.25$ (Ashcroft, 1968), which leads to $T_c > 73$ K. Metallization of hydrogen might provide new insights into the nature of giant planets like Jupiter and Saturn, predominantly composed of hydrogen under extreme conditions. Hydrogen can exist as a low-temperature rotational crystal to very high pressures, $P < 110$ GPa, where the H_2 molecules are orientationally disordered (phase I), the individual molecules executing complete rotational motion in addition to the usual molecular vibrations. Phase II, characterized by a freezing of the H_2 molecules in random orientations in the crystal, is achieved in the $110 < P < 150$ GPa range at temperatures below *c.* 120 K. Above 150 GPa a new phase appears (phase III) where the presence of electric dipole moments is assumed to play a determinant role. The high pressures needed for such experiments can be achieved in a diamond-anvil cell (Mao & Hemley, 1992). Water, one of the most important and ubiquitous substances on Earth, is also an excellent example of the complex formation of different aggregation states to form solids of small molecules. Many different polymorphs of ice are known (up to 13), in addition to many types of amorphous ice.

Table 1.1. *Scheme, acronym and reduced name of the most relevant molecules discussed in the book*

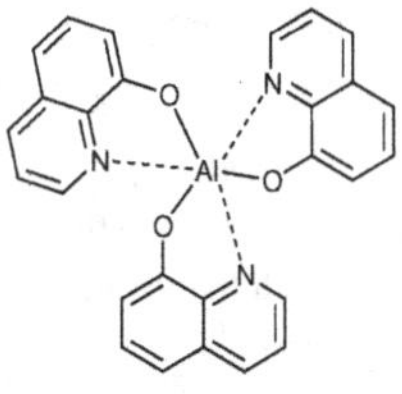

Alq₃
tris(hydroxyquinoline) aluminium

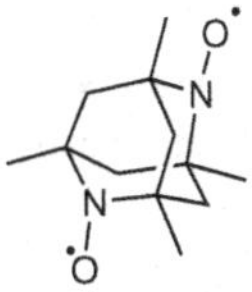

adam(NO)₂
dioxytetramethyl-
diazaadamantane

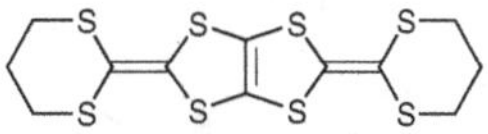

BDA-TTP
bis(dithianylidene)-
tetrathiapentalene

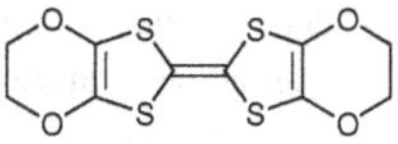

BEDO-TTF
bis(ethylenedioxo)-
tetrathiafulvalene

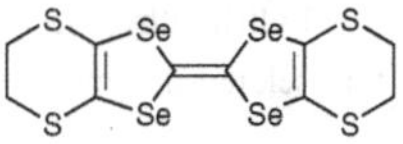

BEDT-TSF
bis(ethylenedithia)-
tetraselenafulvalene

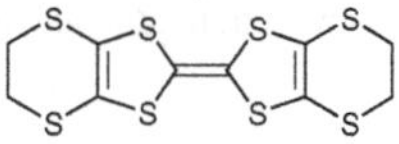

BEDT-TTF
bis(ethylenedithia)-
tetrathiafulvalene

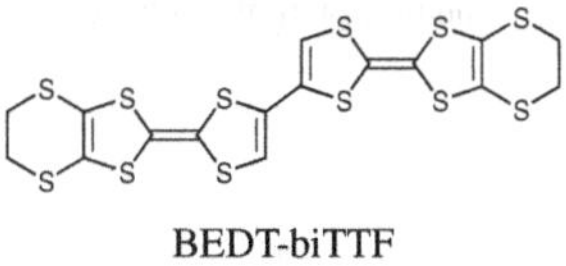

BEDT-biTTF
bis(ethylenedithia)-
bis(tetrathiafulvalene)

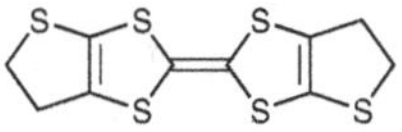

BET-TTF
bis(ethylenethia)-
tetrathiafulvalene

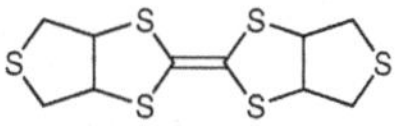

BTDM-TTF
bis(thiodimethylene)-
tetrathiafulvalene

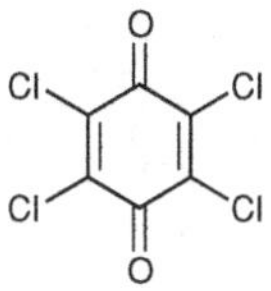

CA
chloranil

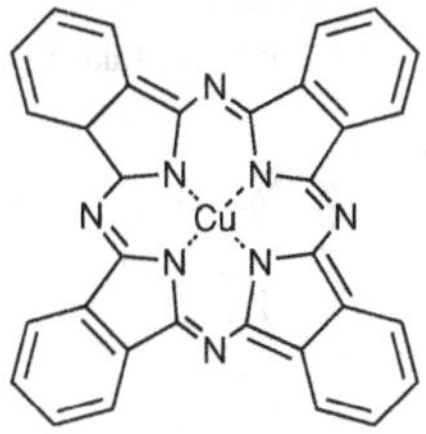

CuPc
copper phthalocyanine

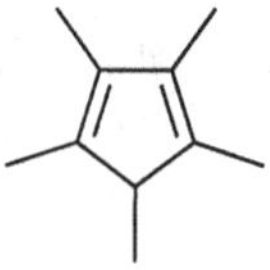

Cp*
pentamethyl-cyclopentadiene

C₆H₆

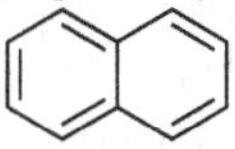

naphthalene

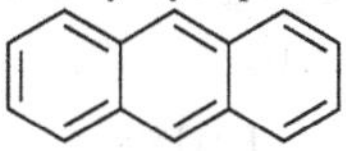

anthracene

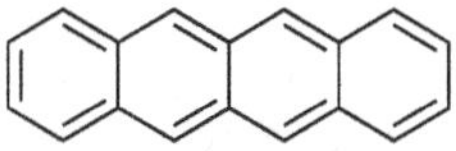

tetracene

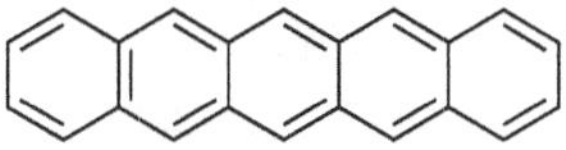

pentacene

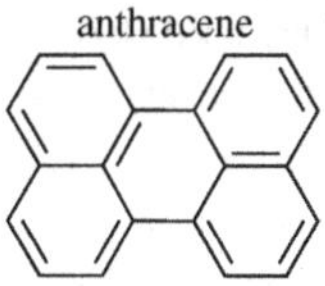

perylene

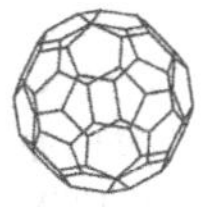

C₆₀
buckminsterfullerene

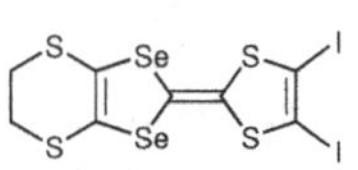

DIETS
diiodo(ethylenedithio)-
diselenadithiafulvalene

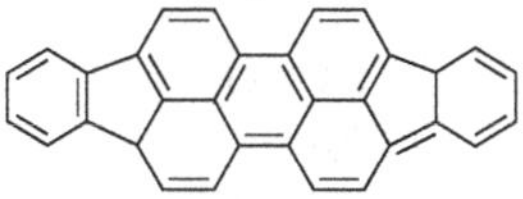

DIP
diindenoperylene

(cont.)

Table 1.1. (*cont.*)

DP5T
dipentyl-quinquethiophene

DMe-DCNQI
dimethyl-dicyanoquinonediimine

DMET
dimethyl(ethylenedithio)-
diselenadithiafulvalene

DMTTF
dimethyl-tetrathiafulvalene

DMtTSF
dimethyltrimethylene-
tetraselenafulvalene

DTEDT
dithiolylidene-ethanediylidene-
dithiole-tetrathiapentalene

DTTTF
dithiophene-tetrathiafulvalene

EC4T
endcapped-quadrithiophene

EDT-TTF
ethylenedithia-tetrathiafulvalene

EDT-TTF-CN
cyano-ethylenedithia-
tetrathiafulvalene

EDT-TTF-(CONHMe)₂
ethylenedithio-
dimethylcarbamoyltetrathiafulvalene

EDT-TTF-COOH
ethylenedithio-tetrathiafulvalene-
carboxylic acid

Fe(S₂CNEt₂)₂Cl
bis(diethyldithiocarbamato)
iron III chloride

FA
fluoranthene

HBC
hexa-peri-hexabenzocoronene

HMTSF
hexamethylene-
tetraselenafulvalene

HMTTF
hexamethylene-
tetrathiafulvalene

MDT-TSF
methylenedithio-
tetraselenafulvalene

MDT-TTF
methylenedithio-
tetrathiafulvalene

MeDC2TNF
dicyanomethylene-trinitrofluorene
carboxylic acid methyl ester

p-NC·C₆F₄·CNSSN
p-cyanotetrafluorophenyl-
dithiadiazolyl

(*cont.*)

Table 1.1. (*cont.*)

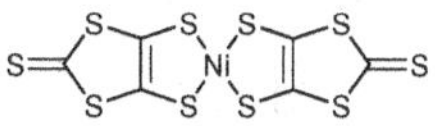

Ni(dmit)$_2$

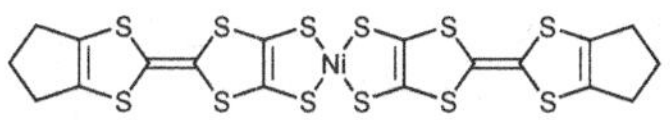

Ni(tmdt)$_2$

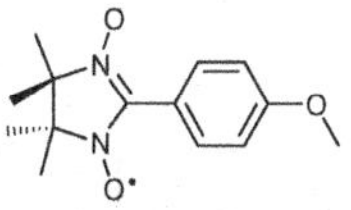

NITPhOMe
methoxyphenyl-
tetramethylimidazoline-oxyloxide

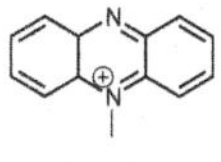

NMP
N-methylphenazinium

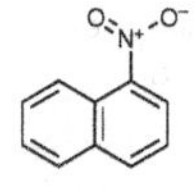

NN
nitronaphthalene

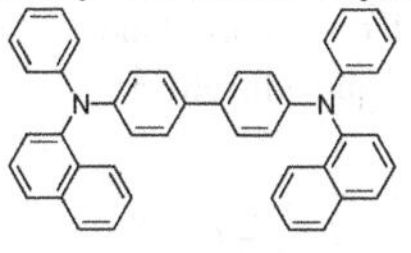

α-NPD
diphenylbisnaphthyl-biphenyldiamine

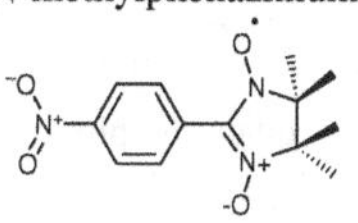

p-NPNN
p-nitrophenyl
nitronyl nitroxide

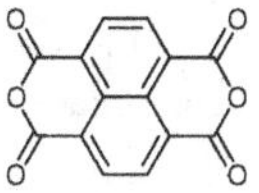

NTCDA
naphthalene
tetracarboxylic dianhydride

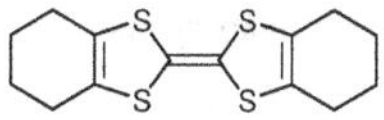

OMTTF
octamethylene-
tetrathiafulvalene

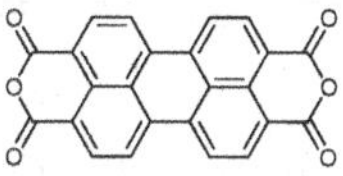

PTCDA
perylene
tetracarboxylic dianhydride

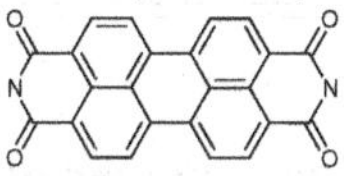

PTCDI
perylene
tetracarboxylic diimide

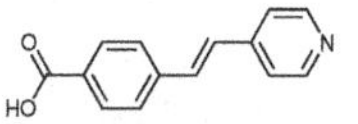

PVBA
trans-pyridylvinyl-
benzoic acid

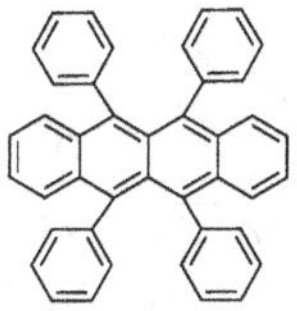

rubrene
tetraphenylnaphthacene

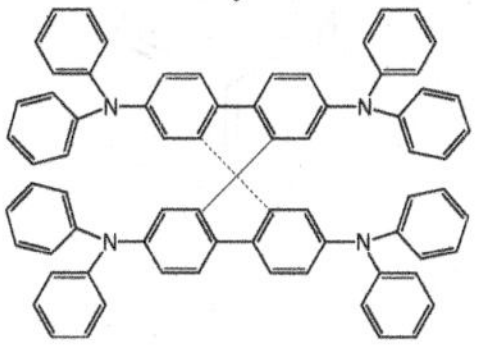

Spiro-TAD
tetrakisdiphenylamino-spirobifluorene

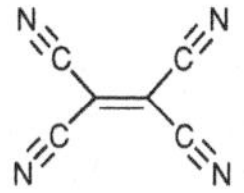

TCNE
tetracyanoethylene

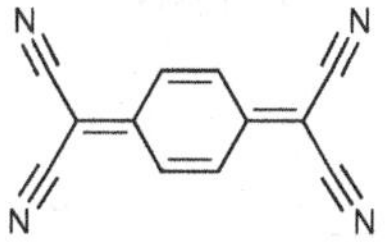

TCNQ
tetracyanoquinodimethane

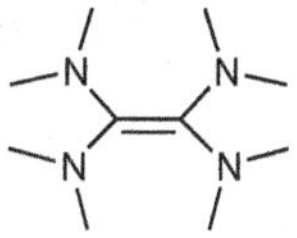

TDAE
tetrakis(dimethylamino)ethylene

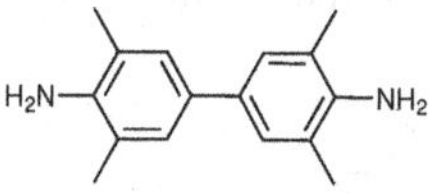

TMB
tetramethylbenzidine

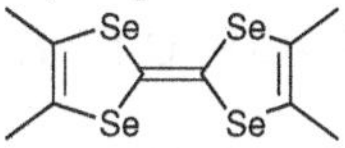

TMTSF
tetramethyl-tetraselenafulvalene

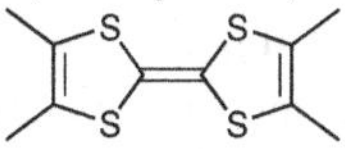

TMTTF
tetramethyl-tetrathiafulvalene

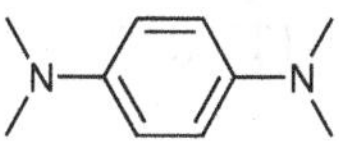

TMPD
tetramethyl-phenylenediamine

(*cont.*)

Table 1.1. (*cont.*)

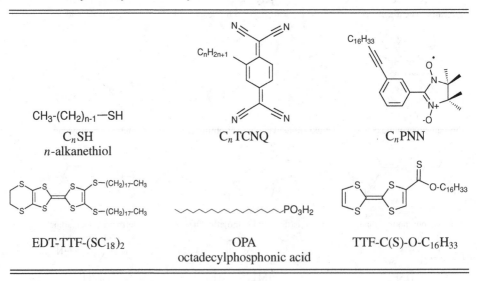

TNAP	**TPD**	**TTF**
tetracyanonaphtho-quinodimethane	diphenyl-methylphenyl-biphenyl-diamine	tetrathiafulvalene
TTP	**TTT**	**α-4T**
tetrathiaperylene	tetrathiatetracene	quaterthiophene
p-6P	**α-6T**	
sexiphenyl	sexithiophene	

Table 1.2. *Scheme, acronym and reduced name of the most relevant molecules used for the formation of LB and SAM films discussed in the book*

$CH_3-(CH_2)_{n-1}-SH$	C_nH_{2n+1}	$C_{16}H_{33}$
C_nSH	**C_nTCNQ**	**C_nPNN**
n-alkanethiol		
EDT-TTF-$(SC_{18})_2$	**OPA**	**TTF-C(S)-O-$C_{16}H_{33}$**
	octadecylphosphonic acid	

according to their degree of symmetry and topology, which ultimately determine the physical and chemical properties of the solids they build.

Electronic configuration

From the electronic point of view the most remarkable property of MOMs is that metals, superconductors and magnets can be built when only s- and p-electrons are involved. For metals this is not surprising since alkali metals are based on s-electrons, and for graphite delocalized p_z-electrons are responsible for its semimetallic character. However, the existence of magnetic order without d-electrons is more intriguing, especially when thinking of classical magnets, known for centuries, which are all based on transition metals. In fact W. Heisenberg concluded, back in 1928, that ferromagnetism could not exist in compounds consisting only of light elements (Heisenberg, 1928). Fortunately he was wrong, otherwise Section 1.5 would make little sense.

As will be discussed later (Section 1.5), molecules containing no metallic elements are able to combine and form materials exhibiting metallic character, e.g., HMTSF-TCNQ, TTF-TCNQ, etc., or even lose any electrical resistance below a given temperature and thus become superconductors, e.g., $(TMTSF)_2ClO_4$. Metal-free molecules can also, in the solid state, show magnetic order, such as p-NPNN and p-NC·C_6F_4·CNSSN, where in the absence of d-electrons the magnetic properties are related to unpaired p-electrons.

Symmetry

Symmetry is one of the most important parameters when characterizing molecules and solids because it permits the rationalization of their physical properties and helps in simplifying the resolution of the Schrödinger equation (see Section 1.7). Following the common practice, Schönflies and short Hermann–Mauguin symbols will be used for point and space groups, respectively. Readers interested in learning more about group theory applied to chemistry are encouraged to read e.g., Cotton, 1971.

Let us start with the neutral molecule TTF, the fundamental building block of an immense pool of CTSs exhibiting magnetic ground states, semiconducting, metallic or superconducting properties that will be extensively discussed in the book. Figure 1.3 shows a TTF molecule oriented in a Cartesian coordinate system, where the long molecular axis is arbitrarily set along the x-axis. We will keep this convention throughout the book.

If we assume that TTF is perfectly flat (contained in the xy-plane), TTF remains invariant under the following symmetry operations:

$$E, C_2(x), C_2(y), C_2(z), \sigma_{xy}, \sigma_{xz}, \sigma_{yz}, i,$$

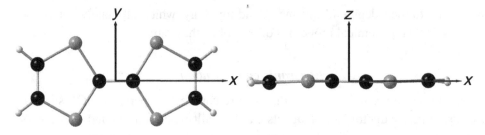

Figure 1.3. Neutral TTF molecule with the Cartesian coordinate system used for the determination of its symmetry operations. C, S and H are represented by black, medium grey and light grey balls, respectively.

where E represents the identity operator (from German *Einheit*), $C_2(x)$, $C_2(y)$, $C_2(z)$ stand for proper rotations of π degrees (rotations of order 2) with respect to the x-, y- and z-axes, respectively, σ_{xy}, σ_{xz}, σ_{yz} are planes of symmetry with respect to the xy-, xz- and yz-planes, respectively (from German *Spiegel*), and i is an inversion operation with its centre at the origin of the coordinate system. The full set of such symmetry operations leads to the D_{2h} point group (see Appendix A for the symmetry operations of other point groups of interest). Table 1.3 classifies some of the molecules from Table 1.1 according to their corresponding point groups. The table is divided into three blocks. The topmost one contains the special groups $C_{\infty v}$ and $D_{\infty h}$, characteristic of asymmetric and symmetric linear molecules, respectively, and the groups containing high-order axes, such as I_h. The second block refers to point groups containing no proper or improper axes of rotation, such as C_s, and finally the third block contains those groups containing proper axes of rotation.

In Table 1.3 we have assumed that molecules belonging to the groups with proper axes of rotation (third block) are planar (σ_{xy} applies). This is not strictly true since molecules tend to deform in space in order to accommodate to the 3D crystal structure. One of the most notable examples is BEDT-TTF, which is illustrated in Fig. 1.4, where the dimer structure of the monoclinic phase of neutral BEDT-TTF is shown. From the figure we observe that the molecules are non-planar, the angles formed between the planes containing the central carbon–carbon bond and the four neighbouring sulfur atoms (TTF core) with the planes containing the four more external sulfur atoms (half TTF core and ethylenedithia groups) are 166.6 and 162.6 degrees, respectively. However, the most remarkable non-planarity is observed for the outer $(CH_2)_2$ groups. Upon oxidation of BEDT-TTF these angles become larger, as will be discussed in Section 1.5.

PTCDA represents a further example of non-planarity. When chemisorbed on an Ag(111) surface the carboxylic oxygen atoms become 0.018 nm closer to the

Table 1.3. *Classification of some relevant molecules discussed in the text according to their point groups*

Point group	Molecule
$C_{\infty v}$	CO
$D_{\infty h}$	H_2, N_2, C_2H_2
I_h	C_{60}
C_s	EDT-TTF-CN, EDT-TTF-COOH
C_{2v}	H_2O, EDT-TTF, α-nT (n odd), DMET, MDT-TSF
C_{2h}	α-nT (n even), DMTTF, BEDT-biTTF
C_{3v}	NH_3
D_{2h}	TTF, TCNQ, TMTTF, TMTSF, BEDT-TTF, PTCDA, Ni(dmit)$_2$, BEDT-TSF, BDA-TTP, pentacene
D_{4h}	CuPc
D_{6h}	C_6H_6

Figure 1.4. BEDT-TTF dimer in the crystal structure of neutral BEDT-TTF. $P2_1/c$, $a = 0.661$ nm, $b = 1.398$ nm, $c = 1.665$ nm, $\beta = 109.55°$. Crystallographic data from Kobayashi *et al.*, 1986a. Dimers are contained in the $(10\bar{2})$ plane. C and S are represented by black and medium grey balls, respectively.

surface than the perylene core (Hauschild *et al.*, 2005). The series of unsubstituted oligothiophenes α-nT exhibits C_{2v} and C_{2h} symmetry for n odd and n even, respectively. However, for the series of end-capped ECnT the symmetry is reduced by the non-planar conformation of the terminal cyclohexene rings. Hence the highest possible symmetries are, depending on the relative conformation of the two cyclohexene rings, C_2 or C_s for n odd and C_2 or C_i for n even. On the other hand we observe that functionalization of TTF may result in the loss of the degree of symmetry, as exemplified in the series TTF (D_{2h}) $\rightarrow$ EDT-TTF (C_{2v}) $\rightarrow$ EDT-TTF-CN (C_s). Functionalization will be discussed in the next chapter.

Table 1.4. C_6H_6 *as a building block*

2D geometry					3D geometry
n	linear	non-linear	non-linear	linear	
1					
2					
3					
4					
5					
6					
7					heptahelicene
...					
∞		graphite			carbon nanotubes

Topology

In a naïve but pedagogical way the molecules we are considering here can be built from a few fundamental building blocks. From a strictly geometrical approach they can be constructed from hexagons, pentagons and functional groups. We can get closer to reality if we allow relaxation of the lengths (bonds) and permit identifying the vertices with any atom. Let us see the amazingly large number of molecules that can be built based exclusively on the aromatic C_6H_6 molecule. A scheme of this LEGO-like classification is given in Table 1.4. Molecules are first distributed according to their spatial conformation. They can be either planar (2D geometry) or non-planar (3D geometry). Planar molecules can be either linear or non-linear and the molecules can attach to each other either sideways or by connecting vertices with bonds. Finally, in the table n indicates the number of C_6H_6 rings, which from the topology point of view corresponds to the genus. The genus of a space corresponds to the number of holes: for a sphere the genus equals 0, for a ring 1, for an eyeglass frame 2 and so on. This is all we need to build any molecule from the table. $n = 1$ is the identity case and for $n = 2$ we can already build two molecules, naphthalene (C_{10}) and biphenyl (C_{12}). For $n = 3$, two non-linear combinations compete with anthracene. As long as $n > 3$ more and more combinations are possible. We plot here just some of them. Row $n = 5$ has two prominent representatives, pentacene and perylene, of great relevance for organic semiconductors, but other combinations are also possible, such as the case of dibenzoanthracene. HBC, with $n = 13$ is already a large molecule (by our standards) and if we continue *ad infinitum* adding C_6H_6 rings in a plane we will obtain a graphene plane, the planes that build graphite. If we build a circular ring with six C_6H_6 rings we get coronene ($n = 6$, 2D), also termed hexabenzobenzene, a molecule that has been observed on the surface of Saturn's moon Titan. Adding an extra ring leads to the opening of the cyclic structure due to steric repulsion and a chiral molecule, heptahelicene ($n = 7$, 3D), is obtained (see Fig. 4.17). Thus, heptahelicene has the π-electron system delocalized over the helical backbone.

It would be interesting to see if closed-loop molecules made exclusively from C_6H_6 rings with Möbius-band shape could be synthesized since they would be 3D *single-faced* molecules, built from two-faced C_6H_6 rings (see discussion in Section 2.1). The Möbius band is constructed by joining the ends of a rectangular strip after having given one end half a twist. Back in 1964 E. Heilbronner predicted that cyclic molecules with the π-orbital topology of a Möbius band should be aromatic if they contain $4m$ π electrons ($m \in \mathbb{N}$) (Heilbronner, 1964). According to the Hückel rule, conventional aromatic hydrocarbon compounds are stabilized by the delocalization of π electrons when they number $4m + 2$, but cyclic compounds with $4m$ π electrons are anti-aromatic and unstable. Twisting cyclic molecules induces destabilization because of the large ring strains and suppresses the overlap of the π orbitals. A Möbius compound stabilized by its extended π system has been only recently synthesized by combining two aromatic structures, with their π orbitals orthogonal and within the ring plane, respectively (Ajami *et al.*, 2003). The molecule results from the combination of tetradehydrodianthracene (in-plane π orbitals) and the annulene *syn*-tricyclooctadiene (out-of-plane π orbitals) by methatesis, where tetradehydrodianthracene plays the role of the rigid structure. Figure 1.5 shows a view of the molecule.

Table 1.4 is completed with the extensively studied carbon nanotubes. Nanotubes are cyclinders that can be thought of as originating from curving a graphene plane and gluing two edges together. Increasing the size of the molecules enables the delocalization of π electrons in a larger area, increasing electrical conductivity. Solid C_6H_6 is an insulator but at the end of the table we have a semimetal (graphite) and a metal or semiconductor (carbon nanotubes).

If we now incorporate functional groups in the C_6H_6-based molecules we will have important molecules such as CA, DMe-DCNQI, TCNQ, TMPD, TNAP and TTT, but the really big jump is encountered when pentagons are allowed to be included. The possibilities to build new molecules become even more numerous than with only C_6H_6 pieces, although we already had an immense number

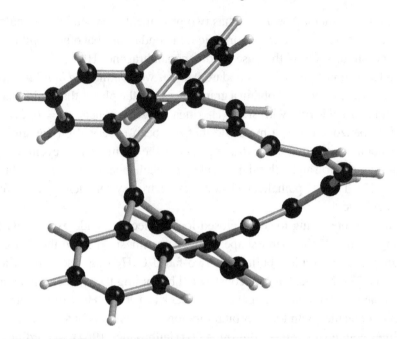

Figure 1.5. View of the Möbius stabilized annulene. Crystallographic data from Ajami, 2003. The tetradehydrodianthracene and twisted structures are shown left and right of the figure, respectively. C and H are represented by black and light grey balls, respectively.

of combinations. Including pentagons leads to the assembling of crucial molecules such as TTF, thiophenes and transition metal dithiolates, as well as TTF-derivatives, FA, DIP, C_{60}, p-NPNN, p-NC·C_6F_4·CNSSN, etc.

Once topologically classified, the molecules can be classified in families. Based on a core molecule such as TTF, a myriad of molecules can be synthesized. We can thus schematize a molecule in three parts: core, functional groups and spacers. For EDT-TTF-CN, TTF is the core and EDT and CN the functional groups. Perylene is the core of molecules such as PTCDA, PTCDI, etc. A spacer separates two parts of a molecule. A carbon–carbon bond in biphenyl is a simple example of a spacer, DTEDT is a further example, but in a large molecule a spacer can be a full TTF molecule. The chemical manipulation of molecules is a key property of MOMs. Think of inorganic materials where the building blocks, atoms, cannot be chemically designed. The intense research on chemical synthesis has led to the preparation of extremely complex molecules. Adding more and more cores, functional groups and spacers leads to very large molecules with open (dendrimers) or closed (macrocycles) structures, which will be discussed in Section 2.5.

1.3 Intermolecular interactions

The molecular packing of MOMs results from a precise and subtle balance of several intermolecular interactions within a narrow cohesion energy range of less than 1 eV molec^{-1}. This is the reason why crystal engineering is so powerful because this balance can be intentionally modified but at the same time it implies that MOMs are soft materials and that polymorphism is favoured. Detailed descriptions on the fundamentals of interatomic and intermolecular interactions can be found in many books (see e.g., Kitaigorodskii, 1961). Here we briefly describe the relevant interactions for MOMs and give a new approach supported on the nanoscience perspective.

The interactions involved in the formation of molecular organic solids are of dipole–dipole, hydrogen bonding and π-overlap types. When a permanent polar molecule induces a dipole in a non-polar (but polarizable) molecule or when charge fluctuation in a non-polar molecule induces a dipole in a non-polar surrounding molecule, there is a weak attractive interaction between them, known as van der Waals–London dispersion forces. The energies involved are typically less than 0.2 eV molec^{-1} and the interaction ranges are short, since forces vary as D^{-6}, where D represents the interatomic/intermolecular distances. The energies involved in permanent dipole–permanent dipole interactions, known as Keeson forces, are ~ 0.5 eV molec^{-1}.[3]

Hydrogen bonding is of electrostatic origin, where there is an attraction between an H atom covalently bound to a donor D and a region with a high electronic density over an atom A (D–H $\cdots$ A–X). Hydrogen-bonded architectures are abundant in biological systems, which has motivated their exploitation in supramolecular chemistry. Hydrogen bonding is responsible e.g., for the surprising physical properties of water and ice and for the DNA double-helix structure, which is held together by N–H groups. Hydrogen bonds of the type (O–H $\cdots$ O) and (N–H $\cdots$ O) are usually termed *strong* (0.2–0.4 eV molec^{-1}) in order to distinguish them from the weaker (C–H $\cdots$ O) and (O–H$\cdots \pi$) interactions (0.02–0.2 eV molec^{-1}). The distinctive geometrical attributes of a hydrogen bond D–H $\cdots$ A–X are the D $\cdots$ A and the H $\cdots$ A lengths, the hydrogen bond angle D–H–A, the H–A–X angle and the planarity of the DHAX system. Typical values for the H $\cdots$ A distance are 0.18–0.20 nm for N–H $\cdots$ O bonds and 0.16–0.18 nm for O–H $\cdots$ O bonds. D–H–A and H–A–X angles range around 150–160° and around 120–130°, respectively.

---

[3] With the exception of He, solid noble gases crystallize in the face-centred-cubic Bravais lattice and their cohesive energies (eV atom^{-1}) are 0.02 for Ne, 0.08 for Ar, 0.11 for Kr and 0.17 for Xe (Ashcroft & Mermin, 1976; p. 401). Van der Waals–London forces are expressed by the Lennard-Jones potential $U_{\text{LJ}} \propto C_R D^{-12} - C_A D^{-6}$, where C_R and C_A stand for two constants multiplying the repulsive and attractive components, respectively.

Overlap between p_π orbitals leads to cohesive energies of typically less than 0.4 eV molec^{-1}. The much stronger ionic and covalent bonding have binding energies of ~ 10 and ~ 3 eV atom^{-1}, respectively. Finally, physisorption is the weakest form of absorption to a solid surface characterized by a lack of a true chemical bond (chemisorption) between substrate and adsorbate and will be discussed in Chapter 4 (see e.g., Zangwill, 1988).

The effect of intermolecular interactions can be readily observed when comparing the absorption spectrum of a molecule in solution to that in the solid state. In solution, where the molecules can be considered as isolated, the spectra are characterized by sharp lines corresponding to absorption bands. However, in the solid, intermolecular interactions cause the formation of exciton bands and splitting of the levels. This phenomenon is often referred to as Davydov splitting. This splitting is thus a measure of the strength of the interactions and for MOMs it can amount to 0.2–0.3 eV.

The binding forces and cohesion energies can be experimentally evaluated with an atomic force microscope (AFM) in some cases. The AFM was developed by G. Binnig, C. F. Quate and Ch. Gerber (Binnig et al., 1986) and its working principle is based on the deflection of a microfabricated cantilever due to repulsive and attractive forces between atoms on the sample surface and atoms at the cantilever tip. The deflection is measured using a laser beam and a photodiode detector and scanning in the x-, y- and z-directions is performed by a piezoelectric translator. For instance, the binding force of an electron donor–acceptor complex has been evaluated as c. 70 pN (Skulason & Frisbie, 2002). These experiments have been performed by measuring pull-off forces between AFM tips and substrates coated with complementary self-assembled monolayers (SAMs) capable of specific chemical binding (see Section 3.2 for the preparation technique leading to SAMs). In a pull-off experiment, typically used to unfold proteins, exposed functional groups on the two SAMs surfaces bind upon tip–substrate contact. The binding is quantified by measuring the pull-off force required to rupture the contact (see Fig. 1.6). If the tip is sufficiently sharp, the contact area is of the order of a few nm^2, so that pull-off involves breaking a small integer number of bonds. In the example discussed here a gold-covered AFM tip has been covered with a SAM consisting of an alkyl linear chain with a thiol end that binds covalently to the gold film, and a TMPD end. Additionally, a flat gold surface has been covered with a SAM again with a thiol end and a TCNQ end.

AFMs have also been used to estimate the cohesive energy of ionic materials with face-centred-cubic structure (Fraxedas et al., 2002a). In these experiments an ultrasharp AFM tip (tip radius $R < 10$ nm) indents a flat surface of a single crystal and the dynamical mechanical response of the surface during indentation is transformed into a force plot (applied force vs. penetration). It turns out that the

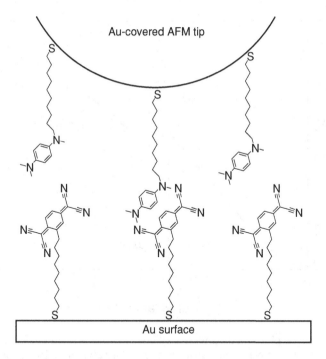

Figure 1.6. Schematic representation of a CT complex formation of a SAM of TMPD on an Au-coated AFM tip and a SAM of TCNQ on an Au substrate. Adapted from Skulason & Frisbie, 2002.

slope of the elastic part is close to the Debye stiffness k_D, defined as $k_D = m\omega_D^2 = m(k_B\Theta_D/\hbar)^2$, where m stands for the mean atomic mass. k_D can be considered as the stiffness of the crystal, because Θ_D is a measure of the temperature above which all vibrational modes begin to be excited and below which modes begin to be frozen out. An approximate estimate of the cohesive energy can thus be obtained. In the case of TTF-TCNQ, force plots obtained with nanoindentation with ultrasharp tips have the shape shown in Fig. 1.7.

Below $\delta_Y \simeq 12.7$ nm, where δ_Y stands for the deformation at the plastic yield point, the material responds elastically and above δ_Y plasticity is evidenced by discrete discontinuities in multiples of $\simeq 0.9$ nm. Such discontinuities correspond to the distance between two adjacent *ab*-planes. From the elastic region we can evaluate the maximum accumulated energy that TTF-TCNQ can withstand without irreversible deformation by simply integrating the force plot for $0 \le \delta \le \delta_Y$. The integral gives $\sim 2.0 \times 10^{-15}$ J. The affected volume is larger than $\pi R^2 \delta_Y$ (~ 3000 nm^3 in this case). Thus, the estimated maximum mean energy per TTF-TCNQ pair is 2.3 eV. This value compares well to the experimentally derived enthalpy of sublimation of TTF-TCNQ, which amounts to 2.7 eV per TTF-TCNQ pair (de Kruif

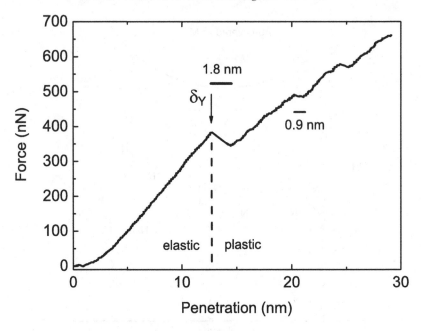

Figure 1.7. Force plot obtained with a silicon microfabricated cantilever with a force constant of $\sim 25\,\mathrm{N\,m^{-1}}$ and an ultrasharp tip of radius $R < 10\,\mathrm{nm}$ on a highly *ab*-oriented thin TTF-TCNQ film.

& Govers, 1980). The 2.3 eV value is lower than the measured cohesive or crystal binding energy of TTF-TCNQ, which is 4.9 eV per TTF-TCNQ pair (Metzger, 1977). The cohesion energy essentially accounts for the Coulomb (Madelung term), polarization (charge-induced dipole), dispersion (van der Waals) and repulsion energies. The fact that nanoindentation is performed on the surface of the material facilitates sublimation of the molecules. Theoretical estimates of the cohesive energy of pentacene using first-principles pseudopotential density functional theory (DFT) give a value of 1.3 eV per molecule (Northrup *et al.*, 2002).

The estimated Θ_D is about 90 K, in agreement with previous heat capacity determinations performed on single crystals (Coleman *et al.*, 1973a). Table 1.5 shows some Θ_D values found in the literature for some MOMs.

1.4 Crystal engineering: synthons

As already mentioned in Section 1.1, chemists regard molecular crystals as supermolecules. This is fully justified since molecules are built by connecting atoms through covalent bonds and crystals are built by connecting molecules with intermolecular interactions. Crystal engineering can be defined as *the understanding of*

Table 1.5. *Debye temperatures of selected MOMs*

Material	Θ_D [K]	Reference
$(TMTTF)_2BF_4$	51	Coulon *et al.*, 1982
$(TMTTF)_2PF_6$	55	Coulon *et al.*, 1982
$(TMTTF)_2ClO_4$	59	Coulon *et al.*, 1982
δ-*p*-NPNN	66	Nakazawa *et al.*, 1992
γ-*p*-NPNN	89	Nakazawa *et al.*, 1992
TTF-TCNQ	90	Coleman *et al.*, 1973a
β-*p*-NPNN	140	Nakazawa *et al.*, 1992
TTF-TCNQ	160	Saito *et al.*, 1999
$(TMTSF)_2ClO_4$	213	Garoche *et al.*, 1982
$(MDT-TSF)(AuI_2)_{0.44}$	300	Kawamoto *et al.*, 2002

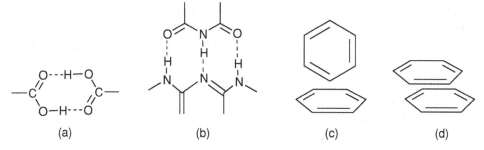

Figure 1.8. Some representative supramolecular synthons.

intermolecular interactions in the context of crystal packing and in the utilization of such understanding in the design of new solids with desired physical and chemical properties (Desiraju, 1989). The major challenge of crystal engineering is that a crystal structure is in fact a compromise between different interactions of varying weak strengths and spatial distributions. Hence considering only one type of interaction, even if it turns out to be the strongest, when predicting crystal structures usually leads to erroneous results, because the contribution of the neglected weaker interactions can dominate.

Interactions can be combined by a designed placement of functional groups in the molecular skeleton to generate supramolecular synthons, which are defined as structural units within supermolecules that can be formed and/or assembled by known or conceivable synthetic operations involving intermolecular interactions (Desiraju, 1995). In other words supramolecular synthons are spatial arrangements of intermolecular interactions.

Figure 1.8 shows a few examples of representative supramolecular synthons. Synthons are derived from designed combinations of interactions and in general are not identical to the interactions. The goal of crystal engineering is to recognize and design synthons that are robust enough to be exchanged from one network structure to another, which ensures generality and predictability. Such structural predictability leads, in turn, to the anticipation of 0D, 1D, 2D and 3D patterns formed with intermolecular interactions. Examples of some simple network architectures are: 0D (polyoxometallate clusters), 1D (ladder and zig-zag chains), 2D (honeycomb and square grids) and 3D (octahedral and hexagonal diamondoid networks) (Moulton & Zawarotko, 2001).

Figures 1.8(a) and (b) represent examples of double and triple hydrogen bonding, respectively. An example of the first will be discussed in Section 5.3 for solid EDT-TTF-COOH and the second is the basis for recognition between nucleotide bases in DNA. Let us now discuss in more detail the synthons represented in Figs. 1.8(c) and (d), which are based on herringbone and stacking modes of association of aromatic rings, respectively. Figures 1.9(a) and (b) show the crystal structures of C_6H_6 and of coronene, respectively. C_6H_6 shows a T-shaped herringbone structure while coronene adopts a slightly modified L-shaped structure, known as the γ-structure (Gavezzoti & Desiraju, 1988). In this case these two similar but clearly differentiated arrangements depend on the relative balance of $C \cdots C$ and $C \cdots H$ intermolecular interactions.

An interesting example is heptafulvalene (Fig. 1.9(c)), a non-planar molecule with an S-type shape, because it exhibits the characteristic herringbone distribution in spite of its non-planarity.

Many crystal structures encountered in the next section will be based on the generalization of the synthon shown in Fig. 1.8(d), incorporating sulfur and selenium atoms to the molecules, representing the proclivity of planar π donors and acceptors to form overlapping layered structures. Neutral TMTTF crystallizes in the $C2/c$ space group and the molecules in the ($\bar{2}02$) plane are arranged according to the γ-structure (see Fig. 1.10(a)) while neutral TMTSF exhibits a pure stacking ordering, with no herringbone arrangement (see Fig. 1.10(b)). The comparison is interesting because the only difference between TMTTF and TMTSF is the presence of sulfur or selenium atoms, respectively. This example leads us to the conclusion that intuition based only on geometrical arrangement of molecules does not always work since the interactions, strictly speaking the synthons, have to be considered. Flat molecules are unlikely to have many polymorphic modifications under normal P and T conditions. However, this phenomenon should be more common among molecules having flexible conformations or different groups capable of hydrogen bonding. We shall see in Section 1.5 the large number of known polymorphs of BEDT-TTF salts and discuss the fascinating phenomenology of polymorphism in Section 1.6.

(a) (b)

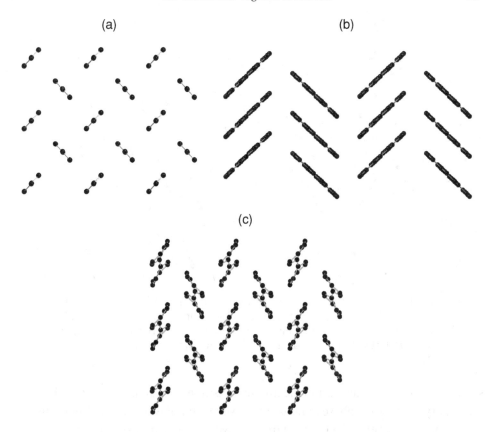

(c)

Figure 1.9. (a) View of the 218 K crystal structure of C_6H_6 along the b-direction, $Pbca$, $a = 0.744$ nm, $b = 0.955$ nm, $c = 0.692$ nm. Crystal structure from Bacon *et al.*, 1964. (b) ab-plane of coronene. $P2_1/a$, $a = 1.610$ nm, $b = 0.469$ nm, $c = 1.015$ nm, $\beta = 110.8°$. Crystal structure from Robertson & White, 1945. (c) bc-plane of heptafulvalene. $P2_1/c$, $a = 0.969$ nm, $b = 0.773$ nm, $c = 0.697$ nm, $\beta = 98.03°$. Crystal structure from Thomas & Coppens, 1972.

1.5 Molecular organic materials

When classifying a given material according to its physical properties, the common tendency is to emphasize the most important property: metallicity, superconductivity, magnetic order, etc. However, strictly speaking one has to specify which point in the phase diagram is being considered, hence indicating the corresponding external variables P, T, B, etc., since the same material can behave differently when spanning the parameter hyperspace. The parameter hyperspace consists of the set of variables and will be defined in detail in Chapter 3. For instance, a given material can be a metal above a given temperature but a semiconductor below

(a)   (b)

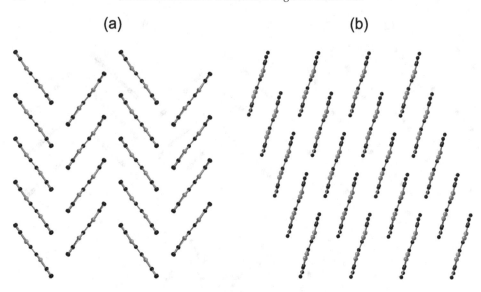

Figure 1.10. (a) ($\bar{2}$02) plane of TMTTF and (b) (100) plane of TMTSF. TMTTF: $C2/c$, $a = 1.614$ nm, $b = 0.606$ nm, $c = 1.428$ nm, $\beta = 119.974°$. TMTSF: $P\bar{1}$, $a = 0.693$ nm, $b = 0.809$ nm, $c = 0.631$ nm, $\alpha = 105.51°$, $\beta = 95.39°$, $\gamma = 108.90°$. Crystallographic data from Kistenmacher *et al.*, 1979.

such temperature, or can exhibit electronic transitions as a function of P. It is thus mandatory to know the phase diagram of a given material when discussing its physical properties. In general, phase diagrams of MOMs are extremely rich but seldom completely explored.

Before introducing some reference materials from the myriad of possibilities, let us first define important energy parameters that are schematized in Fig. 1.11. Case (a) corresponds to a metal, where the valence band is filled with electrons up to the Fermi level E_F. This ideal situation corresponds to the $T = 0$ K limit and the electronic density distribution around E_F at finite temperatures will be discussed in Section 1.7.

The minimum energy required to promote an electron at E_F to the vacuum level E_{vac} is the work function ϕ_M. E_{vac} can be defined as the energy of an electron at rest (zero kinetic energy) just outside the surface of the solid. Figure 1.11(b) represents a semiconductor, characterized by a band gap E_g between the valence band maximum (VBM) and the conduction band minimum (CBM). Hence E_g represents the threshold energy barrier to transitions from occupied to unoccupied states. The electron affinity E_A is defined as the energy required to excite an electron from the CBM to the local E_{vac} and the ionization energy I_E is defined as the energy needed to excite an electron from the VBM to the local E_{vac}. For a metal $I_E = \phi_M$. Finally Fig. 1.11(c) describes the energy band diagram of an isolated molecule,

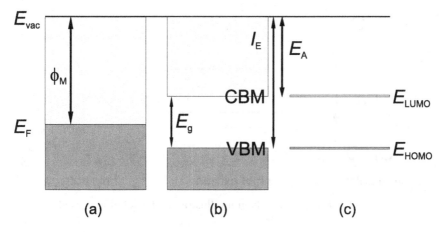

Figure 1.11. Definition of E_F, E_g, E_A, E_{vac}, E_{HOMO}, E_{LUMO}, I_E, W and ϕ_M. For simplicity the unoccupied bands are represented only up to E_{vac}.

with an energy gap E_g between E_{HOMO} and E_{LUMO}, the energies of the highest occupied MO (HOMO) and the lowest unoccupied MO (LUMO), respectively ($E_g = E_{LUMO} - E_{HOMO}$).

For molecules the chemical definition of E_A is the energy gained by an originally neutral molecule when an electron is added to the LUMO of the neutral molecule. Analogously, I_E corresponds to the energy gained by an originally neutral molecule when an electron is ejected from the HOMO of the neutral molecule. MOMs exhibit low bandwidths W, because of the weak intermolecular interactions involved. Hence they can be classified according to Fig. 1.11(b), as narrow W, or to Fig. 1.11(c), with finite width HOMO and LUMO bands. We will use the terminology HOMO and LUMO bands with finite width and therefore the use of terms such as HOMO band maximum or LUMO band minimum are fully justified. Details on the experimental determination of the parameters discussed here will be given in Section 4.2.

Organic semiconductors

According to the definitions given above, semiconductors are characterized by $E_g \neq 0$. Inorganic materials are classified as either semiconductors or insulators if $E_g < 3$ eV or $E_g > 3$ eV, respectively. However, the MOMs scientific community often refers to insulators for $E_g \sim 0.1$–0.2 eV, which could also be defined as narrow gap semiconductors. E_g can be experimentally determined by optical and transport methods. However, the experimental E_g values obtained by optical methods, E_{opt}, e.g., by means of absorption/reflection experiments, may differ from those derived

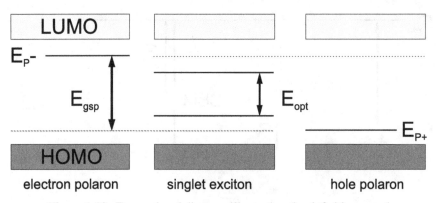

Figure 1.12. Energy band diagram illustrating the definitions used.

from transport methods, E_t, because of the formation of excitons (electron–hole pairs). E_{opt} is defined as the onset for optical absorption while E_t is the minimum energy required for the creation of a separated free electron and hole. Their difference is the exciton binding energy E_{ex} (binding energy between the generated electron and hole):

$$E_{ex} = E_t - E_{opt}. \tag{1.1}$$

In covalently bonded inorganic semiconductors, charge carriers are delocalized, polarization is rather small and carrier screening is efficient, leading to E_{ex} values of the order of a few meV and therefore $E_t \simeq E_{opt}$. However, in organic materials E_{ex} values are far larger owing to the localization of charges on individual molecules and large polarizabilities. For organic semiconductors E_{opt} corresponds to the formation of Frenkel (localized) excitons, with the electron and hole on the same molecule. When the electron and hole are localized at different molecules the bound state is called a localized charge transfer (CT) exciton. On the other hand charge transport is via hopping and involves polarons consisting of the carrier and its polarization cloud.

The combination of a hole polaron and an electron polaron, with binding energies E_{P+} and E_{P-}, respectively, results in the formation of an exciton. Their difference corresponds to E_t and is also referred to as the single particle energy gap E_{gsp}:

$$E_t = E_{gsp} = E_{P-} - E_{P+}. \tag{1.2}$$

Figure 1.12 illustrates these definitions. From the application and scientific points of view it is clear that knowledge of the electronic band diagram of organic semiconductors, and of any MOM in general, is mandatory.

Materials with large E_{ex} values would not be indicated for e.g., photovoltaic devices, since the photogeneration of charge carriers would be inefficient, but could in principle be used as light emitting diodes. Section 4.2 will show examples of the experimental determination of the band diagrams of selected organic semiconductors. E_t is usually obtained from σ vs. T measurements, since $\sigma \propto e^{-E_t/k_B T}$. However, the derived value for E_t is usually termed the activation energy E_a and may contain other contributions in addition to the intrinsic E_t value, such as those arising from the orientation and morphology of the sample. For this reason E_a is most commonly used. In the case of polycrystalline thin films, grain boundaries may have a dominant effect, as will be discussed in Section 6.4, where examples of intrinsically metallic materials behaving as semiconductors will be given.

Organic semiconductors can be arbitrarily divided into single component and multicomponent materials. Table 1.6 provides a list of electronic energies for single component materials in alphabetical order. A detailed analysis of the electronic processes in organic crystals and polymers can be found in Pope and Swenberg, (1999). Examples of binary or ternary component materials that become semiconducting in a given region of their phase-space will be given throughout this section. From the table we observe that the reported values of E_{ex} lie below 1.4 eV. However, some discrepancies are found when describing the same material, as in the case of Alq$_3$, where values as low as 0.2 eV and as large as 1.4 eV have been obtained. Such discrepancies are due to the different experimental methods used for the determination of E_{ex} and to different sample quality depending on the effective order range (degree of crystallinity, morphology, etc.). The presence of defects, especially in films grown from the vapour phase, plays a determinant role, particularly when localized charges are induced. Therefore, comparison of the obtained magnitudes is sometimes difficult. We will insist throughout the book that it is imperative to evaluate the quality of the samples (single crystals or films) before discussing intrinsic physical properties. In addition, the phenomenology of charge generation and transport is complex for organic materials, including polymers, and fundamental research has to be pursued.

Pentacene is the leading representative of the acenes family ($C_{4n+2}H_{2n+4}$), which are planar organic molecules consisting of n aromatic rings arranged linearly, displayed in the second column of Table 1.4. Pentacene exhibits high carrier mobility values, ~ 1 cm^2 s^{-1} V^{-1}, close to the values found for amorphous hydrogenated silicon (a-Si:H), making pentacene a good candidate to compete with silicon microelectronics for particular applications (see Table 6.1). Most of the MOMs exhibit low mobility values, $\sim 10^{-3}$ cm^2 s^{-1} V^{-1}, comparable e.g., to those of α- and β-N$_2$, $\sim 2 \times 10^{-3}$ cm^2 s^{-1} V^{-1} in the range $52 < T < 63$ K (Loveland *et al.*, 1972). The mobility values are reduced by the presence of grain boundaries and

Table 1.6. *Electronic energies of some relevant organic semiconductors and insulators*

Molecule	E_{es}	E_t	E_{opt}	E_{ex}	I_E	E_A
Alq₃	5.4[e]	3.0[l,o], 4.6[e]	2.7[l], 2.8[k], 3.2[e]	0.2[l], 1.4[e,g]	5.6[f], 5.8[d]	3.2
anthracene		4.1	3.9[m], 4.4[n]	1.0	5.8[m]	2.0[m]
BCP	4.7				6.4	
C₆₀	2.3[j]					
CBP	4.0				6.3[d]	
CuPc	3.1[e]	2.3[e], 0.3[a]	1.7[e]	0.6[e]	5.1	
DIP					5.8[b]	
F₄TCNQ					8.3[c]	5.2[c]
F₁₆CuPc		1.5[q]			6.3[q]	4.8
Gaq₃			2.7[o]		5.8[o]	
naphthalene	5.2		4.4[m]		6.8[m]	
NPB		3.3[l]	3.0[l]	0.3[l]	6.4	
α-NPD	5.3[e]	4.5[e]	3.5[e]	1.0[e]	5.4[d]	2.6
pentacene			2.2[m]		4.9[h], 5.1[m]	2.9[m]
perylene			2.9[m]		5.4[m]	2.5[m]
PTCDA	4.0[e]	3.2[e]	2.6[e]	0.6[e]	6.8[d]	4.6
tetracene	2.9		3.1[m]		5.4[m]	2.4[m]
TPD					5.1[f]	
TTT					4.3[f]	
ZnPc	1.9				5.3[c]	3.3[c]
p-6P	3.1				6.1[i]	
α-6T	4.2[e]	3.4[e]	3.0[e]	0.4[e]		

Note: All energies in eV. E_{es} is defined as the HOMO-LUMO peak-to-peak gap as determined by photoemission and inverse photoemission measurements. *References:* a, Alvarado *et al.*, 2001a; b, Dürr *et al.*, 2003a; c, Gao & Kahn, 2001; d, Hill *et al.*, 1998; e, Hill *et al.*, 2000; f, Ishii *et al.*, 1998; g, Knupfer *et al.*, 2001; h, Koch *et al.*, 2002; i, Koch *et al.*, 2003; j, Lof *et al.*, 1992; k, Martin *et al.*, 2000; l, Müller *et al.*, 2001; m, Pope & Swenberg, 1999; n, Sato *et al.*, 1987; o, Schlaf *et al.*, 1999; p, Schwieger *et al.*, 2001; q, Shen & Kahn, 2001.

in the high-temperature limit electrons tend to be driven over the barriers through thermal activation.

The materials p-6P, α-6T, metal Pcs and PTCDA also have interesting optical and transport properties, becoming serious candidates for applications in optical devices such as organic light emitting diodes (OLEDs). In particular α-6T has been used as the active material for organic field-effect transistors (OFETs) and will be discussed in Section 6.3. Pcs show intense absorption in the red and their appearance in the solid state ranges from dark blue to metallic bronze to green, depending on the central metal, the crystalline form and the particle size. They are widely used as colourants for plastics, inks and automobile paints. In addition, they have also been found to exhibit interesting optical, magnetic, catalytic, semiconductive and photoconductive properties (Law, 1993; McKeown, 1998). PTCDA is an electrically neutral pigment and consists of a perylene core with a delocalized π-electron system and two anhydride endgroups. The endgroups give rise to a permanent quadrupole moment with the positive charge situated around the centre of the molecule, and the negative charge around the endgroups.

Let us now briefly examine some aspects of their crystal structures. The crystal structure of pentacene grown from solution is shown in Figs. 1.13(a) and (b). This structure is triclinic and the projections along the c- and b-axes reveal the modified herringbone structure and the 2D molecular plane distribution, respectively. Vapour-grown pentacene exhibits a slightly different crystal structure and higher mobility values have been reported for this polymorph (Siegrist *et al.*, 2001).

Projections of the monoclinic β-phase of *p*-6P are shown in Figs. 1.13(c) and (d). As depicted in Fig. 1.13(c), the molecules are organized forming a herringbone pattern while the layered arrangement is shown in Fig. 1.13(d). The long axes of the planar molecules are aligned parallel to each other and tilted about 17 degrees relative to the layer normal.

The monoclinic arrangement of α-6T is very similar to that of *p*-6P, as observed from Figs. 1.13(e) and (f). In α-6T the molecules are also planar and the molecular long axis is at an angle of $23.5°$ with the a-axis. The most salient difference is the more pronounced dimerized character in α-6T.

C_{60}, originally termed buckminsterfullerene and nowadays called colloquially buckyballs or even footballene, is an extremely appealing and interesting molecule with truncated icosahedral structure (I_h point group), readily viewed as a football or a soccerball by replacing each vertex of such a ball by a carbon atom (Kroto *et al.*, 1985). Solid C_{60} is a semiconductor with a face-centred-cubic structure (space group $Fm\bar{3}m$) at RT, in which the molecules exhibit dynamic orientational disorder. C_{60} opened two main avenues of research soon after its discovery: doping-induced superconductivity and carbon nanotubes. Superconductivity will be discussed later and carbon nanotubes are a matter of intense research because of their extraordinary properties: metallic/semiconductor depending on helicity, larger stiffness, flexibility, chemical inertness, etc. (Dresselhaus *et al.*, 1996). Such nanotubes have allowed the preparation of single-molecule transistors, since nanotubes are very long 1D molecules (Postma *et al.*, 2001).

Our last example is $Mo_3S_7[(dmit)_2]_3$, a single-component material based on the trinuclear Mo_3S_7 cluster (with idealized C_{3v} symmetry) coordinated to three dithiolate ligands dmit (see Fig. 1.14(a)) (Llusar *et al.*, 2004). This solid exhibits sizeable intermolecular magnetic interactions together with a significant electron delocalization and contains, in addition, large open channels. Single crystals and powder samples of $Mo_3S_7[(dmit)_2]_3$ show $\sigma_{RT} \sim 20$ and $\sigma_{RT} \sim 1$ Ω^{-1} cm^{-1},

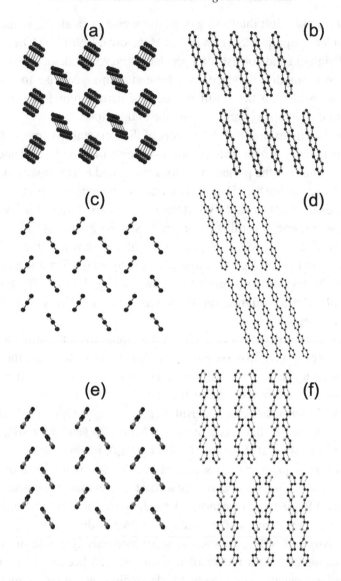

Figure 1.13. Crystal structure of solution-grown pentacene along (a) the c-axis and (b) the b-axis. $P\bar{1}$, $a = 0.790$ nm, $b = 0.606$ nm, $c = 1.601$ nm, $\alpha = 101.9°$, $\beta = 112.6°$, $\gamma = 85.8°$. Crystallographic data from Campbell *et al.*, 1961. See Table 1.10 for comments on the values of the lattice parameters. Crystal structure of *p*-6P: (c) *bc*-planes along their long molecular axis and (d) along the b-axis. $P2_1/c$, $a = 2.624$ nm, $b = 0.557$ nm, $c = 0.809$ nm, $\beta = 98.17°$. Crystallographic data from Baker *et al.*, 1993. Crystal structure of α-6T: (e) *bc*-plane and (f) projection along the c-axis. $P2_1/n$, $a = 4.471$ nm, $b = 0.785$ nm, $c = 0.603$ nm, $\beta = 90.76°$. Crystallographic data from Horowitz *et al.*, 1995.

(a) (b)

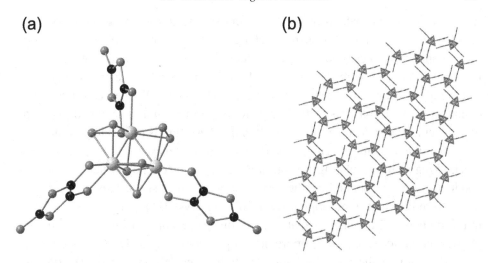

Figure 1.14. (a) $Mo_3S_7[(dmit)_2]_3$ cluster and (b) view of the crystal structure of the $Mo_3S_7[(dmit)_2]_3$ solid parallel to the (001) plane. C, S and Mo atoms are represented by black, medium grey and light grey balls, respectively. $P\bar{3}$, $a = 1.944$ nm, $c = 0.656$ nm. Crystallographic data from Llusar *et al.*, 2004.

respectively, and $E_a \sim 10$ and $E_a \sim 30$ meV, respectively. The crystal structure of $Mo_3S_7[(dmit)_2]_3$ is hexagonal and is shown in Fig. 1.14(b). The 3D packing results in cavities with a diameter of 1 nm. This new semiconductor material leads to the possibility of obtaining a series of new materials by chemically modifying the cluster complex, which are also nanostructured, thus enabling in principle the introduction of other molecules (guest) in their structure (host).

Organic metals

In 1911 H. N. McCoy and W. C. Moore predicted *that it is possible to prepare composite metallic substances from non-metallic constituent elements*, inspired by their results on the preparation of tetramethylammonium-mercury amalgams by cold electrolysis (McCoy & Moore, 1911). The prediction has come true, as will be evident in this section, but it was not until 1954 that a relatively stable perylene-bromine complex with conductivities up to 1 Ω^{-1} cm^{-1} was synthesized (Akamatu *et al.*, 1954). Although the material reported in this work behaved as a semiconductor with $E_a = 0.055$ eV, it can certainly be considered as the starting point of an extremely productive field leading to a wealth of conducting organic materials exhibiting remarkable results of fundamental interest. Since then, many perylene-based conductors have been prepared, which are exhaustively reviewed in Almeida and Henriques, 1997. The host of different conducting materials that have been synthesized essentially retain the structure of a purely organic molecule (perylene or others), which builds insulating solids, and an inorganic or organic counterpart, which induces transfer of charge and thus enables electrical transport. The organic molecules should possess delocalized electrons and the natural choice is planar molecules with p_π orbitals. The resulting materials can conduct electricity only if charge is transferred between both entities. We can distinguish CT complexes, where charge is partially transferred from a donor molecule to an acceptor molecule and radical cation salts (RCSs), where anions have integer oxidation states but donors behave formally as having mixed valency.

Thinking of the insulating neutral organic material (e.g., a perylene crystal), a conductor can be obtained by the introduction of other substances in the lattice (e.g., bromine), that is by doping the insulating crystal. Doping is the key issue in the synthesis of conductors that may become superconductors under particular physical conditions, hence its importance. Superconductors will be discussed in the next subsection. In the case of polymers, metallicity can also be obtained by doping. At this point we can differentiate four ways of doping: disordered, ordered, self (or internal) and interfacial. By disordered doping we mean that dopants do not form a regular lattice in the newly formed organic–inorganic hybrid, in analogy to doped inorganic semiconductors (Si, Ge, GaAs, etc.), while ordered doping will be used for hybrid materials consisting of two sublattices: one corresponding to the organic part and one corresponding to the inorganic part. Both lattices can have different lattice parameters and can be either commensurate or non-commensurate. Internal doping is a subtle concept and will be discussed below. Finally, interfacial doping corresponds to the direct injection of electrons or holes at an organic/insulator interface and will be discussed in Section 6.3. Later on in this chapter, in Section 1.7, we will face the problem of dimensionality vs. doping, that is how a quasi-1D metal can become 2D or vice versa.

The stoichiometric combination of TTF with TCNQ led to the discovery in 1973 of TTF-TCNQ (Ferraris *et al.*, 1973), the most representative organic conductor, which continues to be intensively studied as a reference system. TTF-TCNQ is a charge transfer salt (CTS) with $\varrho = 0.59$ electrons molec^{-1} exhibiting monoclinic crystal structure built up from parallel, segregated chains of TTF and TCNQ (see Fig. 1.15). A partial CT from the TTF HOMO to the TCNQ LUMO and the subsequent charge delocalization within the stacks lead to metallic conductivity. Within these segregated donor and acceptor columns the molecules do not lie directly on top of each other: there is a lateral displacement giving rise to a ring-over-bond overlap, where the exocyclic carbon–carbon double bond lies over the ring of the molecule adjacent to it in the stack. π-overlap along the stack (b-axis) is the origin of its 1D metallic behaviour. σ is highly anisotropic: $\sigma_b/\sigma_a \sim 10^3$ at RT, where σ_a and σ_b represent the conductivity along the a- and b-directions, respectively. σ_b rises from ~ 1000 Ω^{-1} cm^{-1} at RT to $>10^4$ at 60 K, but on further cooling a

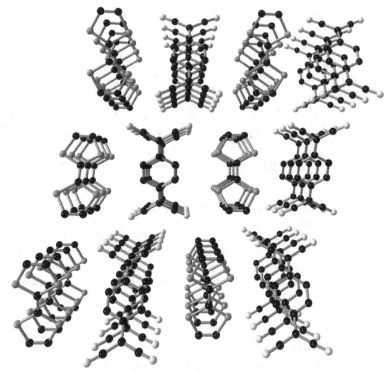

Figure 1.15. Crystal structure of TTF-TCNQ: perspective view along the stacking *b*-axis. $P2_1/c$, $a = 1.230$ nm, $b = 0.382$ nm, $c = 1.847$ nm, $\beta = 104.46°$. C, S and N atoms are represented by black, medium grey and light grey balls, respectively. H atoms are not represented for clarity. Crystallographic data from Kistenmacher *et al.*, 1974.

metal–semiconductor (also known as metal–insulator) transition occurs at *c.* 54 K, decreasing σ_b, which is interpreted in terms of a Peierls transition.

It soon became apparent that high conductivity is associated with crystal structures in which the donor and acceptor molecules form seggregated stacks, with considerable π-overlap and delocalization along the stacks. Therefore, research was focused on the synthesis of new donor and acceptor molecules with planar or near-planar geometries. For the TMTSF-TCNQ salt two crystallographic phases are known, with either mixed or segregated donor–acceptor stacks (see Fig. 1.16). The first one is semiconducting and the second is metallic, exhibiting a metal–insulator transition at 57 K (Jacobsen *et al.*, 1978), thus closely related to TTF-TCNQ.

TTF-TCNQ continues to be a unique compound because large single crystals can be grown, the crystals are stable in air and only one crystallographic phase is known (monoclinic). Concerning TTF and TCNQ molecules, both have similar size and identical weight (204 amu, TTF being $C_6S_4H_4$ and TCNQ, $C_{12}N_4H_4$), belong to the same point group D_{2h} and their ionization potential/electron affinity values

(a)   (b)

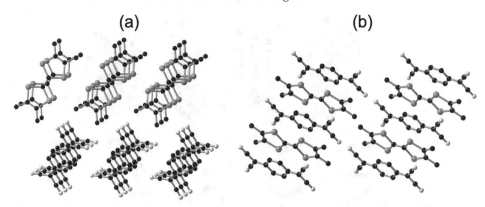

Figure 1.16. Crystal structures of TMTSF-TCNQ. (a) Black phase, $P\bar{1}$, $a =$ 0.388 nm, $b = 0.764$ nm, $c = 1.885$ nm, $\alpha = 77.34°$, $\beta = 89.67°$, $\gamma = 94.63°$. Crystallographic data from Bechgaard *et al.*, 1977. (b) Red phase, $P\bar{1}$, $a = 0.810$ nm, $b = 1.046$ nm, $c = 0.700$ nm, $\alpha = 103.78°$, $\beta = 98.49°$, $\gamma = 94.91°$. Crystallographic data from Kistenmacher *et al.*, 1982. C, S and N are represented by black, medium grey and light grey balls, respectively. H atoms are not represented for clarity.

favour incomplete CT. The TTF molecule is a good donor because its first I_E is low ($\simeq 7.0$ eV), identical to heptafulvalene, and TCNQ a good acceptor because of its low E_A value ($\simeq 2.8$ eV) (Klots *et al.*, 1974).

Table 1.7 summarizes some relevant organic conductors ordered by their σ_{RT} values at ambient pressure. HMTSF-TNAP exhibits to date the highest σ_{RT} value followed by HMTSF-TCNQ. HMTSF-TCNQ is isomorphous with HMTSF-TCNQF$_4$. However, HMTSF-TCNQF$_4$ is an insulator, with $\sigma_{RT} \sim 10^{-6}$ Ω^{-1} cm^{-1}, almost nine orders of magnitude lower that HMTSF-TCNQ. This clearly differentiated behaviour is related to the corresponding ϱ values. For HMTSF-TCNQ $\varrho = 0.74$, while for HMTSF-TCNQF$_4$ $\varrho = 1$, and therefore CT is complete, because TCNQF$_4$ is a far stronger electron acceptor than TCNQ, resulting in a fully ionic salt. For comparison let us mention that for *trans*-polyacetylene doped with AsF$_5$ $\sigma_{RT} \simeq$ 200 Ω^{-1} cm^{-1} (Chiang *et al.*, 1977).

It is interesting to note that the only *single component* molecular metal known to date is Ni(tmdt)$_2$, with $\sigma_{RT} \simeq 400$ Ω^{-1} cm^{-1} (Tanaka *et al.*, 2001). The possibility of obtaining such metals was predicted by theoretical work on the so-called two-bands systems by showing that electron transfer could be induced internally between two types of bands of the same component and that metal bis(dithiolene) molecules could lead to single-component molecular metals (Canadell, 1997). This is an example of internal doping.

(FA)$_2$PF$_6$ is also an interesting material because its structure strongly resembles that of the Bechgaard–Fabre salts (BFS), as will discussed next. It is composed of slightly dimerized donor FA stacks piled in a zig-zag manner along the *a*-direction

Table 1.7. *Material and σ_{RT} of selected highly conducting molecular organic metals*

Material	σ_{RT} [Ω^{-1} cm^{-1}]	Reference
HMTSF-TNAP	2400	Bechgaard *et al.*, 1978
HMTSF-TCNQ	1400–2200	Bloch *et al.*, 1975
(DMtTSF)$_2$AsF$_6$	1200–1500	Delhaes *et al.*, 1985
TTT-I$_{1.5}$	600–1200	Buravov *et al.*, 1976
TMTSF-TCNQ	1000	Jacobsen *et al.*, 1978
(FA)$_2$PF$_6$	1000	Ilakovac *et al.*, 1993
TTF-TCNQ	1900	Cohen *et al.*, 1974
(DTEDT)[ClO$_4$]$_{0.67}$	830	Misaki *et al.*, 1995
TSF-TCNQ	800	Bloch *et al.*, 1977
(BEDT-biTTF)-AuI$_2$	800	Iyoda *et al.*, 1999
TTP-I$_{1.28}$	800	Hilti *et al.*, 1981
(TMTSF)$_2$X	400–800	Bechgaard *et al.*, 1980
(perylene)$_2$Au(mnt)$_2$	700	Almeida & Henriques, 1997
TMTSF-DMTCNQ	400–600	Jacobsen *et al.*, 1978
NiPcI	500	Martinsen *et al.*, 1984
TTF-Br$_{0.79}$	400	Torrance *et al.*, 1979
Ni(tmdt)$_2$	400	Tanaka *et al.*, 2001
NMP-TCNQ	380	Coleman *et al.*, 1973b
θ-(BET-TTF)$_2$Br.3H$_2$O	350	Laukhina *et al.*, 2000
TTF[Ni(dmit)$_2$]$_2$	300	Brossard *et al.*, 1986
(TMTSF)$_2$ReO$_4$	300	Jacobsen *et al.*, 1982
(TMTTF)$_2$Br	240	Coulon *et al.*, 1982
K$_x$C$_{60}$	200	Hebard *et al.*, 1991
BTDMTTF-TCNQ	130	Rovira *et al.*, 1995
(TMTTF)$_2$PF$_6$	40	Coulon *et al.*, 1982
(TMTTF)$_2$ReO$_4$	33	Kobayashi *et al.*, 1984

Note: For copper $\sigma_{RT} \sim 6 \times 10^5$ Ω^{-1} cm^{-1}. For (TMTSF)$_2$X, X = PF$_6^-$, AsF$_6^-$, SbF$_6^-$, NO$_3^-$ and BF$_4^-$.

as shown in Fig. 1.17. (FA)$_2$PF$_6$ is a highly anisotropic 1D metal and exhibits a metal–insulator phase transition at about 180 K.

Organic superconductors

W. A. Little proposed in 1964 the possibility of synthesizing organic materials exhibiting no electrical resistance (Little, 1964). Based on the BCS theory, Little extended the electron pairing mechanism to describe the electrons moving along an organic polymer with highly polarizable side chains. Superconductivity has not yet been observed in doped conjugated polymers at least in part because the degree of crystallinity is not high enough and scattering through defects dominates. The

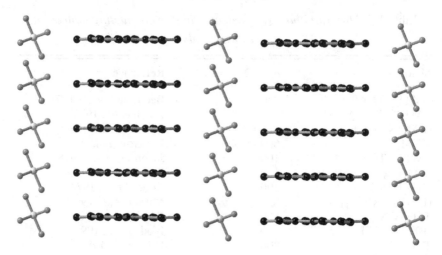

Figure 1.17. RT crystal structure of $(FA)_2PF_6$, $A2/m$, $a = 0.661$ nm, $b = 1.257$ nm, $c = 1.477$ nm, $\beta = 104.0°$. Crystallographic data from Enkelmann *et al.*, 1982. C, F and P atoms are represented by black, medium grey and light grey balls, respectively. H atoms are not represented for clarity.

only known polymeric system exhibiting a superconductor transition is crystalline $(SN)_x$ with a transition temperature $T_c = 0.26$ K (Greene *et al.*, 1975).

The possibility of growing sufficiently large highly pure and crystalline single crystals, e.g., by the well-known and widely used electrocrystallization (EC) technique (see Section 3.1), of MOMs based on small molecules has permitted the preparation of a long list of superconductors with $T_c < 15$ K. Table 1.8 summarizes some selected organic compounds exhibiting superconductivity. An exhaustive list of materials discovered up to 1998 can be found in Ishiguro *et al.*, 1998. Table 1.8 includes some of the most recent materials.

The first compound exhibiting a metal–superconductor transition was $(TMTSF)_2PF_6$, with $T_c \simeq 0.9$ K above $P \simeq 1.2$ GPa (Jérome *et al.*, 1980). In $(TMTSF)_2PF_6$ a sharp metal–insulator transition occurs at 15 K at ambient pressure, indicative of a Peierls transition (Bechgaard *et al.*, 1980). The structure of $(TMTSF)_2PF_6$ consists of nearly uniform stacks of TMTSF molecules ordered in sheets separated by anion sheets (see Fig. 1.18). $(TMTSF)_2PF_6$ is a representative of the more general isostructural $(TMTSF)_2X$ family of quasi-1D conductors, known as the Bechgaard salts, where X stands for monovalent anions such as AsF_6^-, ClO_4^-, ReO_4^-, etc. As mentioned above this structure is quite similar to that of $(FA)_2PF_6$ (see Fig. 1.17).

$(TMTSF)_2ClO_4$ exhibits $T_c \simeq 1.4$ K at $P = 0$ GPa, hence becoming the first organic superconductor at ambient pressure (Bechgaard *et al.*, 1981). The tetrahedral

Table 1.8. *Material, P and T_c of selected molecular organic superconductors classified according to the point group of the donor*

Material	P [GPa]	T_c [K]	Reference
D_{2h}			
$(TMTSF)_2PF_6$	1.2	0.9	Jérome *et al.*, 1980
$(TMTSF)_2ClO_4$	0	1.3	Bechgaard *et al.*, 1981
$(TMTTF)_2Br$	2.6	0.8	Balicas *et al.*, 1994
$(TMTTF)_2PF_6$	5.3	1.6	Adachi *et al.*, 2000
$(TMTTF)_2BF_4$	3.7	1.4	Auban-Senzier *et al.*, 2003
$(ET)_4(ReO_4)_2$	0.4	2.0	Parkin *et al.*, 1983a
α-$(ET)_2NH_4Hg(SCN)_4$	0	0.8	Wang *et al.*, 1990
β-$(ET)_2IBr_2$	0	2.7	Williams *et al.*, 1984
β_L-$(ET)_2I_3$	0	1.5	Yagubskii *et al.*, 1984
β_H-$(ET)_2I_3$	0	8.1	Creuzet *et al.*, 1985
β''-$(ET)_4Fe(C_2O_4)_3H_2O \cdot C_6H_5CN$	0	7.0	Kurmoo *et al.*, 1995
γ-$(ET)_3(I_3)_{2.5}$	0	2.5	Shibaeva *et al.*, 1985
κ-$(ET)_2Cu[N(CN)_2]Br$	0	11.6	Kini *et al.*, 1990
κ-$(ET)_2Cu[N(CN)_2]Cl$	0.03	12.8	Williams *et al.*, 1990
θ-$(ET)_2(I_3)_{1-x}(AuI_2)_x$ $(x < 0.02)$	0	3.6	Kobayashi *et al.*, 1986b
λ-$(BETS)_2GaCl_4$	0	6	Kobayashi *et al.*, 1997
λ-$(BETS)_2FeCl_4$	0	0.1	Uji *et al.*, 2001
κ-$(BETS)_2FeBr_4$	0	1.1	Ojima *et al.*, 1999
β-$(BDA-TTP)_2SbF_6$	0	7.5	Yamada *et al.*, 2001
β-$(BDA-TTP)_2AsF_6$	0	5.8	Yamada *et al.*, 2001
α-$TTF[Ni(dmit)_2]_2$	0.7	1.6	Brossard *et al.*, 1986
α'-$TTF[Pd(dmit)_2]_2$	2	6.5	Brossard *et al.*, 1988
C_{2v}			
$(DMET)_2AuCl_2$	0	0.8	Kikuchi *et al.*, 1987
$(DMET)_2AuI_2$	0.5	0.5	Kikuchi *et al.*, 1987
κ-$(MDT-TTF)_2AuI_2$	0	3.5	Papavassiliou *et al.*, 1988
$(MDT-TSF)(AuI_2)_{0.44}$	0	4.5	Kawamoto *et al.*, 2002
α-$(EDT-TTF)[Ni(dmit)_2]_2$	0	1.3	Inokuchi *et al.*, 1996
θ-$(DIETS)_2[Au(CN)_4]$	1.0	8.6	Imakubo *et al.*, 2002
C_s			
$(DTEDT)[Au(CN)_2]_{0.4}$	0	4.0	Misaki *et al.*, 1995

Note: The resistivity measurements for $\lambda - (BETS)_2FeCl_4$ were performed at $\boldsymbol{B} > 17\,T$, and for $\theta - (DIETS)_2[Au(CN)_4]$ were performed under uniaxial strain parallel to the crystallographic *c*-axis.

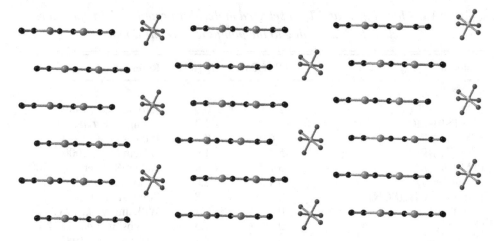

Figure 1.18. Crystal structure of the Bechgaard salt $(TMTSF)_2PF_6$. $P\bar{1}$, $a = 0.730$ nm, $b = 0.771$ nm, $c = 1.352$ nm, $\alpha = 83.39°$, $\beta = 86.27°$, $\gamma = 71.01°$. C, Se, F and P atoms are represented by black, medium grey, dark grey and light grey balls, respectively. H atoms are omitted. Crystallographic data from Thorup *et al.*, 1981.

ClO_4^- ion being smaller than the octahedral PF_6^- ion, the TMTSF sheets become closer to each other, *chemically* inducing the same effect as the *physical* hydrostatic pressure. When replacing TMTSF by TMTTF one obtains the isostructural Fabre salts.

Materials with the general formula $(TMTCF)_2X$, where C stands for chalcogens sulfur and selenium and X for monovalent anions, are known as Bechgaard–Fabre salts (BFS). Because of the extremely high quality of the BFS that can be achieved, these are, together with TTF-TCNQ and BEDT-TTF salts, the most extensively studied crystalline MOMs and a matter of intensive research.

The generic *T–P* phase diagram of the BFS is shown in Fig. 1.19. The pressure axis describes both the externally applied hydrostatic pressure P as well as the internal *chemical* pressure, where the selenium-based materials (TMTSF) would exhibit a larger pressure than the sulfur-based counterparts (TMTTF). The choice of the origin on the pressure axis, $(TMTTF)_2PF_6$, will become evident below. Let us start our discussion with the $(TMTTF)_2X$ materials, which are Mott-Hubbard (MH) insulators (semiconductors) at RT and ambient pressure. Increasing pressure at RT leads to a MH–metal (M) transition, thus in general $(TMTSF)_2X$ materials are metals at 300 K. The pressure axis can also be viewed as a decreasing electron localization axis.

We saw in Section 1.1 for N_2 that electron localization implies insulating ground states and that localization can be reduced by applying external pressure. When reducing temperature from RT down to about 20 K, $(TMTTF)_2X$ and $(TMTSF)_2X$

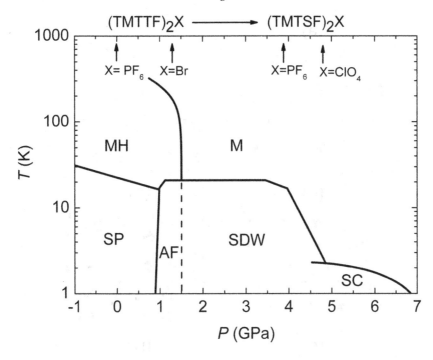

Figure 1.19. Generic *T–P* phase diagram for BFS. The origin on the pressure axis is arbitrarily set for (TMTTF)$_2$PF$_6$. MH, Mott–Hubbard; M, Metal; SP, Spin-Peierls; AF, Antiferromagnetic; SDW, Spin-Density-Wave; SC, Superconductor. Adapted from Auban-Senzier & Jérome, 2003.

salts undergo MH–Spin-Peierls (SP) and M–spin-density-wave (SDW) transitions. In both cases $E_t \neq 0$, which becomes more evident for the (TMTSF)$_2$X salts because of the metal to semiconductor transition. The most salient feature of this generic phase diagram is the existence of a wide variety of ground states below *c.* 20 K. (TMTTF)$_2$PF$_6$ is the only known system that can be driven through the entire series of ground states: SP, AF, SDW and SC by applying an external pressure, deserving selection as origin of the pressure axis in Fig. 1.19. The case of (TMTSF)$_2$ClO$_4$ is also important since it is the only material that becomes a superconductor at ambient pressure. The generic M region should in fact be divided into Fermi liquid and Luttinger liquid regions, a point that will be briefly discussed in Sections 1.7 and 6.1.

The series of 2D superconductors based on the BEDT-TTF molecule is extremely rich because of the large number of polymorphs and because, to date, they exhibit the highest T_c values with $T_c \simeq 12.8$ K for κ-(BEDT-TTF)$_2$Cu[N(CN)$_2$]Cl (see Table 1.8). Figure 1.20 shows different arrangements of the BEDT-TTF layers for the α, β, β', β'', γ, δ, κ and θ crystallographic phases. The α-phase exhibits a herringbone arrangement, similar to the γ-structure found in fused-ring aromatic hydrocarbons, as discussed in Section 1.4. The θ-phase can be regarded as

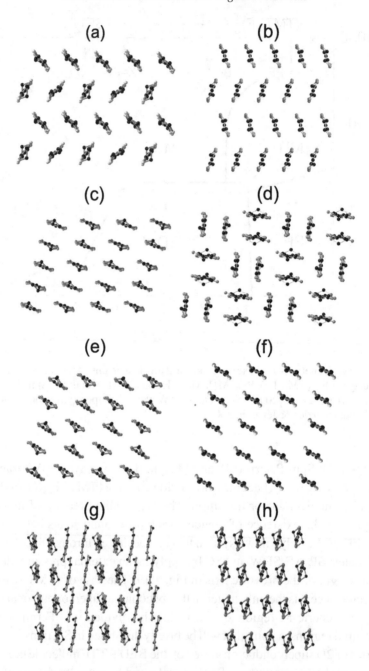

Figure 1.20. Perspective views of the BEDT-TTF layers for different crystallographic phases: (a) *ac*-plane of α-phase, (b) *ac*-plane of θ-phase, (c) *ab*-plane of β-phase, (d) *ac*-plane of κ-phase, (e) *ab*-plane of β'-phase, (f) *ab*-plane of β''-phase, (g) *ab*-plane of γ-phase and (h) *ac*-plane of δ-phase. C and S atoms are represented by black and medium grey balls, respectively.

a modification of the α-phase. In the β-phase the BEDT-TTF are packed face-to-face in a linear stack alignment while the β' and β'' modifications essentially differ from the β-structure in the inclination of the BEDT-TTF molecules with respect to the stacking axis. A slight dimerization is found in the β'-phase. While the γ and δ-phases are more exotic, the relevant κ-phase has a rather differentiated structure, formed by almost orthogonal dimers. The large conformational freedom of the ethylenedithio groups in BEDT-TTF favours lateral interactions between molecules and enhances polymorphism.

In contrast to the apparently endless list of polymorphs of the BEDT-TTF salts, only recently have the first polymorphs of the BFS been obtained with confined electrocrystallization (CEC), a modification of the EC technique. The origin of this remarkable difference is related to the topology of the TMTTF and BEDT-TTF molecules, with four and eight sulfur atoms per molecule, which favours the competing differences of the donor–donor and donor–anion intermolecular interactions, and the large conformational freedom of the ethylene groups together with the rather flexible molecular framework. The experimental conditions are so crucial, due to the weak interactions involved, that different T_c values can be obtained for the same highly perfect materials depending on e.g., the cooling procedure. For β-(BEDT-TTF)$_2$I$_3$ the L and H subscripts indicate low and high T_c phases, respectively. When samples are cooled from RT under atmospheric pressure the β_L phase is obtained. However, if a small hydrostatic pressure is applied during cooling down to about 10 K and then the subsequent cooling down to liquid helium temperatures is performed under atmospheric pressure, the β_H phase is obtained.

The first salt of the series was (BEDT-TTF)$_4$(ReO$_4$)$_2$, with $T_c \simeq 2$ K for $P > 0.4$ GPa (Parkin *et al.*, 1983a) and superconductivity under ambient pressure was first achieved for the β-(BEDT-TTF)$_2$I$_3$ compound with $T_c \simeq 1.4$ K (Yagubskii *et al.*, 1984). Figure 1.21(a) shows the crystal structure of β-(BEDT-TTF)$_2$IBr$_2$, where the BEDT-TTF molecules are aligned in a zig-zag stack forming sheets parallel to the *ab*-plane with the IBr$_2$ anions separating them by forming an insulating layer (Williams *et al.*, 1984). As discussed above, the molecules are non-planar, the angles formed between the planes containing the central carbon–carbon bond and the four neighbouring sulfur atoms (TTF-core) with the plane containing the four more external sulfur atoms (half TTF-core and ethylenedithia groups) are 168.2 and 173.1 degrees, respectively, thus larger than for the neutral BEDT-TTF crystals. BEDT-TTF molecules with oxidation state +1 exhibit a nearly flat conformation. The example of κ-(BEDT-TTF)$_4$PtCl$_6\cdot$C$_6$H$_5$CN is extremely interesting since in the RT structure BEDT-TTF is found in three different oxidation states: 0, +1/2 and +1 (Doublet *et al.*, 1994). In the case of κ-(BEDT-TTF)$_2$Cu[N(CN)$_2$]Br (see Fig. 1.21(b)) two BEDT-TTF molecules are paired with their central tetrathioethylene planes almost parallel, and adjacent pairs are almost perpendicular to one

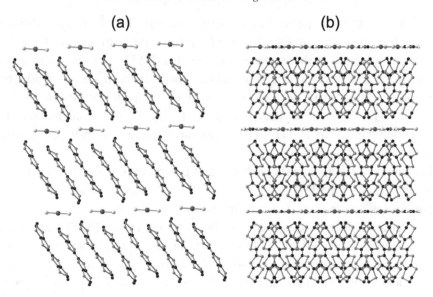

Figure 1.21. (a) Crystal structure of the β-(BEDT-TTF)$_2$IBr$_2$ salt, $P\bar{1}$, $a =$ 0.659 nm, $b = 0.897$ nm, $c = 1.509$ nm, $\alpha = 93.79°$, $\beta = 94.97°$, $\gamma = 110.54°$. Crystallographic data from Williams *et al.*, 1984. (b) Crystal structure of the κ-(BEDT-TTF)$_2$Cu[N(CN)$_2$]Br salt, *Pnma*, $a = 1.287$ nm, $b = 2.954$ nm, $c = 0.847$ nm. Crystallographic data from Geiser *et al.*, 1991. C and S atoms are represented by black and medium grey balls, respectively.

another in the *ac*-plane, resulting in a donor sheet. The insulating layer is formed by arrays of Cu[N(CN)$_2$]Br anions exhibiting a planar polymer-like structure (Geiser *et al.*, 1991). From these figures it can be again observed that the crystals are composed of well-defined organic/inorganic interfaces. Such interfaces can be chemically manipulated by the introduction of neutral guest molecules thus modifying the electronic structure, phase diagram and electronic instabilities (Deluzet *et al.*, 2002a) and will be discussed in Section 6.4.

Nowadays much effort is dedicated to the study of superconductors under higher hydrostatic pressures and under high magnetic fields. Above 5 GPa, the Fabre salt (TMTTF)$_2$PF$_6$ becomes a superconductor with $T_c \simeq 1.4–1.8$ K (Adachi *et al.*, 2000), being the first example of a superconducting BFS sharing the same anion. The case of λ-(BEDT-TSF)$_2$FeCl$_4$ is quite remarkable: upon application of magnetic fields exactly parallel to the conducting layers superconductivity is induced for fields $\boldsymbol{B} > 17$ T with $T_c \simeq 0.1$ K (Uji *et al.*, 2001). This experimentally observed field-induced superconducting state survives between 18 and 41 T (Balicas *et al.*, 2001), so that the well-known fact that the superconducting state is destroyed for sufficiently strong magnetic fields, the so-called critical field $\boldsymbol{B}_c$, is not in question. Let us conclude this description of relevant superconductors with the series of ambient-pressure superconductors β-(BDA-TTP)$_2$X, with X = SbF$_6^-$ and AsF$_6^-$, whose discovery has been crucial since it demonstrates that the TTF core is not essential for the production of organic superconductors (Yamada *et al.*, 2001).

Those readers not familiar with superconductivity in organic materials may find the T_c values rather low. However, they are comparable to values for inorganic metallic elements. Here is a list of some selected examples: $T_c(\text{Nb}) = 9.25$ K, $T_c(\text{Pb}) = 7.20$ K, $T_c(\alpha\text{-Hg}) = 4.15$ K, $T_c(\text{Sn}) = 3.72$ K, $T_c(\text{Al}) = 1.17$ K, $T_c(\text{Ti}) = 0.40$ K, etc. It is interesting to note that copper does not exhibit a superconducting transition. The highest known T_c values of any material correspond to the copper-oxide series with $T_c \simeq 138$ K as the absolute record for the thallium-doped mercury-cuprate compound.

These copper-oxide compounds crystallize in the perovskite structure and super-conductivity is based on the (hole or electron) doping in the copper-oxide planes. This is the reason why these materials can be regarded as being 2D. The first compound of the family was La$_{2-x}$Sr$_x$CuO$_4$ with $T_c \simeq 38$ K, which soon led to YBa$_2$Cu$_3$O$_{7-\delta}$ with $T_c \simeq 92$ K for $\delta < 1$ (Burns, 1993). The non-copper oxide electron-doped perovskite Ba$_{1-x}$K$_x$BiO$_3$ exhibits superconductivity near 30 K for $0.3 < x < 0.5$ (Cava *et al.*, 1988).

Solid C$_{60}$ is a semiconductor, as previously discussed, and becomes metallic under doping. Electron-doped Cs$_x$Rb$_y$C$_{60}$ has T_c as high as 33 K (Tanigaki *et al.*, 1991). The recently discovered MgB$_2$ superconductor, with $T_c \simeq 39$ K, deserves special attention (Nagamatsu *et al.*, 2001). MgB$_2$ is structurally and electronically related to graphite. The crystal structure of MgB$_2$ is shown in Figs. 1.22(a) and (b) and consists of alternating sheets of honeycomb boron layers and hexagonal magnesium layers.

Magnesium atoms donate electrons to the conduction band where no d-electrons are involved. The honeycomb planes of boron determine the electronic process. We thus have the formation of σ and π bands, as in graphite. The structure of graphite is depicted in Fig. 1.22(c) and consists of stacked planar sheets, in which the carbon atoms are covalently bound (σ, sp^2-hybridization) into a honeycomb lattice and the sheets are weakly bound through van der Waals interactions (the interlayer distance is $c/2 = 0.335$ nm). Perpendicular to the basal planes p_π-electrons are delocalized forming a conducting 2D system. As a result of its electronic structure, graphite is a semimetal exhibiting highly anisotropic conductivity. The high T_c value of MgB$_2$ is ascribed to the strong coupling between electrons in the 2D σ band and the optical E_{2g} phonon associated to in-plane motion of boron atoms. The energy of this phonon is $\hbar\omega_D \simeq 570$ meV and hence the high T_c value, since $T_c \propto \hbar\omega_D$.

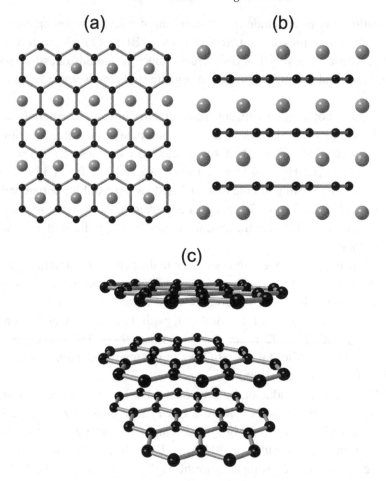

(a) **(b)**

(c)

Figure 1.22. View along the (a) *c*-axis and (b) *b*-axis of MgB_2. $P6/mmm$, $a = 0.309$ nm, $c = 0.352$ nm. B and Mg atoms are represented by black and medium grey balls, respectively. (c) Perspective view of graphite. $P6_3mc$, $a = 0.246$ nm, $c = 0.671$ nm.

Band structure calculations of MgB_2 show that the σ states are unfilled and hence metallic, whereas for graphite σ states are completely filled (An & Pickett, 2001). It is sometimes useful to consider the geometrical and electronic structure of graphite when studying MOMs, because it summarizes the main features: carbon-based, 2D and electrical conductivity given by π-electrons.

Graphite also becomes a superconductor when doped with alkali metals (Al-Jishi, 1983). The obtained materials are called graphite intercalation compounds (GICs), and C_8K, C_8Rb, C_8Cs and C_8RbHg, with $T_c < 2$ K, are some examples. All this recalls the crystal structures of the Bechgaard and BEDT-TTF-based salts, with alternating organic–inorganic planes and with CT. Therefore, the existence of organic superconductors should not surprise us. However, one has to be careful since the segregated layered arrangement does not necessarily imply that the electronic structures are equivalent.

Organic magnets

Under the general term molecule-based magnets we include solids exhibiting spontaneous magnetic ordering (i.e., for $B = 0$) below a critical temperature and consisting of molecular units (building blocks) bound together by ionic, covalent or van der Waals interactions. When such materials possess unpaired electron spins residing in p orbitals they are called organic molecule-based magnets or simply organic magnets. Organic magnets possessing only electron spins residing in p orbitals are called purely organic (metal-free) magnets. Note that classical inorganic magnets are based on unpaired spins localized in metal d or f orbitals. In this section we mention the most representative reported molecule-based magnets. Some recommended general references on this subject are Kahn, 1993; Gatteschi, 1994; Miller & Epstein, 1994; Day, 2002; Blundell & Pratt, 2004 and the series initiated by Miller & Drillon, 2001.

At this point one can argue whether single-molecule magnets, individual molecules with high spin values ($S \geq 10$), should be considered as molecule-based magnets. Such molecules are chemically synthesized and obtained in the form of single crystals and their properties derived from this aggregated state. The molecules are intentionally designed with a metallic core (e.g., Mn) surrounded by inert molecules, strongly reducing magnetic intermolecular interactions. In this sense, such solids are built from weakly interacting molecules, and cannot be considered as molecule-based magnets, because no collective effects are found (no spontaneous magnetic ordering); their properties are related to the individual molecules and not to the solid. Magnetic clusters such as $Mn_{12}O_{12}(CH_3COO)_{16}(H_2O)_4$ compounds, known as Mn_{12} clusters, exhibit a high-spin ground-state value ($S = 10$) because eight of the manganese ions are in the $+3$ oxidation state ($S = 2$, up) and four are in the $+4$ state ($S = 3/2$, down). These clusters undergo a very slow relaxation of the magnetization below the so-called blocking temperature (~ 4 K). This is different from paramagnetism, where the magnetization vanishes with the applied field (Sessoli *et al.*, 1993).

Let us briefly explore MOMs exhibiting transitions to a ferromagnetic state. The first genuine examples of bulk ferromagnetism in molecule-based crystals were reported for transition-metal cyanide complexes of the Prussian Blue type (Holden *et al.*, 1956). Prussian Blue, $Fe_4^{+3}[Fe^{+2}(CN)_6]_3 \cdot 14H_2O$, is itself a ferromagnet with $T_C \simeq 5.5$ K and it builds a 3D cubic network of alternating Fe^{+2} and Fe^{+3} bridged by CN groups. Prussian Blue is the first example of mixed-valence compounds. Following the definition of molecule-based magnets given above, Prussian Blue is clearly a member of this group. What may be disturbing is perhaps the term *molecule*, which tends to suggest that the building blocks should only be bound by weak van der Waals interactions. It is simply a semantic question. $[Cr_5(CN)_{12}]10H_2O$, which also belongs to the Prussian Blue family, containing three Cr^{+2} ($S = 2$) and two Cr^{+3} ($S = 3/2$) ions, orders as a bulk ferromagnet at 240 K (Mallah *et al.*, 1993).

The first molecule-based magnet, built from weakly interacting molecules, was $Fe^{+3}(S_2CNEt_2)_2Cl$, with $T_C \simeq 2.5$ K (Wickman *et al.*, 1967). $[Fe^{+3}(Cp^*)_2]$ [TCNE] is a ferromagnet with $T_C \simeq 4.8$ K (Miller *et al.*, 1988) while the 1D $[Fe(Cp^*)_2]TCNQ$ is a metamagnet. A metamagnet exhibits either antiferromagnetic or ferromagnetic behaviour depending on the magnitude of the applied external field B. Antiferromagnetic or ferromagnetic behaviour is found below 2.5 K for $B < 1500$ or $B > 1500$ Oe, respectively (Candela *et al.*, 1979).

As purely organic ferromagnets we cite the nitroxide radical materials *p*-NPNN and adam$(NO)_2$, the sulfur-nitrogen based radical *p*-NC·C_6F_4·CNSSN and finally TDAE-C_{60}. Solid *p*-NPNN exhibits four polymorphs: α, β, γ and δ. Projections of their crystal structures are presented in Fig. 1.23. The thermodynamically most stable β-phase undergoes a ferromagnetic transition ($T_C \simeq 0.6$ K) (Tamura *et al.*, 1991), while the γ-phase orders antiferromagnetically at $T_N \simeq 0.65$ K (Kinoshita, 1994). β-*p*-NPNN was the first purely organic material exhibiting bulk ferromagnetism. The paramagnetic α- and δ-phases show weak antiferromagnetic and sizable ferromagnetic intermolecular couplings, respectively. No magnetic ordering has been observed for either phase. The radical adam$(NO)_2$ orders ferromagnetically in the solid state below 1.48 K (Chiarelli *et al.*, 1993) and β-*p*-NC·C_6F_4·CNSSN shows weak ferromagnetism due to spin canting (non-collinear antiferromagnetism) at 35.5 K (Banister *et al.*, 1996). The fullerene-based CT compound TDAE-C_{60} has been described as a ferromagnet below 16 K (Allemand *et al.*, 1991).

The general definition of molecule-based magnets permits the inclusion of complex hybrid organic–inorganic materials, e.g., the layered compound [BEDT-TTF]$_3$[MnCr$(C_2O_4)_3$] (Coronado *et al.*, 2000), where ferromagnetism and metallic conductivity coexist induced by the well-differentiated organic and hybrid organic–inorganic sublattices. This material will be discussed in detail in Section 6.5 as an example of a material exhibiting two different physical properties.

Weak ferromagnetic order can also be obtained by what is termed spin frustration. Spin frustration can be understood by comparing the behaviour of magnetic spins located at the vertices of a square lattice with those on a triangular lattice. In antiferromagnetic materials magnetic spins tend to align antiparallel with their neighbours, which is possible if the spins are in a square lattice, but impossible for a triangular lattice because one of the spins in each triangle cannot align simultaneously with both its neighbours. In a Kagomé lattice (see Fig. 1.24) some of the spins remain frustrated and the material can thus develop weak long-range ferromagnetic

(a) (b)

(c) (d)

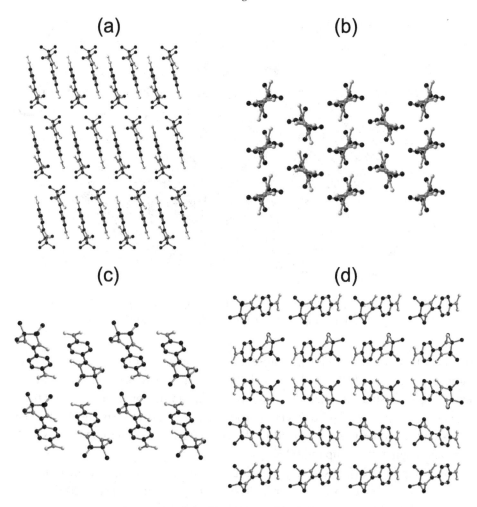

Figure 1.23. Crystal structure of *p*-NPNN. (a) α-phase projected along the *b*-axis, $P2_1/c$, $a = 0.730$ nm, $b = 0.762$ nm, $c = 2.468$ nm, $\beta = 93.62°$. Crystallographic data from Tamura *et al.*, 2003. (b) β-phase projected along the *c*-axis, $Fdd2$, $a = 1.096$ nm, $b = 1.935$ nm, $c = 1.235$ nm. Crystallographic data from Awaga *et al.*, 1989. (c) γ-phase projected along the *c*-axis, $P\bar{1}$, $a = 0.919$ nm, $b = 1.210$ nm, $c = 0.647$ nm, $\alpha = 97.35°$, $\beta = 104.44°$, $\gamma = 82.22°$. Crystallographic data from Turek *et al.*, 1991. (d) δ-phase projected along the *c*-axis, $P2_1/c$, $a = 0.896$ nm, $b = 2.380$ nm, $c = 0.673$ nm, $\beta = 104.25°$. Crystallographic data from Tamura *et al.*, 2003. C, N and O atoms are represented by black, medium grey and light grey balls, respectively. H atoms are not represented for clarity.

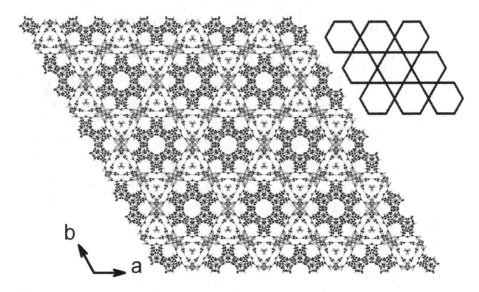

Figure 1.24. *ab*-plane of $[(Cu_2(C_5H_5N)_2(bdc)_2)_3]_n$. $P\bar{3}c1$, $a = b = 1.865$ nm, $c = 1.983$ nm. Crystallographic data from Moulton *et al.*, 2002.

order. $[(Cu_2(C_5H_5N)_2(bdc)_2)_3]_n$ represents the first example of a nanoscale Kagomé lattice (Moulton *et al.*, 2002). Kagomé lattices are attractive because they naturally lead to spin frustration when the system contains antiferromagnetic interactions. The crystal structure of $[(Cu_2(C_5H_5N)_2(bdc)_2)_3]_n$ is illustrated in Fig. 1.24 where the Kagomé lattice is schematized to the right for clarity. The material exhibits remnant magnetization up to 300 K.

An example of a MOM showing ordering to a ferrimagnetic state, with a transition temperature of about 2.5 K, is $[MnCp_2^*][Ni(dmit)_2]$ (Faulmann *et al.*, 2003).

Some MOMs containing transition metals exhibit the interesting property called spin crossover. In a spin crossover transition the central $3d$ atom (Cr, Mn, Fe, Co and Ni) changes from a high spin to a low spin state when the temperature is lowered. In addition to this thermally induced spin transition, light can also induce switching of one state to the other. This change in spin multiplicity is accompanied by modifications in the magnetic and optical properties as well as in the structure (Gütlich *et al.*, 1994). Obviously, the possibility of switching between two phases suggests potential applications in optical information technology.

1.6 Polymorphism

Polymorphism can be defined as the ability of an element, molecule or compound to crystallize in more than one distinct crystal structure. It is out of the scope of this book to give a lengthy description of polymorphism and in this section we will deal with a few concepts intimately related to molecular organic crystals. Those willing to learn more about this intriguing phenomenon are recommended to read Bernstein, 2002.

Polymorphism is very important for MOMs because of the weak intermolecular interactions involved. This implies that molecular crystals exhibit low cohesion energies, as previously discussed, and molecules may adopt different shapes, resulting in what is termed conformational polymorphism (Bernstein & Hagler, 1978). Torsion about single bonds defines the molecular conformation. The energy involved for torsion is *c.* 0.05 eV molec^{-1}, comparable to the energy difference between different crystalline forms. It thus becomes clear that for molecules which possess torsional degrees of freedom, various polymorphs are expected. The conformational degrees of freedom are thus intrinsic factors. The previously discussed cases of BEDT-TTF and *p*-NPNN molecules are good examples of conformational polymorphism because of the presence of mobile ethylenedithio groups and the torsion of the carbon–carbon bond connecting the nitrophenyl and nitronyl nitroxide groups, respectively. BEDT-TTF salts and *p*-NPNN exhibit several polymorphs, but for the vast majority of MOMs only one or two crystallographic phases have been *observed.*

In fact, the number of polymorphs of a given material can be larger than the number established to date. It mainly depends on the chosen synthesis route. Thus, many polymorphs are still waiting to be discovered! Extrinsic factors influencing the observation of polymorphs are more difficult to define and refer to the dimensionality of the parameter hyperspace. The mysterious phenomenon of disappearing polymorphs should serve as an example (Dunitz & Bernstein, 1995). Polymorphs of several materials have been *unambiguously* prepared and characterized, and for some unknown and uncontrollable factors have never been found again, despite longstanding efforts. Logic tells us that if they have been prepared at some stage it should always be possible to obtain them again, provided that the right experimental conditions are met again. It is of course a matter of time, dedication and systematic work, sometimes incompatible with the ever decreasing scientific time scale, urged by the absurd competition towards fast reporting of results. It is clear that the larger the dimensionality of the parameter hyperspace (the relevant external parameters involved), the larger the time needed for successful synthesis. This is further complicated by the small energy barriers between polymorphs, which means that small variations in e.g., temperature by few degrees makes the target synthetic route unviable. On the other hand, a given material, because of its extraordinary properties, becomes synthesized in a single way after its discovery leading to a unique crystallographic phase. However, when synthesized differently, new phases can be obtained. This is the case for the BFS. They have usually been obtained by standard EC in many laboratories around the world, leading to a unique stable crystallographic form (triclinic). However, using the CEC technique, a new crystallographic phase has been obtained (monoclinic), a subject that will be developed in detail in Section 6.4.

A delicate question directly concerns the phase diagrams of MOMs. As discussed in Section 1.5 these low cohesion energy solids are subject to quite extreme external conditions (high hydrostatic pressures, cryogenic temperatures, high magnetic fields, etc.). It would seem reasonable to think that the initial crystallographic phase (at ambient conditions) transforms, hopefully reversibly, into unknown phases during the measurements, since they are extremely sensitive to such changes. This point is, however, not always considered, in part because of the difficulty of obtaining the crystallographic structures under such external conditions. There are examples of experimentally determined crystallographic modifications as a function of temperature, e.g., TTF-CA that will be discussed in Section 6.4, illustrating the relationship between structure and physical property. Nowadays some research groups have become increasingly interested in determining the crystallographic phases of MOMs as a function of applied pressure.

Tables 1.9, 1.10 and 1.11 summarize some of the known polymorphs of selected MOMs, those discussed with a higher degree of dedication throughout this book. Some polymorphs may fail, but one of the stringent selection criteria has been the existence of reliable published crystallographic data. From Table 1.9 we observe that TTF exhibits two polymorphs, despite the relative rigidity of its molecular framework. However, for the closely related TMTTF molecule only one crystallographic phase is known. In Section 5.6 experimental hints of structural modifications of TMTTF, obtained as thin films, will be given. In spite of the two TTF modifications only one crystallographic phase is known for TTF-TCNQ, corresponding to segregated molecular stacking. However, under particular experimental conditions (as thin films grown from the vapour phase), TTF-TCNQ seems to order in an additional slightly different arrangement. Again, this will be discussed in Section 5.6.

Apparently, TCNQ seems to prefer to order in a single crystallographic phase but, for instance, the closely related terephthalic acid, consisting of a central C_6H_6 ring and two carboxylic acid groups in *para* positions, crystallizes in two distinct phases (Davey *et al.*, 1994). Surprisingly, while for neutral TMTSF only one phase is known, two phases have been determined for TMTSF-TCNQ, corresponding to different molecular stackings (segregated and mixed).

Two modifications, α and β, are known for PTCDA. Figure 1.25 shows a projection onto the (102) plane of α-PTCDA. The major difference between both crystallographic phases is the angle between the b-axis and the longer axis of the

Table 1.9. *Crystallographic phases of selected TTF- and TSF-based MOMs*

Material	Phase	Space group	a (nm)	b (nm)	c (nm)	α (°)	β (°)	γ (°)	Reference
TTF	α	$P2_1/c$	0.736	0.402	1.392		101.42		Cooper et al., 1971
	β	$P\bar{1}$	0.838	1.291	0.814	98.91	96.62	100.44	Ellern et al., 1994
TTF-CA	N	$P2_1/n$	0.740	0.762	1.459		99.10		Le Cointe et al., 1995
	I	Pn	0.719	0.754	1.444		98.60		Le Cointe et al., 1995
TTF-TCNE	α	$P2_1/n$	1.331	1.547	0.686		101.18		Clemente & Marzotto, 1996
	β	$P2_1/n$	1.543	0.663	1.416		104.32		Clemente & Marzotto, 1996
TTF-MeDC2TNF	α	$P2_1/n$	0.678	3.057	1.244		94.30		Salmerón-Valverde et al., 2003
	β	$C2/c$	1.525	1.269	1.379		104.35		Salmerón-Valverde et al., 2003
(TMTTF)$_2$IO$_4$		$P\bar{1}$	0.723	0.766	1.323	81.99	96.54	106.68	Yakushi et al., 1986
	μ'	$C2/c$	1.304	0.880	2.399		98.03		Perruchas et al., 2005
(TMTTF)$_2$ReO$_4$		$P\bar{1}$	0.717	0.762	1.323	86.62	95.56	108.20	Kobayashi et al., 1984
	μ'	$C2/c$	1.303	0.879	2.386		97.94		Perruchas et al., 2005
TMTSF-TCNQ	black	$P\bar{1}$	0.388	0.764	1.885	77.34	89.67	94.63	Bechgaard et al., 1977
	red	$P\bar{1}$	0.810	1.046	0.700	103.78	98.49	94.91	Kistenmacher et al., 1982
(TMTSF)$_2$ClO$_4$		$P\bar{1}$	0.727	0.768	1.327	84.58	86.73	70.43	Rindorf et al., 1982
	μ''	$C2/c$	1.206	0.879	2.639		95.30		Perruchas et al., 2005
(BEDT-TTF)$_2$I$_3$	α	$P\bar{1}$	0.921	1.085	1.749	96.95	97.97	90.75	Bender et al., 1984
	β	$P\bar{1}$	0.661	0.910	1.529	94.35	95.55	109.75	Hennig et al., 1985
	θ	$Pnma$	1.008	3.385	0.496				Kobayashi et al., 1986b
	κ	$P2_1/c$	1.639	0.847	1.283		108.56		Kobayashi et al., 1987

Note: N and I stand for neutral and ionic, respectively.

Table 1.10. *Crystallographic phases of selected semiconductor MOMs*

Material	Phase	Space group	a (nm)	b (nm)	c (nm)	α (°)	β (°)	γ (°)	Reference
Alq₃	α	P1̄	0.626	1.291	1.474	109.66	89.66	97.68	Brinkmann et al., 2000
	β	P1̄	0.844	1.025	1.317	108.58	97.06	89.74	Brinkmann et al., 2000
	γ	P3̄1c	1.441		0.622				Brinkmann et al., 2000
pentacene	δ	P1̄	1.324	1.442	0.618	88.55	95.92	113.93	Cölle et al., 2002
	S	P1̄	0.790	0.606	1.601	101.90	112.60	85.80	Campbell et al., 1961
	V	P1̄	0.625	0.779	1.451	76.65	87.50	84.60	Siegrist et al., 2001
perylene	α	P2₁/a	1.135	1.087	1.031		100.80		Donaldson et al., 1953
	β	P2₁/a	1.127	0.588	0.965		92.10		Tanaka, 1963
p-6P	β	P2₁/c	2.624	0.557	0.809		98.17		Baker et al., 1993
α-4T	LT	P2₁/c	0.608	0.786	3.048		91.81		Siegrist et al., 1998
	HT	P2₁/a	0.893	0.575	1.434		97.22		Siegrist et al., 1998
α-6T	LT	P2₁/n	4.471	0.785	0.603		90.76		Horowitz et al., 1995
	HT	P2₁/a	0.914	0.568	2.067		97.78		Siegrist et al., 1995
PTCDA	α	P2₁/c	0.374	1.196	1.734		98.80		Ogawa et al., 1999
	β	P2₁/c	0.378	1.930	1.077		83.60		Ogawa et al., 1999
H₂Pc	α	C2/c	2.614	0.381	2.397		91.10		Ashida et al., 1966
	β	P2₁/a	1.985	0.472	1.480		122.25		Robertson, 1936
	X	P2₁/a	1.063	2.315	0.489		95.90		Hammond et al., 1966
CuPc	α	P1̄	1.289	0.377	1.206	96.22	90.62	90.32	Hoshino et al., 2003
	β	P2₁/a	1.941	0.479	1.463		120.00		Brown, 1968
TiOPc	α (II)	P1̄	1.217	1.258	0.864	96.28	95.03	67.86	Hiller et al., 1982
	β (I)	P2₁/c	1.341	1.323	1.381		103.72		Hiller et al., 1982
	Y	P2₁/c	1.385	1.392	1.514		120.20		Brinkmann et al., 2002a

Note: S, solution-grown; V, vapour-grown; LT, low temperature and HT, high temperature. The reduced unit cell of S-pentacene is $a = 0.606$ nm, $b = 0.790$ nm, $c = 1.488$ nm, $\alpha = 96.74°$, $\beta = 100.54°$, $\gamma = 94.2°$ to be compared to the crystal structure of V-pentacene. *p*-6P exhibits a monoclinic modification at 110 K with $a = 2.628$ nm, $b = 1.100$ nm, $c = 1.600$ nm, $\beta = 99.79°$ (Baker *et al.*, 1993).

Table 1.11. *Crystallographic phases of selected MOMs exhibiting magnetic order*

Material	Phase	Space group	a (nm)	b (nm)	c (nm)	α (°)	β (°)	γ (°)	Reference
p-NPNN	α	$P2_1/c$	0.730	0.762	2.468		93.62		Tamura et al., 2003
	β	$Fdd2$	1.096	1.935	1.235				Awaga et al., 1989
	γ	$P\bar{1}$	0.919	1.210	0.647	97.35	104.44	82.22	Turek et al., 1991
	δ	$P2_1/c$	0.896	2.380	0.673		104.25		Tamura et al., 2003
p-NCC$_6$F$_4$CNSSN	α	$P\bar{1}$	0.757	0.806	0.951	65.73	69.17	67.52	Banister et al., 1995
	β	$Fdd2$	1.510	1.083	1.193				Banister et al., 1996
adam(NO)$_2$	α	$C2/c$	0.838	1.449	1.035		105.35		Dromzee et al., 1996
	β	$Pbcn$	1.116	1.157	0.968				Dromzee et al., 1996
TTTA	HT	$P2_1/c$	0.940	0.371	1.506		104.66		Fujita & Awaga, 1999
	LT	$P\bar{1}$	0.752	1.003	0.702	100.57	97.01	77.58	Fujita & Awaga, 1999

Note: LT and HT stand for low and high temperature, respectively.

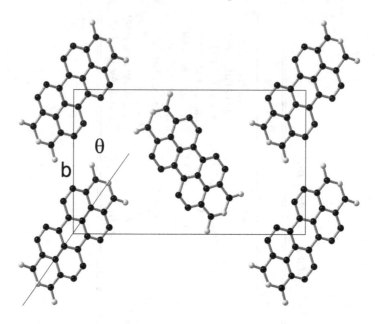

Figure 1.25. Projection onto the (102) plane of α-PTCDA. C and O atoms are represented by black and medium grey balls, respectively. H atoms are omitted for clarity. Crystallographic data from Ogawa *et al.*, 1999.

molecules. This angle is of 42 and 38 degrees for the α- and β-phases, respectively. Note that the molecules are slightly inclined and do not lie flat on the (102) plane.

Polymorphism is also a common phenomenon in Pcs. Among them, CuPc has been the most studied compound because of its stability and its extensive use in the dye industry. Five polymorphs, namely the α-, β-, γ-, δ- and ε-forms of CuPc are known. β-CuPc is the thermodynamically stable polymorph while the α-, γ-, δ- and ε-CuPc modifications are metastable and can be converted to β-CuPc upon heating or solvent recrystallization.

The stacking arrangements of CuPc molecules in the α- and β-polymorphs are depicted in Fig. 1.26. Two major differences distinguish both phases. First of all β-CuPc exhibits the γ- or L-shaped herringbone structure in clear opposition to the α-phase. On the other hand the inclination of the plane of the molecules is larger for the β-phase. Details of the stacking arrangements of δ-, γ-, and ε-CuPc are not available. In Section 3.2 we shall see the case of the M-CuPc polymorph, obtained as a thin film grown under microgravity.

1.7 Orbitals, bonds and bands

When studying the electronic structure of crystalline solids, physicists tend to think in terms of the mathematical concept of electronic bands, the so-called dispersion

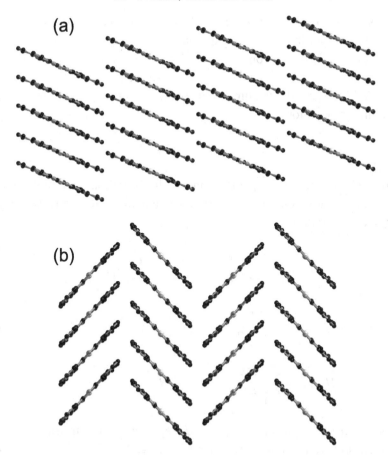

Figure 1.26. Details of the stacking of (a) α-CuPc (crystallographic data from Hoshino *et al.*, 2003) and (b) β-CuPc (crystallographic data as supplied by the Cambridge Crystallographic Data Centre). C atoms are represented by black balls.

relations, that is energy E as a function of the wave vector k: $E(k)$. Chemists, however, try to retain the orbital and bond signatures to the band structure. In the case of MOMs the chemists' approach seems more appropriate since MOs are only slightly perturbed by the formation of the solid, hence the electronic origin of the band can be unambiguously described in terms of MOs. This is the reason for the sometimes confusing assignation of point groups to bands, which should strictly be classified according to the space group. In what follows we try to set a common theoretical ground, based on the minimum required quantum-mechanical concepts, to describe the electronic structure of molecules and solids. In spite of the simplicity of the model, several fundamental concepts applicable to MOMs will be derived, so let me suggest, in particular to those not familiar with the field, a careful reading of this section.

Let us start by considering the general many-electron problem of N_e valence electrons, which contribute to chemical bonding, and N_{ion} ions, which contain the nuclei and the tightly bound core electrons. The positions of the electrons and ions are given by r_i and R_j, respectively, referred to the same arbitrary origin. This problem can be described quantum-mechanically, in the absence of external fields, by the Hamilton operator H_0:

$$H_0 = H_{ee} + H_{ion-ion} + H_{e-ion}, \qquad (1.3)$$

where H_{ee}, $H_{ion-ion}$ and H_{e-ion} correspond to the Hamilton operators concerning electron–electron, ion–ion and electron–ion interactions, respectively, which are given by the expressions:

$$H_{ee} = -\sum_{i=1}^{N_e} \frac{\hbar^2}{2m_e}\nabla_i^2 + \sum_{i>j} \frac{e^2}{|r_i - r_j|}, \qquad (1.4a)$$

$$H_{ion-ion} = -\sum_{j=1}^{N_{ion}} \frac{\hbar^2}{2M_j}\nabla_j^2 + \sum_{i>j} V_{ion-ion}(R_i - R_j), \qquad (1.4b)$$

$$H_{e-ion} = \sum_{i=1}^{N_e}\sum_{j=1}^{N_{ion}} V_{e-ion}(r_i - R_j). \qquad (1.4c)$$

In Eqs. (1.4a), (1.4b) and (1.4c) m_e and M_j represent the electron and ion masses, respectively, and ∇_i the Laplacian operator ($\nabla^2 \equiv \nabla \cdot \nabla = \partial^2/\partial x^2 + \partial^2/\partial y^2 + \partial^2/\partial z^2$). In Eq. (1.4a) we have inserted a Coulomb term for the electron–electron repulsive interactions and for $V_{ion-ion}$ and V_{e-ion}, the ion–ion and electron–ion interaction potentials, we leave open their explicit form but we assume that they can be described as sums over two-particle interactions.

Since electrons are much faster than nuclei, owing to $m_e \ll M_j$, ions can be considered as fixed and one can thus neglect the $H_{ion-ion}$ contribution (formally $H_{ion-ion} \ll H_{ee}$, where $V_{ion-ion}$ is a constant). This first approximation, as formulated by N. E. Born and J. R. Oppenheimer, reflects the instantaneous adaptation of electrons to atomic vibrations thus discarding any electron–phonon effects. Electron–phonon interactions can be a-posteriori included as a perturbation of the zero-order Hamiltonian H_0. This is particularly evident in the photoemission spectra of molecules in the gas phase, as already discussed in Section 1.1 for N_2^+, where the π_u^- state exhibits several lines separated by a constant quantized energy.

Equation (1.3) simply transforms to:

$$H_0 \simeq H_{ee} + H_{e-ion}. \qquad (1.5)$$

We should thus solve the Schrödinger equation:

$$H_0|\Psi^{Ne}\rangle = E|\Psi^{Ne}\rangle, \tag{1.6}$$

where $|\Psi^{Ne}\rangle$ and E represent the N_e-electron wave function (eigenfunction) and energy, respectively.

The second important approximation that enables the resolution of Eq. (1.6) consists in considering that every electron is subject to an effective interaction potential $V(r_i)$, which takes into account the full attractive electron–ion interactions as well as somehow a part of the repulsive electron–electron interactions. Ideally we would like to express H_0 in the form:

$$H_0 \simeq \sum_{i=1}^{N_e} \left\{ -\frac{\hbar^2}{2m_e}\nabla_i^2 + V(r_i) \right\} = \sum_{i=1}^{N_e} H_i^{1e}. \tag{1.7}$$

This is the *one-electron approximation*, also called the independent electron approximation and hence the 1e superscript, where a Hamiltonian H_0 of an N_e-electron system can be expressed as the sum of N_e one-electron H_i^{1e} Hamiltonians and the Schrödinger equation to be solved becomes:

$$H_i^{1e}|\Psi_i^{1e}\rangle = E|\Psi_i^{1e}\rangle. \tag{1.8}$$

Transforming Eq. (1.4a), which exhibits a $|r_i - r_j|$ dependence, at least partially into a $|r_i|$ dependence is not obvious and deserves special attention for, a priori, electron Coulomb repulsion cannot be ignored. The energy contribution from the repulsive Coulombic term will be represented by U. In transition metals and their oxides, electrons experience strong Coulombic repulsion due to spatial confinement in d and f orbitals. Spatial confinement and electronic correlations are closely related and because of the localization of electrons materials may become insulators.

One of the simplest models describing correlated electrons is the Hubbard model, which describes electrons with spins up and down moving between localized states at lattice sites i and j. In this model, electrons are allowed to interact only when they meet on the same lattice site i, keeping in mind that the Pauli exclusion principle requires them to have opposite sign. The kinetic energy and the interaction energy are characterized by the hopping term $t = W/4$ and U, respectively. These two terms compete because the kinetic part favours the electrons being as mobile as possible, while the interaction energy is minimal when electrons stay apart from each other, that is, they are localized on different atomic sites. The parameters that determine the properties described by the Hubbard model are the U/W ratio, the temperature and the doping. For $U/W \ll 1$ the kinetic energy term dominates and electrons are delocalized, representing highly conducting materials, while for $U/W > 1$ electronic correlations dominate leading to localized states and thus to insulating materials. Solids described by the latter case are known as Mott–Hubbard insulators. Equation (1.7) is fully justified in the $U/W \ll 1$ limit but when U becomes dominant, that is in the case of strongly correlated systems, other approximations have to be found. In what follows we will stay within the framework of the one-electron approximation.

One-electron approximation

A first attempt consists of assuming that each electron feels a smooth distribution of negative charge with a charge density ρ_c arising from the remaining $N_e - 1$ electrons. In this case Eq. (1.4a) transforms into:

$$H_{ee} = -\sum_{i=1}^{N_e} \frac{\hbar^2}{2m_e} \nabla_i^2 - e \int dr' \frac{\rho_c(r')}{|r - r'|}, \tag{1.9}$$

where $\rho_c(r) = -e \sum_{i=1}^{N_e} \langle \Psi_i^{1e}(r)|\Psi_i^{1e}(r)\rangle$. H_i^{1e} from Eq. (1.8) thus becomes:

$$H_i^{1e} = -\frac{\hbar^2}{2m_e} \nabla_i^2 + \sum_{j=1}^{N_{ion}} V_{e-ion}(r_i - R_j)$$

$$+ \left\{ e^2 \sum_{j=1}^{N_e} \int dr' \frac{\langle \Psi_j^{1e}(r')|\Psi_j^{1e}(r')\rangle}{|r - r'|} \right\}. \tag{1.10}$$

Equation (1.10) represents the Hartree Hamiltonian and Eq. (1.8) has to be solved by iteration, in the sense that a guessed trial wave function $|\Psi_i^{1e}\rangle$ is introduced in Eq. (1.10) and the Schrödinger equation Eq. (1.8) solved. The resulting wave function is again introduced in Eq. (1.10) and Eq. (1.8) is again solved until self-consistency is achieved.

When $|\Psi^{Ne}\rangle$ is expressed as the antisymmetric combination given by the Slater determinant:

$$|\Psi^{Ne}\rangle = \begin{vmatrix} |\Psi_1^{1e}(r_1)\rangle & |\Psi_1^{1e}(r_2)\rangle & \cdots & |\Psi_1^{1e}(r_{N_e})\rangle \\ |\Psi_2^{1e}(r_1)\rangle & |\Psi_2^{1e}(r_2)\rangle & \cdots & |\Psi_2^{1e}(r_{N_e})\rangle \\ \vdots & \vdots & \vdots & \vdots \\ |\Psi_{N_e}^{1e}(r_1)\rangle & |\Psi_{N_e}^{1e}(r_2)\rangle & \cdots & |\Psi_{N_e}^{1e}(r_{N_e})\rangle \end{vmatrix}, \tag{1.11}$$

an extra term is added to Eq. (1.10), known as the exchange term. The Schrödinger equation thus becomes:

$$H_i^{1e}|\Psi_i^{1e}(r)\rangle = \left\{ -\frac{\hbar^2}{2m_e}\nabla_i^2 + \sum_{j=1}^{N_{ion}} V_{e-ion}(r_i - R_j) \right\} |\Psi_i^{1e}(r)\rangle$$
$$+ \left\{ e^2 \sum_{j=1}^{N_e} \int dr' \frac{\langle \Psi_j^{1e}(r')|\Psi_j^{1e}(r')\rangle}{|r - r'|} \right\} |\Psi_i^{1e}(r)\rangle$$
$$- e^2 \sum_{j=1}^{N_e} \int dr' \frac{\langle \Psi_j^{1e}(r')|\Psi_j^{1e}(r)\rangle}{|r - r'|} |\Psi_i^{1e}(r')\rangle. \qquad (1.12)$$

Note that the exchange term is of the form $\int V(r, r')\Psi(r')dr'$ instead of the $V(r)\Psi(r)$ type. Equation (1.12), known as the Hartree–Fock equation, is intractable except for the free-electron gas case. Hence the interest in sticking to the conceptually simple free-electron case as the basis for solving the more realistic case of electrons in periodic potentials. The question is how far can this approximation be driven. Landau's approach, known as the Fermi liquid theory, establishes that the electron–electron interactions do not appear to invalidate the one-electron picture, even when such interactions are strong, provided that the levels involved are located within $k_B T$ of E_F. For metals, electrons are distributed close to E_F according to the Fermi function $f(E)$:

$$f(E) = \frac{1}{1 + e^{(E-E_F)/k_B T}}. \qquad (1.13)$$

A plot of $f(E)$ is shown in Fig. 1.27. E_F is usually defined as the chemical potential in the $T \to 0$ limit.

However, if we consider strong electronic correlations it looks doubtful that the simplified one-electron approximation can be applied at all. To circumvent this conceptual problem Landau proposed the bright idea of substituting electrons by something closely related, quasi-electrons (in general particles will be substituted by quasi-particles). In practice, this theoretical idea, so common in particle physics, transforms the energy expression of a free electron, $E = \hbar^2 k^2/2m_e$ into $E = \hbar^2 k^2/2m_e^*$, where m_e^* represents the effective mass, which is a function of k: $m_e^*(k)$. In other words, the non-interacting electron case is maintained but the energy scale is *renormalized*. The main goal is to preserve the free-electron model because we can handle it analytically. This approach will be limited to the case where correlations can be treated as perturbations.

So far we have studied the general many-electron problem of N_e electrons and N_{ion} fixed or very slow ions. Let us now apply the above-developed formalism to molecules and solids.

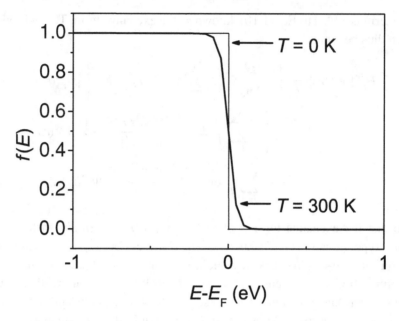

Figure 1.27. Plot of the Fermi distribution function $f(E)$ for $T = 0$ and $T = 300$ K.

Molecules

Let us consider the case of molecules with N_{ion} small (N_{ion} typically less than 100) and where there is no long range order in the sense that any eventual periodicity would be of the order of the molecular shape. In the case of molecules, the one-electron wave functions $|\Psi^{1e}\rangle$ represent MOs and can be simply expressed as a linear combination of N_{at} atomic orbitals $|\psi^{at}\rangle$ (LCAO) of different atoms forming the molecule:

$$|\Psi^{1e}\rangle = \sum_{j=1}^{N_{at}} c_j |\psi_j^{at}\rangle, \qquad (1.14)$$

where the coefficients c_j have to be determined. Introducing Eq. (1.14) into Eq. (1.8) gives rise to:

$$\sum_{j=1}^{N_{at}} c_j \left\{ H^{1e}|\psi_j^{at}\rangle - E|\psi_j^{at}\rangle \right\} = 0, \qquad (1.15)$$

and multiplying Eq. (1.15) to the left by $\langle\psi_i^{at}|$, the complex conjugate of $|\psi_i^{at}\rangle$, we obtain:

$$\sum_{j=1}^{N_{at}} c_j \left\{ \langle\psi_i^{at}|H^{1e}|\psi_j^{at}\rangle - E\langle\psi_i^{at}|\psi_j^{at}\rangle \right\} = 0. \qquad (1.16)$$

By defining:

$$H_{ij} = \langle \psi_i^{at} | H^{le} | \psi_j^{at} \rangle, \tag{1.17a}$$

$$S_{ij} = \langle \psi_i^{at} | \psi_j^{at} \rangle, \tag{1.17b}$$

the secular Eq. (1.16) is simplified to the expression:

$$\sum_{j=1}^{N_{at}} c_j \{ H_{ij} - E S_{ij} \} = 0. \tag{1.18}$$

Therefore, with the LCAO approximation, Eq. (1.8) transforms to a system of N_{at} equations with N_{at} unknown parameters c_j. The resolution of Eq. (1.18) implies that the determinant of the $\{ H_{ij} - E S_{ij} \}$ matrix has to be zero, otherwise we would obtain the trivial solution $c_j = 0 \; \forall j$, which obviously has no physical meaning. Therefore,

$$\begin{vmatrix} H_{11} - E S_{11} & H_{12} - E S_{12} & \cdots & H_{1N_{at}} - E S_{1N_{at}} \\ H_{21} - E S_{21} & H_{22} - E S_{22} & \cdots & H_{2N_{at}} - E S_{2N_{at}} \\ \vdots & \vdots & \vdots & \vdots \\ H_{N_{at}1} - E S_{N_{at}1} & H_{N_{at}2} - E S_{N_{at}2} & \cdots & H_{N_{at}N_{at}} - E S_{N_{at}N_{at}} \end{vmatrix} = 0. \tag{1.19}$$

In conclusion, the energies E that satisfy Eq. (1.19) are associated to molecular electronic states. Since Eq. (1.19) is an equation of N_{at} order, we obtain N_{at} energy values E_l ($l = 1, \ldots, N_{at}$), that is, as many molecular levels as atomic orbitals. In the simple example of H_2 discussed in Section 1.1, $N_{at} = 2$ and both $1s$ atomic orbitals combine to form bonding σ_g and antibonding σ_u MOs. In the case of N_2 (see Fig. 1.1), neglecting $1s$ core electrons, the combination of two sp and one p_z atomic orbitals per N atom leads to six MOs.

The easiest way to calculate the terms H_{ij} and S_{ij} is within the simple Hückel approximation, where it is assumed that:

$$H_{ii} = (E_\alpha)_i, \tag{1.20a}$$

$$H_{ij} = (E_\beta)_{ij} \qquad j = i \pm 1, \tag{1.20b}$$

$$S_{ij} = \delta_{ij}, \tag{1.20c}$$

where δ_{ij} stands for the Kronecker delta function ($\delta = 1$ for $i = j$ and $\delta = 0$ for $i \neq j$). Note that for $i \neq j$ $H_{ij} = 0$ unless the ith and jth orbitals are on adjacent atoms (nearest neighbours). The term $(E_\alpha)_i$ corresponds to the energy of the atomic orbital $|\psi_i^{at}\rangle$ and $(E_\beta)_{ij}$ represents the nearest-neighbour resonance integrals. This extremely simplified method works surprisingly well for small systems and has the advantage that it permits the resolution of the secular equation by hand and that enables a realistic evaluation of more complex systems. A more elaborated approach is called the extended Hückel method, which allows the determination of H_{ij} for $i \neq j$ from the following expression:

$$H_{ij} = 1.75 S_{ij} \frac{H_{ii} + H_{jj}}{2}. \tag{1.21}$$

Some relevant references concerning the application of the Hückel method are Hoffmann, 1963; Cotton, 1971; Iung & Canadell, 1997.

Periodic solids

In principle we could apply the methodology described in the previous section for very large molecules and solids, where $N_{at} \rightarrow \infty$, but in this case the resolution of the secular determinant given in Eq. (1.19) would be impossible. Some boundaries have to be given in order to make the problem manageable and the most elegant way is by taking symmetry into account. An infinite solid can be quite simply described if it exhibits real-space long-range periodicity (period negligible compared with the crystal dimensions), remaining invariant under primitive or lattice translations T_{R_n} of vectors $R_n = n_1 a_1 + n_2 a_2 + n_3 a_3$, where n_1, n_2 and $n_3 \in \mathbb{Z}$ and a_1, a_2 and a_3 are non-coplanar basis vectors. The set of all R_n vectors leads to all equivalent points in the lattice. In an infinite crystal with a real-space period R_n, the application of T_{R_n} would transform the crystal into itself, so that any physical property should be invariant under such operation. The effective interaction potential $V(r_i)$ must thus be periodic:

$$T_{R_n} V(r_i) = V(r_i + R_n) = V(r_i), \tag{1.22}$$

and as a consequence the Hamilton and the lattice translation operator T_{R_n} commute:

$$[H^{le}, T_{R_n}] = H^{le} T_{R_n} - T_{R_n} H^{le} = 0. \tag{1.23}$$

Hence, applying the T_{R_n} operator to the left of Eq. (1.8) we obtain:

$$T_{R_n} H^{le} |\Psi^{le}\rangle = T_{R_n} E |\Psi^{le}\rangle = H^{le} T_{R_n} |\Psi^{le}\rangle = E T_{R_n} |\Psi^{le}\rangle. \tag{1.24}$$

In conclusion, the invariance under lattice translations means that if $|\Psi^{le}\rangle$ is a wave function of the Hamilton operator, $T_{R_n} |\Psi^{le}\rangle$ is a solution as well. It can be shown that if two operators commute they have a common orthonormal basis of wave functions. In the case of the Hamilton and lattice translation operator this common base is given by the Bloch functions $|\phi(k, r)\rangle$:

$$T_{R_n} |\phi(k, r)\rangle = |\phi(k, r + R_n)\rangle = e^{ikR_n} |\phi(k, r)\rangle. \tag{1.25}$$

Bloch's theorem establishes that the eigenstates of the one-electron Hamiltonian H^{le}, where $V(r_i) = V(r_i + R_n)$ for all R_n in a Bravais lattice, can be chosen to have the form of a plane wave times a function, u, with the periodicity of the Bravais lattice:

$$|\phi(\boldsymbol{k}, \boldsymbol{r})\rangle = e^{i\boldsymbol{k}\boldsymbol{r}} u(\boldsymbol{k}, \boldsymbol{r}), \tag{1.26}$$

where $u(\boldsymbol{k}, \boldsymbol{r} + \boldsymbol{R}_n) = u(\boldsymbol{k}, \boldsymbol{r})$ for all $\boldsymbol{R}_n$. Equation (1.25) is readily obtained from Eq. (1.26) since $T_{\boldsymbol{R}_n}|\phi(\boldsymbol{k}, \boldsymbol{r})\rangle = |\phi(\boldsymbol{k}, \boldsymbol{r} + \boldsymbol{R}_n)\rangle = e^{i\boldsymbol{k}(\boldsymbol{r}+\boldsymbol{R}_n)} u(\boldsymbol{k}, \boldsymbol{r} + \boldsymbol{R}_n) = e^{i\boldsymbol{k}\boldsymbol{r}} u(\boldsymbol{k}, \boldsymbol{r}) e^{i\boldsymbol{k}\boldsymbol{R}_n} = e^{i\boldsymbol{k}\boldsymbol{R}_n}|\phi(\boldsymbol{k}, \boldsymbol{r})\rangle$. For a free electron, $V = 0$, $|\phi(\boldsymbol{k}, \boldsymbol{r})\rangle = e^{i\boldsymbol{k}\boldsymbol{r}}$ and $E = \hbar^2 k^2 / 2m_e$.

If we assume that the Bloch functions have periodic boundary conditions:

$$\left|\phi\left(\boldsymbol{k}, \boldsymbol{r} - \frac{\boldsymbol{R}_N}{2}\right)\right\rangle = \left|\phi\left(\boldsymbol{k}, \boldsymbol{r} + \frac{\boldsymbol{R}_N}{2}\right)\right\rangle, \tag{1.27}$$

where $\boldsymbol{R}_N = N_1\boldsymbol{a}_1 + N_2\boldsymbol{a}_2 + N_3\boldsymbol{a}_3$ represents the dimensions of the crystal modelled by a $N_1\boldsymbol{a}_1 \times N_2\boldsymbol{a}_2 \times N_3\boldsymbol{a}_3$ parallelepiped with the lattice origin in its centre (N_1, N_2 and N_3 are large numbers). It can be readily shown by using Eq. (1.26) that $N_1 k_1 a_1 + N_2 k_2 a_2 + N_3 k_3 a_3 = 2\pi p$, where $p \in \mathbb{Z}$ ($p = 0, \pm 1, \pm 2, \ldots, \pm N_1 N_2 N_3 / 2$). For a 1D system (e.g., $N_1 > 1$, $N_2 = N_3 = 1$), $k_1 = (2p\pi)/(a_1 N_1)$, where $p = 0, \pm 1, \pm 2, \ldots, \pm N_1/2$. The k_i values are confined to the $[-\pi/a_i, \pi/a_i]$ interval. $k = \pm\pi/a_i$ are called the zone boundaries or critical points. Within the periodic boundary conditions, known as the Born–von Karman approximation, the k values are discrete but the separation between two consecutive values is so small that it can be considered as continuous. For a 1D system the Born–von Karman approximation makes a periodic linear chain equivalent to a cyclic system. This finite and unbounded 1D system is said to be topologically connected and compact. A translation of a vector $p\boldsymbol{a}_1$ is equivalent to a rotation of angle $2\pi p/N_1$ in the interval $[-\pi, \pi]$.

We further proceed, as previously, by building a one-electron wave function from the associated atomic orbitals, Eq. (1.14), but imposing the Bloch condition expressed in Eq. (1.25). This is the well-known TB approximation (Ashcroft & Mermin, 1976). The resulting wave function is often called the crystal orbital and has the form:

$$|\Phi^{1e}(\boldsymbol{k}, \boldsymbol{r})\rangle = \sum_{\mu=1}^{N_{at}} c_\mu |\phi_\mu^{1e}(\boldsymbol{k}, \boldsymbol{r})\rangle, \tag{1.28}$$

where

$$|\phi_\mu^{1e}(\boldsymbol{k}, \boldsymbol{r})\rangle = \frac{1}{\sqrt{N_{ion}}} \sum_{j=1}^{N_{ion}} e^{i\boldsymbol{k}\boldsymbol{R}_j} |\psi_\mu^{at}(\boldsymbol{r} - \boldsymbol{R}_j)\rangle. \tag{1.29}$$

N_{at} stands for the number of atomic orbitals for each lattice. The resolution of the Schrödinger equation and the derivation of the secular determinant can be done as previously simply by substituting $|\psi^{at}\rangle$ by $|\phi^{le}(k)\rangle$ and $|\Psi^{le}\rangle$ by $|\Phi^{le}(k)\rangle$. Equations (1.17a) and (1.17b) thus become:

$$H_{\mu\nu} = \langle \phi_{\mu}^{le}(k)|H^{le}|\phi_{\nu}^{le}(k)\rangle, \tag{1.30a}$$

$$S_{\mu\nu} = \langle \phi_{\mu}^{le}(k)|\phi_{\nu}^{le}(k)\rangle, \tag{1.30b}$$

respectively, where $S_{\mu\nu} = \delta_{\mu\nu}$ and the determinant of Eq. (1.19) has exactly the same expression. The initial utopic problem of solving a system of equations of $N_{ion} \times N_{ion}$ order has reduced to an $N_{at} \times N_{at}$ system. Again, since Eq. (1.19) is an equation of N_{at} order, we obtain N_{at} energy values E_l ($l = 1, \ldots, N_{at}$), as many as atomic orbitals in the unit cell.

Note the k-dependence of Eqs. (1.30a) and (1.30b) with respect to Eqs. (1.17a) and (1.17b) providing an extra index to l for the allowed energies. For each l value the energy will be a function of k, $E_l(k)$, and thus has the possibility of exhibiting different values. The dispersion relation $E_l(k)$ is called the electronic band structure. It can be shown (Madelung, 1978) that the $E_l(k)$ function is continuous and differentiable in k, symmetric with respect to $k = 0$ and, finally, $E_l(k) = E_l(k + G)$.[4] These properties imply that a given E_l should have a maximum and a minimum value and that $\partial E_l / \partial k = 0$ for some k, at least at zone boundaries. All this is telling us that periodicity has generated energy bands, which should have near-sinusoidal shape, since Bloch functions are built as Fourier series.[5] Because of the last two properties $E_l(k)$ may be restricted to the positive part of the first Brillouin zone $[0, \pi/a_i]$.

Let us see all the information we can extract from the most simple infinite periodic system: a linear chain of equally spaced atoms, separated by a distance a (see Fig. 1.28(a)). Let us further assume for simplicity that each atom contributes one orbital ($N_{at} = 1$). Equation (1.28) becomes:

$$|\Phi^{le}(k, x)\rangle = |\phi^{le}(k, x)\rangle = \frac{1}{\sqrt{N_{ion}}} \sum_{n=1}^{N_{ion}} e^{ikna}|\psi^{at}(x - (n-1)a)\rangle. \tag{1.31}$$

The analytical expression of the energy can be calculated from Eq. (1.19). Since $N_{at} = 1$ we readily obtain $E(k) = H_{11}$ since $S_{11} = 1$ because of the orthonormality

---

[4] G is defined in reciprocal space as $G_m = m_1\mathbf{b}_1 + m_2\mathbf{b}_2 + m_3\mathbf{b}_3$, where m_1, m_2 and $m_3 \in \mathbb{Z}$ and $\mathbf{b}_1$, $\mathbf{b}_2$ and $\mathbf{b}_3$ are related to $\mathbf{a}_1$, $\mathbf{a}_2$ and $\mathbf{a}_3$ by the expression $\mathbf{a}_i \cdot \mathbf{b}_j = 2\pi\delta_{ij}$. The two lattices, the point lattice of $\mathbf{R}_n$ and $\mathbf{G}_m$ are reciprocal and $\mathbf{R}_n \cdot \mathbf{G}_m = 2\pi(n_1m_1 + n_2m_2 + n_3m_3) = 2\pi N$, where N is an integer.
[5] A 1D function $f(x)$ exhibiting a period L can be expressed as a Fourier series as $f(x) = f(x + L) = \sum_n A_n e^{i2\pi nx/L}$, where A_n are coefficients and $e^{i2\pi nx/L} = \cos(2\pi nx/L) + i\sin(2\pi nx/L)$.

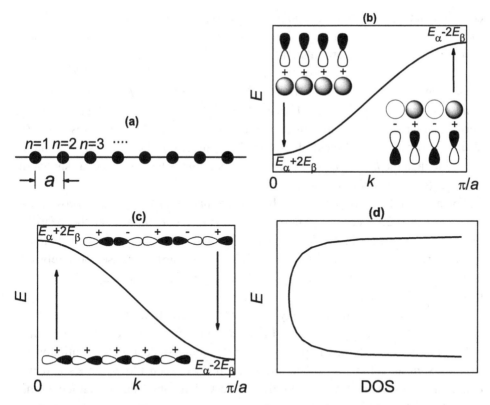

Figure 1.28. (a) Infinite periodic linear chain of atoms separated by the distance a, (b) band structure for $1s$ and p_π atomic orbitals pointing perpendicularly to the chain direction, (c) band structure for p_π atomic orbitals aligned along the chain direction and (d) DOS of a 1D infinite periodic linear chain.

of Bloch's functions. H_{11} is obtained from Eqs. (1.30a) and (1.31):

$$H_{11} = \langle \phi_1^{le}(k)|H^{le}|\phi_1^{le}(k)\rangle$$

$$= \frac{1}{N_{ion}} \sum_{n=1}^{N_{ion}} \sum_{m=1}^{N_{ion}} e^{i(n-m)ka} \langle \psi_m^{at}|H^{le}|\psi_n^{at}\rangle. \tag{1.32}$$

Within the simple Hückel approximation $\langle \psi_m^{at}|H^{le}|\psi_n^{at}\rangle$ equals E_α for $m = n$ and E_β for $m = n \pm 1$ according to Eqs. (1.20a) and (1.20b), respectively, and we obtain:

$$E(k) = \frac{1}{N_{ion}} N_{ion} \left\{ E_\alpha + E_\beta(e^{ika} + e^{-ika}) \right\} = E_\alpha + 2E_\beta \cos ka, \tag{1.33}$$

taking into account the periodic boundary conditions mentioned above (Eq. (1.27)). $E(k)$ satisfies the conditions pointed out earlier, symmetry with respect to $k = 0$, continuity and differentiability and periodicity on $G = 2\pi/a$. For $k = 0$, we obtain $|\Phi^{1e}\rangle = |\psi_1^{at}\rangle + |\psi_2^{at}\rangle + |\psi_3^{at}\rangle + \cdots$. For $1s$ orbitals this would imply the mostly bonding state, since all orbitals add in phase, and this would result in the most stable (lowest energy) state. For $k = \pi/a$, $|\Phi^{1e}\rangle = -|\psi_1^{at}\rangle + |\psi_2^{at}\rangle - |\psi_3^{at}\rangle + \cdots$, which represents the less bound state. This is schematically represented in the band structure of Fig. 1.28(b). The same applies to p_π and d_{z^2} orbitals pointing perpendicularly to the chain direction (see Fig. 1.28(b)). In both cases $E_\beta < 0$. If the p_π orbitals are aligned along the chain direction, the $k = 0$ state is more energetic than the $k = \pi/a$ state because the bonding is less effective, as shown in Fig. 1.28(c). In this case $E_\beta > 0$.

For a general 3D system it can be shown that $E(\mathbf{k}) = E_\alpha + 2E_{\beta a} \cos k_a a + 2E_{\beta b} \cos k_b b + 2E_{\beta c} \cos k_c c$, where $E_{\beta i}$ for $i = a, b, c$ represents the nearest-neighbours resonance integral along the a-, b- and c-directions, respectively, and W is defined as $W = 4(|E_{\beta a}| + |E_{\beta b}| + |E_{\beta c}|)$. The particular cases of purely 1D and 2D systems are represented by $E_{\beta a} \neq 0$, $E_{\beta b} = E_{\beta c} = 0$ and $E_{\beta a} \neq 0$, $E_{\beta b} \neq 0$, $E_{\beta c} = 0$, respectively.

We thus observe that the topology of the orbital interactions determines in which way the band runs, that is the sign of the near-neighbours interaction (Hoffmann, 1987). For p_π orbitals this is particularly interesting since E_β can continuously vary its sign as a function of the angle formed between the p_π lobes and the chain directions. Although we are only at a very elemental stage for an ideal (and thus unrealistic) model system, we can sense that the magnitude of the nearest-neighbours resonance integrals can be modulated if we are able to control the above-mentioned angle. This is an extremely important point, which is the basis of crystal engineering involving planar molecules with p_π orbitals.

Defining the group velocity v as $v \propto \partial E/\partial k$, we obtain $v \propto E_\beta \sin ka$. For $k \neq 0$, $v \neq 0$ if $|E_\beta| \neq 0$, so that electrons can propagate only in the presence of overlap. The case $E_\beta = 0$ thus corresponds to $v = 0$, where electrons cannot move because they are localized, which is the case for core electrons. The density of states (DOS) can be approximated by the inverse of the group velocity. For 1D systems the DOS has the shape shown in Fig. 1.28(d) and has maximum values at the bottom and top of the band because there are more states when approaching both points (v reduces).

The bandwidth W is simply defined as $W = 4|E_\beta|$. The greater the overlap between neighbours, the greater the bandwidth. For RISs, e.g., BEDT-TTF, TMTTF and TMTSF salts formed with inorganic anions, one can choose the shape of the ion in order to intentionally introduce smaller or larger donor interactions. Smaller/larger anions will induce larger/weaker donor molecule overlap, hence larger/smaller $|E_\beta|$ values. Larger $|E_\beta|$ values, or equivalently larger bandwidths W, result in higher σ values because the intermolecular hopping of electrons is enhanced. Smaller $|E_\beta|$ (smaller W) values tend to reduce σ leading to Mott insulators because of electron localization (small v values). This effect has been termed chemical pressure.

Each atomic orbital admits two electrons (spins up and down), so that for an N_{ion}-system each band would be able to allocate $2N_{ion}$ electrons. In our example, if each atomic orbital has only one electron, the band would have N_{ion} electrons, which means it is half-filled, while if there were two electrons per atomic orbital the band would be completely filled. Half-filled bands correspond to the metallic state, and indeed there are 1D materials that are metals, such as TTF-TCNQ and some Bechgaard salts as discussed in Section 1.5 (see also the discussion on page 281). An illustrative example of the formation of bands from MOs has been experimentally obtained for n-phenylene linear molecules ($n = 1, \ldots, 6$) (Seki *et al.*, 1984) and n-alkanes C_nH_{2n+2} (Grobman & Koch, 1979), where ultraviolet photoemission spectroscopy (UPS) has been systematically performed in the gas phase for increasing n-values. When n increases the MO bands broaden due to the increasing contribution of the discrete MOs. In the case of n-alkanes the comparison of gas phase and solid phase spectra reveals that $n = 13$ already provides a convenient finite model for the band structure of an infinite linear polymer. This comparison also shows that the influence of intermolecular interactions on the band structure is rather small, the most significant difference between spectra from the two phases being a slight increase in peak widths in the solid phase.

Within the limit of negligible electronic correlations the filling of the band N_{at} and k_F are related through:

$$N_{at}\frac{2\pi}{a} = 4k_F. \tag{1.34}$$

For $N_{at} = 1$, corresponding to a half-filled band, $k_F = \pi/2a$ and for $N_{at} = 2$, corresponding to a filled band, $k_F = \pi/a$ (see Fig. 1.28). In fact for CTSs $N_{at} = \varrho$.

Let us consider briefly the rather simple polyacetylene molecule, $-(CH)_n$, in which each carbon is σ bonded to only two neighbouring carbons and one hydrogen atom with one π electron on each carbon (the p_π orbital pointing perpendicularly to the chain direction). If the carbon bond lengths were equal, with one π electron per formula unit, it would imply a metallic state ($E_\beta < 0$) as discussed above. However, neutral polyacetylene is a semiconductor with an energy gap of approximately 1.5 eV. The reason for this discrepancy is discussed next.[6]

Let us imagine now that for some reason, either intrinsic or extrinsic, the linear chain from Fig. 1.28(a) dimerizes but preserves its 1D structure. The new distribution is illustrated in Fig. 1.29(a). The first consequence is that the period

---

[6] The bandwidth of linear polymers is of the order of 10 eV, thus sensibly larger than the typical values found for MOMs, which lie in the $\simeq 0.5$ eV range. This difference is due to the fact that the π-overlap is intramolecular and intermolecular for polymers and MOMs, respectively.

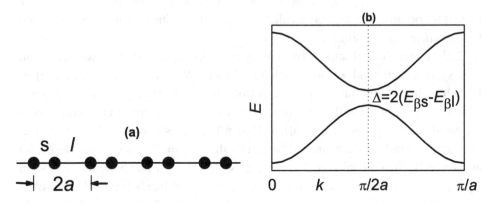

Figure 1.29. (a) Dimerized infinite linear chain, (b) band structure of a dimerized infinite linear chain.

has doubled from a to $2a$ and that the near-neighbour interactions have split into the terms $E_{\beta s}$ and $E_{\beta l}$ referring to short and long distances, respectively. Since we remain with the case of $1s$ orbitals, $E_{\beta s} < 0$ and $E_{\beta l} < 0$ and the asymmetry in length imposes that $|E_{\beta s}| > |E_{\beta l}|$. By doubling the period it follows that $N_{at} = 2$. We thus proceed as for the former case for lattice constant a but for $N_{at} = 2$.

Equation (1.19) becomes:

$$\begin{vmatrix} H_{11} - E & H_{12} \\ H_{21} & H_{22} - E \end{vmatrix} = 0, \tag{1.35}$$

where we recall that $S_{\mu\nu} = \delta_{\mu\nu}$ because of orthonormality. The Bloch functions have the expression:

$$|\phi_\mu^{le}(k, x)\rangle = \frac{1}{\sqrt{N_{ion}/2}} \sum_{n=1}^{N_{ion}/2} e^{ink2a} |\psi_{\mu n}^{at}(x - (n-1)a)\rangle, \tag{1.36}$$

where $\mu = 1, 2$ for both $1s$ orbitals and because of the two inequivalent ions in the unit cell ($2a$) the sum extends to $N_{ion}/2$ instead of N_{ion} for each value of μ. Introducing Eq. (1.36) into Eq. (1.30a) we obtain:

$$\begin{aligned} H_{\mu\nu} &= \langle \phi_\mu^{le}(k)|H^{le}|\phi_\nu^{le}(k)\rangle \\ &= \frac{1}{N_{ion}/2} \sum_{n=1}^{N_{ion}/2} \sum_{m=1}^{N_{ion}/2} e^{i(n-m)k2a} \langle \psi_{\mu m}^{at}|H^{le}|\psi_{\nu n}^{at}\rangle. \end{aligned} \tag{1.37}$$

Within the simple Hückel approximation one obtains $H_{11} = H_{22} = E_\alpha$, $H_{12} = E_{\beta s} + E_{\beta l}e^{ik2a}$ and $H_{21} = E_{\beta s} + E_{\beta l}e^{-ik2a}$. Introducing these expressions into

Eq. (1.35) we obtain:

$$E(k) = E_\alpha \pm \sqrt{E_{\beta s}^2 + E_{\beta l}^2 + 2E_{\beta s} E_{\beta l} \cos k2a}. \tag{1.38}$$

The band structure expressed by Eq. (1.38) is represented in Fig. 1.29(b). Note that at $k = \pi/2a$ a gap $E_{zb} = 2|E_{\beta s} - E_{\beta l}|$ opens as a consequence of the dimerization. If we come back to the initial situation of Fig. 1.28 (period a) or if we arbitrarily assume that the linear chain from Fig. 1.28(a) has a larger period $2a$, $E_{zb} = 0$ in both cases because $E_{\beta s} = E_{\beta l}$. The opening of a gap at the zone boundary is a direct consequence of the period increase. Since $N_{at} = 2$ we should have two bands, as observed, and each band can accommodate $2N_{ion}$ electrons (total of $4N_{ion}$). Since we have assumed one electron per $1s$ orbital and two orbitals only the first band will be completely filled, making the structure semiconducting (or insulating) upon doubling the lattice parameter. Dimerization may arise because it optimizes the intermolecular bonding rendering the dimerized structure more stable than the non-dimerized one, or because of the formation of superstructures induced by dopants (anions in CTSs) or by ordering of dopants. The eventual ordering of spins would also lead to an increase in the period. In all cases, the formation of superstructures will result in the formation of band gaps. The stabilization by structural distortion is usually discussed in terms of structural instabilities. The molecular chemistry concept corresponds to the Jahn–Teller distortions while in solid-state physics structural modulations are discussed in terms of the Peierls transition. Both concepts essentially describe the same phenomenon, the distortion-induced reduction of the total energy (Canadell & Whangbo, 1991). Structural instabilities are associated with $2k_F \equiv \pm k_F$ modulations in the limit of negligible electronic modulations (Conwell, 1988).

In solid-state physics the opening of a gap at the zone boundary is usually studied in the free electron approximation, where the application of e.g., a 1D weak periodic potential V, with period a $[V(x) = V(x + a)]$, opens an energy gap at π/a (Madelung, 1978; Zangwill, 1988). $E(k)$ splits up at the Brillouin zone boundaries, where Bragg conditions are satisfied. Let us consider the Bloch function from Eq. (1.28) in 1D expressed as a linear combination of plane waves:

$$|\Phi^{le}(k, x)\rangle = e^{ikx} u(x) = \sum_{n=-N}^{N} c_{k-nG} e^{i(k-nG)x}, \tag{1.39}$$

where $G = 2\pi/a$ and N is a sufficiently large value. For simplicity we only take two plane waves, so that $c_{k-nG} \neq 0$ for $n = 0, 1$ and zero otherwise. Equation (1.39) then transforms to:

$$|\Phi^{le}(k, x)\rangle = c_k e^{ikx} + c_{k-G} e^{i(k-G)x}. \tag{1.40}$$

$V(x)$ can also be expressed as a Fourier series:

$$V(x) = \sum_{n=-N}^{N} V_{nG}e^{inGx}, \tag{1.41}$$

where V_{nG} are the Fourier coefficients. Introducing Eqs. (1.40) and (1.41) in the Schrödinger equation one obtains the Bethe equation (Bethe, 1928):

$$\left[\frac{\hbar^2 k^2}{2m_e} - E\right]c_k + \sum_{n=-N}^{N} V_{nG}c_{k-nG} = 0, \tag{1.42}$$

which leads to the equation system:

$$\left[\frac{\hbar^2 k^2}{2m_e} - E + V_0\right]c_k + V_G c_{k-G} = 0, \tag{1.43a}$$

$$\left[\frac{\hbar^2 (k-G)^2}{2m_e} - E + V_0\right]c_{k-G} + V_G c_k = 0. \tag{1.43b}$$

The energy E is obtained by setting to zero the determinant from Eq. (1.43):

$$\begin{vmatrix} \dfrac{\hbar^2 k^2}{2m_e} - E + V_0 & V_G \\[2ex] V_G & \dfrac{\hbar^2 (k-G)^2}{2m_e} - E + V_0 \end{vmatrix} = 0, \tag{1.44}$$

which results in:

$$E = V_0 + \frac{\hbar^2}{2m_e}\left[\eta^2 + \left(\frac{1}{2}G\right)^2\right] \pm \sqrt{\left(\frac{\hbar^2}{2m_e}\right)^2 G^2\eta^2 + V_G^2}, \tag{1.45}$$

using $\eta = k - \frac{1}{2}G$. At $k = \pi/a$, $\eta = 0$ and Eq. (1.45) reduces to:

$$E = V_0 + \frac{\hbar^2}{2m_e}\left(\frac{1}{2}G\right)^2 \pm V_G. \tag{1.46}$$

Equation (1.46) shows that a gap $E_{zb} = 2V_G$ opens at $k = \pi/a$ induced by the weak periodic potential. Following the same arguments given above, doubling a to $2a$ opens a gap at $\pi/2a$. We thus see that both the chemical and physical approaches lead to the same conclusions as expected.

Let us now consider the ideal case of the BFS (see Fig. 1.18). In the absence of anions and neglecting sulfur- or selenium-induced lateral molecule–molecule interactions, the organic molecules can be modelled as forming an ideal 1D chain with p_π orbitals aligned in the direction of the chain with a lattice constant a given

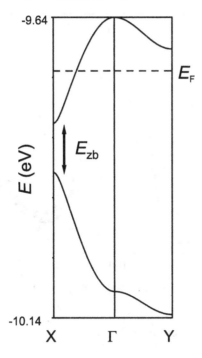

Figure 1.30. Theoretical band structure of $(TMTTF)_2ReO_4$ along the $\Gamma-X$ and $\Gamma-Y$ directions. Courtesy of Dr E. Canadell.

by the separation between two adjacent molecules, as represented in Fig. 1.28(c). In this case the materials should behave as metals as previously discussed. The presence of anions (which ultimately induce the circulation of charge) doubles the lattice parameter, so that one would expect two bands. Figure 1.30 shows the band structure of $(TMTTF)_2ReO_4$ as calculated with the extended Hückel method.

We clearly identify the two bands expected due to the doubling of the lattice parameter as well as the dimerization gap along the $\Gamma-X$ direction, which corresponds to the real-space a-direction, the stacking direction. Because of the orbital overlap along a, the dominant band dispersion is found along the $\Gamma-X$ direction. However, we also observe some dispersion along the $\Gamma-Y$ direction, which means that the system is not purely 1D but slightly 2D. This small degree of two-dimensionality prevents the Peierls transition and permits the BFS to become superconductors. Since the BFS are mixed valence materials (formally one charge to be distributed for two organic molecules) we have in fact $3N_{ion}$ electrons to be accommodated in two bands, with $4N_{ion}$ electrons allowed (hence the term three-quarter filled bands), which predicts metallic behaviour. The real structure is 3D because of lateral interactions and slight displacements of molecules with regard to each other, which means that not all BFS are metals, but the important point is that we summarize most of the dominating physics with an extremely simple model. As will be discussed

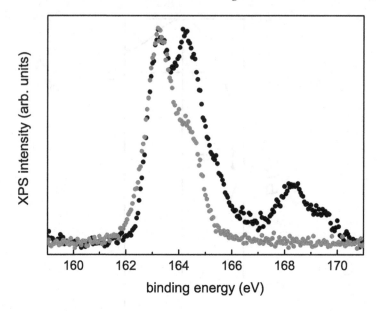

Figure 1.31. XPS spectra of $(TMTTF)_2PF_6$ (black dots) and BEDT-TTF (grey dots) taken at RT on as-received single crystals.

in Chapter 4, non-centrosymmetric anions order below a given temperature within the BFS structure and this leads to metal–insulator transitions. This can be easily understood here because of the doubling of the lattice period.

One way of experimentally exploring the electronic structure of solids is by means of photoemission spectroscopies such as UPS and X-ray photoelectron spectroscopy (XPS), where photoexcited electrons are analyzed dispersively as a function of their kinetic energy. The electronic structure of the reference material TTF-TCNQ will be extensively discussed in Section 6.1. Figure 1.31 shows the XPS spectra of the S$2p$ core line for $(TMTTF)_2PF_6$ (black dots) and BEDT-TTF (grey dots).

The chemical environment of sulfur atoms is identical for both donor molecules TMTTF and BEDT-TTF, formed by carbon–sulfur–carbon bonds (see Table 1.1), so that the XPS spectra should reveal the different charge states involved. Neutral BEDT-TTF shows a single component with a spin-orbit doublet with binding energies 163.2 and 164.5 eV for the S$2p_{3/2}$ and S$2p_{1/2}$ lines, respectively, and with a branching ratio, defined as the intensity ratio between the S$2p_{1/2}$ and S$2p_{3/2}$ features, of 0.46. In the case of $(TMTTF)_2PF_6$ we observe two equally intense main lines, which readily rules them out as originating from a single spin-orbit doublet, and a weaker structure at ∼169 eV, which arises from surface charging effects but is not relevant for the present discussion.

The $(TMTTF)_2PF_6$ spectra can be satisfactorily decomposed into two main lines, each with their corresponding spin-orbit doublet, the lower energy line coinciding with the neutral state. We thus observe that the formal assessment of mixed valency in the BFS, e.g., TMTTF^0 and TMTTF^{+1}, in a discrete boolean sense (0,1) seems to be proven experimentally because photoemission is intrinsically a rapid process. Analogous S2p spectra have been obtained for TTF-TCNQ (Sing *et al.*, 2003a).

In (perylene)$_2$PF$_6$ the coexistence of neutral and charged perylene molecules has been proved by high-resolution NMR measurements (Fischer & Dormann, 1998).

In real *trans*-polyacetylene, the structure is dimerized with two carbon atoms in the repeat unit. Thus the π band is divided into occupied π and unoccupied π^* bands. The bond-alternated structure of polyacetylene is characteristic of conjugated polymers. Consequently, since there are no partially filled bands, conjugated polymers are expected to be semiconductors, as pointed out earlier. However, for conducting polymers the interconnection of chemical and electronic structure is much more complex because of the relevance of non-linear excitations such as solitons (Heeger, 2001).

Spin lattices

Although the spin degrees of freedom have already been considered (spin 1/2 (up) or $-1/2$ (down) for electrons), the explicit contribution to the Hamiltonian has been avoided in order to keep the discussion as simplified as possible. The spin terms can be introduced by assuming a spin-independent Hamiltonian commuting with the spin operator. Hence the global wave function can be expressed as the product of $|\Psi^{1e}\rangle$ by a spin function $|S\rangle$. Keeping both electronic and spin contributions in the equations makes the resolution of the Schrödinger equation almost impossible, so that the spin character is only considered when exploring the magnetic configuration of a given structure. In this case a spin is assigned to each lattice point and the interaction among spins is considered. For simplicity let us imagine the infinite linear chain from Fig. 1.28(a), where each lattice point has either spin $S_n = 1$ or $S_n = -1$. This 1D approximation is referred to as the Ising model (Ising, 1925).

The simplest Hamiltonian expressing the magnetic energy of the system is built by considering only the nearest-neighbours interactions, following the simple Hückel approximation discussed before. The Hamiltonian H_{spin}, termed the Heisenberg Hamiltonian, is given by the expression:

$$H_{\text{spin}} = -J \sum_{n=1}^{N_{\text{ion}}} S_n S_{n+1}, \qquad (1.47)$$

where we consider cyclic boundary conditions and where J stands for the London–Heitler exchange coupling. If J is negative the energy is minimal if the spins on two neighbours tend to point in opposite directions, thus $J < 0$ favours antiferromagnetic order. On the contrary, $J > 0$ favours the spins being oriented ferromagnetically. The case of a 1D antiferromagnetic spin-1/2 lattice was solved exactly by H. Bethe with coupling only between nearest neighbours (Bethe, 1931).

In the presence of an external magnetic field B, Eq. (1.47) becomes:

$$H_{\mathrm{spin}} = -J \sum_{n=1}^{N_{\mathrm{ion}}} S_n S_{n+1} - \mu_{\mathrm{m}} B \sum_{n=1}^{N_{\mathrm{ion}}} S_n, \qquad (1.48)$$

where $\mu_{\mathrm{m}} = g\mu_{\mathrm{B}}$ ($\mu_{\mathrm{B}} = e\hbar/2m_{\mathrm{e}}$). The total magnetization $\mathcal{M}(= \chi B)$ for this 1D chain is given by the expression (Ising, 1925):

$$\mathcal{M} = N_{\mathrm{ion}}\mu_{\mathrm{m}} \frac{\sinh\left(\frac{\mu_{\mathrm{m}}B}{k_{\mathrm{B}}T}\right)}{\sqrt{\sinh\left(\frac{\mu_{\mathrm{m}}B}{k_{\mathrm{B}}T}\right)^2 + e^{\frac{-4J}{k_{\mathrm{B}}T}}}}. \qquad (1.49)$$

One of the striking consequences of the Ising model, pointed out by E. Ising himself, is the absence of permanent magnetization since the total magnetization $\mathcal{M}$ vanishes for $B = 0$ (Ising, 1925). This behaviour is a characteristic property of 1D magnetic systems and will be discussed in Section 6.2 with some examples.

Spin frustration in a linear antiferromagnetic chain can be achieved by adding a second-neighbour antiferromagnetic exchange, since nearest-neighbour antiferromagnetic exchange favours second-neighbour ferromagnetic ordering. On the other hand, if J is modulated the system exhibits spin waves (magnons) and results in spin-Peierls effects in analogy to the phonon-mediated Peierls instabilities discussed before.

Equation (1.47) can be generalized to the 2D case of an $(n \times m)$ lattice. The expression then becomes, again only valid for near-neighbour interactions but without cyclic boundary conditions:

$$H_{\mathrm{spin}} = - \sum_{m=1}^{N_{\mathrm{ion}}-1} \sum_{n=1}^{N_{\mathrm{ion}}-1} \{J_{n,n+1,m} S_{n,m} S_{n+1,m} + J_{n,m,m+1} S_{n,m} S_{n,m+1}\}. \qquad (1.50)$$

Equation (1.50) can be further simplified assuming that $J_{n,n+1,m} \equiv J_{\parallel}$ and $J_{n,m,m+1} \equiv J_{\perp}$. If $J_{\parallel} \gg J_{\perp}$ or $J_{\parallel} \ll J_{\perp}$ we fall into the 1D case of independent 1D chains. However, if $J_{\parallel} < J_{\perp}$, we have parallel chains with some degree of interchain interaction. In this case we have arbitrarily set that $J_{\parallel}$ and $J_{\perp}$ represent the inter- and intrachain interactions, respectively. If $n_{\mathrm{max}} = 2$ we have a two-leg ladder with N_{ion} rungs. More complex spin ladders can be built for $n > 2$. Sr_2CuO_3, whose crystal structure consists of linear chains of corner-sharing CuO_4 squares, is an inorganic example of a spin-1/2 antiferromagnetic Heisenberg chain. (DTTTF)$_2$[Au(mnt)$_2$], a mixed-valence salt, formed by segregated DTTTF and [Au(mnt)$_2$] stacks with a herringbone arrangement, is an example of a two-leg spin-ladder (Rovira *et al.*, 1997).

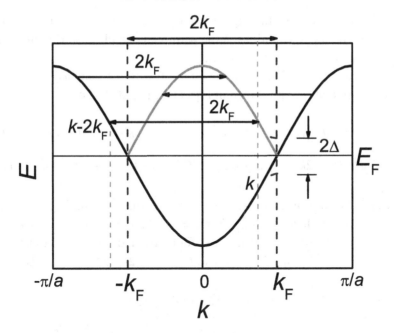

Figure 1.32. Band structure of a linear chain of period a at half filling.

Orbital mixing

Let us now study the effect of including a perturbation H_{per} to the zero-order Hamiltonian H_0, so that the new Hamilton operator H will be given by $H = H_0 + H_{per}$. We further proceed with our example of the linear chain of period a with $N_{at} = 1$, for the sake of simplicity, and continue to extract fundamental information. Figure 1.32 shows the band dispersion of this half-filled system.

Let us consider a wave vector k of an occupied state close to k_F and its equivalent $k - 2k_F$ as indicated in Fig. 1.32. The Bloch functions of both states will be given by Eq. (1.31):

$$|\phi^{1e}(k, x)\rangle = \frac{1}{\sqrt{N_{ion}}} \sum_{n=1}^{N_{ion}} e^{ikna} |\psi^{at}(x - (n - 1)a)\rangle, \qquad (1.51a)$$

$$|\phi^{1e}(k - 2k_F, x)\rangle = \frac{1}{\sqrt{N_{ion}}} \sum_{n=1}^{N_{ion}} e^{i(k-2k_F)na} |\psi^{at}(x - (n - 1)a)\rangle. \qquad (1.51b)$$

The contribution of the perturbation to the total energy of the system is such that

$$\langle \phi^{1e}(k - 2k_F, x)|H_{per}|\phi^{1e}(k, x)\rangle \neq 0,$$

thus enabling interaction or mixing between states or orbitals. Equations (1.51a) and (1.51b) are no longer solutions of the perturbed one-electron Hamiltonian and new Bloch functions have to be built. The most simple way consists in expressing the new wave functions as linear combinations of $|\phi^{1e}(k, x)\rangle$ and $|\phi^{1e}(k - 2k_F, x)\rangle$. Such linear combinations are shown in Eqs. (1.52a) and (1.52b):

$$|\phi_{mix}^{1e}(k, x)\rangle \propto |\phi^{1e}(k, x)\rangle + \varphi e^{-i\delta}|\phi^{1e}(k - 2k_F, x)\rangle, \quad (1.52a)$$

$$|\phi_{mix}^{1e}(k - 2k_F, x)\rangle \propto -\varphi e^{i\delta}|\phi^{1e}(k, x)\rangle + |\phi^{1e}(k - 2k_F, x)\rangle, \quad (1.52b)$$

where δ stands for an arbitrary phase. For $\varphi = 0$ we recuperate the $H_{per} = 0$ situation developed above and therefore $|\phi_{mix}^{1e}(k, x)\rangle = |\phi^{1e}(k, x)\rangle$ and $|\phi_{mix}^{1e}(k - 2k_F, x)\rangle = |\phi^{1e}(k - 2k_F, x)\rangle$. The electronic densities for both k and $k - 2k_F$ states are respectively given by $\langle \phi_{mix}^{1e}(k, x)|\phi_{mix}^{1e}(k, x)\rangle$ and $\langle \phi_{mix}^{1e}(k - 2k_F, x)|\phi_{mix}^{1e}(k - 2k_F, x)\rangle$:

$$\begin{aligned}
\langle \phi_{mix}^{1e}(k, x)|\phi_{mix}^{1e}(k, x)\rangle \propto \ & \langle \phi^{1e}(k, x)|\phi^{1e}(k, x)\rangle \\
& + \varphi^2 \langle \phi^{1e}(k - 2k_F, x)|\phi^{1e}(k - 2k_F, x)\rangle \\
& + \varphi e^{-i\delta} \langle \phi^{1e}(k, x)|\phi^{1e}(k - 2k_F, x)\rangle \\
& + \varphi e^{i\delta} \langle \phi^{1e}(k - 2k_F, x)|\phi^{1e}(k, x)\rangle. \quad (1.53)
\end{aligned}$$

The terms $\langle \phi^{1e}(k, x)|\phi^{1e}(k, x)\rangle$, $\langle \phi^{1e}(k - 2k_F, x)|\phi^{1e}(k - 2k_F, x)\rangle$, $\langle \phi^{1e}(k, x)|\phi^{1e}(k - 2k_F, x)\rangle$ and $\langle \phi^{1e}(k - 2k_F, x)|\phi^{1e}(k, x)\rangle$ are given by the expressions:

$$\langle \phi^{1e}(k, x)|\phi^{1e}(k, x)\rangle = \frac{1}{N_{ion}} \sum_{m=1}^{N_{ion}} \sum_{n=1}^{N_{ion}} e^{ik(n-m)a} \langle \psi_m^{at}|\psi_n^{at}\rangle, \quad (1.54a)$$

$$\langle \phi^{1e}(k - 2k_F, x)|\phi^{1e}(k - 2k_F, x)\rangle = \frac{1}{N_{ion}} \sum_{m=1}^{N_{ion}} \sum_{n=1}^{N_{ion}} e^{i(k-2k_F)(n-m)a} \langle \psi_m^{at}|\psi_n^{at}\rangle,$$

$$(1.54b)$$

$$\langle \phi^{1e}(k, x)|\phi^{1e}(k - 2k_F, x)\rangle = \frac{1}{N_{ion}} \sum_{m=1}^{N_{ion}} \sum_{n=1}^{N_{ion}} e^{i[(k-2k_F)n-km]a} \langle \psi_m^{at}|\psi_n^{at}\rangle,$$

$$(1.54c)$$

$$\langle \phi^{1e}(k - 2k_F, x)|\phi^{1e}(k, x)\rangle = \frac{1}{N_{ion}} \sum_{m=1}^{N_{ion}} \sum_{n=1}^{N_{ion}} e^{i[kn-(k-2k_F)m]a} \langle \psi_m^{at}|\psi_n^{at}\rangle.$$

$$(1.54d)$$

Assuming that $\langle \psi_m^{at} | \psi_n^{at} \rangle \propto \delta_{mn}$ we obtain:

$$\langle \phi_{mix}^{le}(k, x) | \phi_{mix}^{le}(k, x) \rangle \propto$$

$$\frac{1}{N_{ion}} \sum_{n=1}^{N_{ion}} \{1 + \varphi^2 + \varphi e^{-i(2k_F an + \delta)} + \varphi e^{i(2k_F an + \delta)}\} \langle \psi_n^{at} | \psi_n^{at} \rangle$$

$$= \frac{1}{N_{ion}} \sum_{n=1}^{N_{ion}} \{1 + \varphi^2 + 2\varphi \cos(2k_F an + \delta)\} \langle \psi_n^{at} | \psi_n^{at} \rangle, \quad (1.55a)$$

$$\langle \phi_{mix}^{le}(k - 2k_F, x) | \phi_{mix}^{le}(k - 2k_F, x) \rangle \propto$$

$$\frac{1}{N_{ion}} \sum_{n=1}^{N_{ion}} \{1 + \varphi^2 - \varphi e^{-i(2k_F an + \delta)} - \varphi e^{i(2k_F an + \delta)}\} \langle \psi_n^{at} | \psi_n^{at} \rangle$$

$$= \frac{1}{N_{ion}} \sum_{n=1}^{N_{ion}} \{1 + \varphi^2 - 2\varphi \cos(2k_F an + \delta)\} \langle \psi_n^{at} | \psi_n^{at} \rangle. \quad (1.55b)$$

For $\varphi = 0$ Eqs. (1.55) reduce to:

$$\langle \phi^{le}(k, x) | \phi^{le}(k, x) \rangle = \langle \phi^{le}(k - 2k_F, x) | \phi^{le}(k - 2k_F, x) \rangle$$

$$\propto \frac{1}{N_{ion}} \sum_{n=1}^{N_{ion}} \langle \psi_n^{at} | \psi_n^{at} \rangle, \quad (1.56)$$

where the electronic density is modulated by the lattice periodicity a. However, for $\varphi \neq 0$ an extra modulation of period $2a$ is introduced associated with $2k_F$. Figure 1.33 exemplifies the electronic densities for both $\varphi = 0$ and $\varphi \neq 0$, respectively.

In conclusion, allowing orbital mixing or interaction for states close to k_F leads to a modulation of the electronic density with a period given by $2a$. As a result of the mixing the system becomes stabilized by shifting their energies and opening an energy gap $2E_{zb}$ at k_F given by the expression:

$$E_{zb} = \langle \phi^{le}(k - 2k_F, x) | H_{per} | \phi^{le}(k, x) \rangle. \quad (1.57)$$

The situation discussed here is equivalent to a periodic distortion of the lattice with a period $2a$, as developed above. When the perturbation H_{per} is given by lattice vibrations, that is mediated by electron–phonon interactions, the electronic density modulation is expressed in terms of a charge-density wave (CDW), while when electron–electron repulsions dominate the modulation is induced by SDWs (Canadell & Whangbo, 1991).

The opening of a band gap at E_F in the superconducting state can also be interpreted within this framework. The charge carriers of a superconducting state are not individual electrons as in the normal metallic state but electrons coupled in pairs, the so-called Cooper pairs, having opposite wave vectors. Hence, Cooper pairs are described by product functions $|\phi(k)\phi(-k)\rangle$. Following the discussion given above,

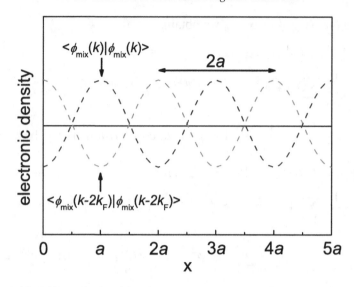

Figure 1.33. Electronic densities for $\varphi = 0$ (continuous straight line) and $\varphi \neq 0$ (discontinuous lines) for an infinite linear chain of period a. For simplicity it has been assumed that $\delta = 0$ and $\langle \psi_n^{at} | \psi_n^{at} \rangle = 1$.

the interaction of an occupied pair function $|\phi(k)\phi(-k)\rangle$ with an unoccupied pair function $|\phi(k')\phi(-k')\rangle$ leads to an energy gap related to

$$\langle \phi(k)\phi(-k)|H_{per}|\phi(k')\phi(-k')\rangle,$$

where H_{per} stands for the electron–phonon interaction in the traditional BCS scenario. A superconducting energy gap prevents Cooper pairs from breaking up when there is no excitation energy greater than the gap.[7]

Fermi surface

As already discussed above, bands may be partially filled and when this occurs the energy of the highest occupied level, the Fermi energy E_F, lies within the energy range of one (or more) bands. For each partially filled band there will be a surface in k-space separating the occupied from the unoccupied levels. The set of all such surfaces is known as the Fermi surface (FS) and is analytically defined by $E(k) = E_F$. The concept of FS plays a key role not only in understanding the dimensionality of the transport properties of metals but also in explaining the electronic instabilities of partially filled band systems.

Let us start by considering a 2D lattice for which the band dispersion is given by $E(k) = E_\alpha + 2E_{\beta a} \cos k_a a + 2E_{\beta b} \cos k_b b$. Figure 1.34 shows 3D band structure

---

[7] The energy gap at $T = 0$ K is given by $\simeq 3.5 k_B T_c$ within the BCS scenario and vanishes at $T = T_c$.

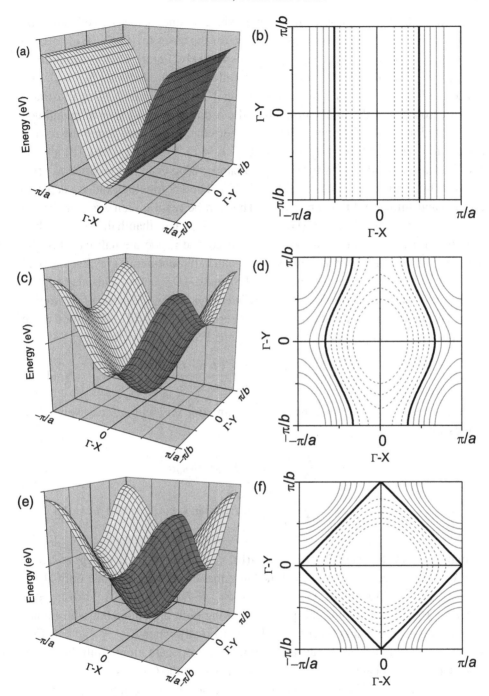

Figure 1.34. $E_{\beta a} < E_{\beta b} = 0$: (a) 3D band structure and (b) contour plot. $E_{\beta a} < E_{\beta b} < 0$: (c) 3D band structure and (d) contour plot. $E_{\beta a} = E_{\beta b} < 0$: (e) 3D band structure and (f) contour plot. FSs of half-filled systems are represented by black lines while FSs of lower and higher band fillings by short dashed and continuous grey lines, respectively.

representations $E(k_a, k_b)$ of a single band, where k_a and k_b are defined along the $\Gamma \to X$ and $\Gamma \to Y$ directions, respectively, for the cases (a) $E_{\beta a} < E_{\beta b} = 0$, (c) $E_{\beta a} < E_{\beta b} < 0$ and (e) $E_{\beta a} = E_{\beta b} < 0$, respectively. Figure 1.34(a) exemplifies the case of a 1D metal with no interaction along the b-direction, $E_{\beta b} = 0$, and hence the band exhibits no dispersion along $\Gamma \to Y$. Figure 1.34(c) illustrates the situation where the interaction along the b-direction is weaker in magnitude than that along the a-direction. The case where the interactions along the a- and b-directions are identical is depicted in Fig. 1.34(e). Figures 1.34(b), (d) and (f) show the corresponding FSs for half-filled bands, which can be classified as open (Figs. 1.34(b) and (d)) or closed (Fig 1.34(f)). The FS topology indeed depends on the band filling and thus on doping. This can be readily seen e.g., for the 2D $E_{\beta a} = E_{\beta b} < 0$ case (Fig. 1.34(f)). For band fillings less than half (short dashed grey lines) the FS would correspond to a closed line inside the half case FS (2D system) and, conversely, band fillings larger than half (continuous grey lines) would derive in open lines outside the half case FS (quasi-1D system). This also applies to the Fig. 1.34(d) case. Hence *electronic dimensionality is a function of the doping level*. The dimensionality of the transport properties can be readily understood when considering the group velocity v confined to the FS. v is then defined as:

$$v = \frac{1}{\hbar} \nabla_k E(k)|_{E_F} = \frac{1}{\hbar} \left(\frac{\partial E}{\partial k_a}, \frac{\partial E}{\partial k_b}, \frac{\partial E}{\partial k_c} \right). \tag{1.58}$$

The components of v along the k_a and k_b directions, v_a and v_b, are given by the expressions:

$$v_a = \frac{1}{\hbar} \frac{\partial E}{\partial k_a} = -2 E_{\beta a} a \sin k_a a, \tag{1.59a}$$

$$v_b = \frac{1}{\hbar} \frac{\partial E}{\partial k_b} = -2 E_{\beta b} b \sin k_b b. \tag{1.59b}$$

In the case of the open FS of Fig. 1.34(b), $v_a \neq 0$ and $v_b = 0$, except at $k_a a = 2\pi n$ ($n \in \mathbb{Z}$), since $E_{\beta a} < E_{\beta b} = 0$ meaning that electrons propagate only along the real-space a-direction (1D case). When $E_{\beta a} < E_{\beta b} < 0$, Fig. 1.34(d), $v_a \neq 0$ and $v_b \neq 0$, hence electrons can propagate along both directions but with higher velocities along the a-direction (quasi-1D case). Finally, when $E_{\beta a} = E_{\beta b} < 0$, Fig. 1.34(f), $v_a \neq 0$ and $v_b \neq 0$, hence electrons can propagate along both directions since k can point in any direction with similar magnitude (2D case).

When a piece of a FS can be translated by a vector q and superimposed on another piece of the FS, the FS is said to be nested by the vector q. This concept is only meaningful for open FSs. In the 1D case of $E_{\beta a} < E_{\beta b} = 0$ (Fig. 1.34(b)) there is an infinite number of nesting vectors q, as shown in Fig. 1.35(a), while in the

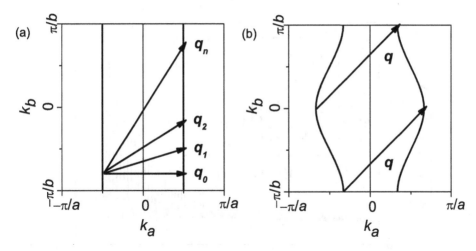

Figure 1.35. Nesting vectors for: (a) $E_{\beta a} < E_{\beta b} = 0$ and (b) $E_{\beta a} < E_{\beta b} < 0$.

quasi-1D case of $E_{\beta a} < E_{\beta b} < 0$ (Fig. 1.34(d)) nesting is only possible through the vector q shown in Fig. 1.35(b). Metallic systems with nested FS are electronically unstable and therefore are likely to undergo a metal–insulator phase transition. In the insulating phase a gap opens at E_F thus destroying the FS.

Finally, let us consider the effect of applying an external magnetic field B (Kartsovnik, 2004). In a magnetic field, the motion of quasi-particles becomes partially quantized according to the expression:

$$E(B, k_z, l) = \frac{\hbar e |B|}{m^*}\left(l + \frac{1}{2}\right) + E(k_z), \qquad (1.60)$$

where $E(k_z)$ is the energy at zero magnetic field of the motion parallel to B, l is a quantum number ($l \in \mathbb{Z}$) and m^* is an averaged effective mass.

The magnetic field quantizes the motion of the quasi-particles in the plane perpendicular to B: the resulting levels are known as Landau levels, and the phenomenon is called Landau quantization. The Landau-level energy separation is given by $\hbar$ multiplied by the angular frequency $\omega_c = e|B|/m^*$, known as the cyclotron frequency, which corresponds to the semiclassical frequency at which the quasi-particle orbits the FS. Magnetic quantum oscillations are caused by the Landau levels passing through E_F. This results in an oscillation of the electronic properties of the system, periodic in $1/|B|$. Experimentally, the oscillations are usually measured in the magnetization or the resistivity, effects known as de Haas–van Alphen or Shubnikov–de Haas, respectively.

Landau quantization only occurs for sections of FS corresponding to closed k-space orbits in the plane perpendicular to B. The frequency of the oscillation (in tesla) is given by $(\hbar/2\pi e)A$, where A is the cross-sectional k-space area of

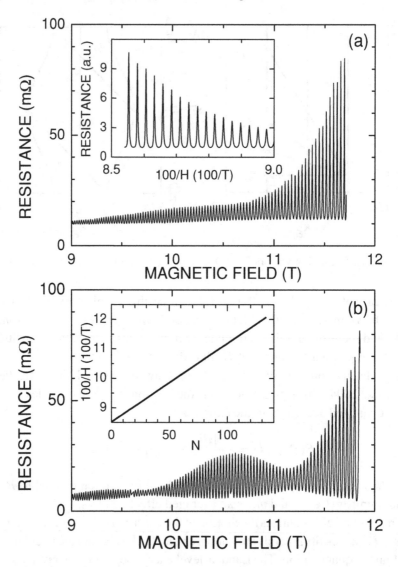

Figure 1.36. Magnetoresistance measurements performed between 9 and 12 T at 380 mK on two high-quality β_H-(BEDT-TTF)$_2$I$_3$ single crystals (a) and (b). In the inset in (a) the high-field magnetoresistance at 380 mK is shown on an enlarged scale versus H^{-1} (H represents $\boldsymbol{B}$) and clearly reveals an enhanced anharmonicity at increased fields. The inset in (b) shows the inverse field values of 133 oscillation peaks vs. integral numbers. The almost perfect linearity is a test for both the origin of the oscillations and the quality of the data. The slope provides a fundamental field of 3730 T. Reprinted with permission from W. Kang, G. Montambaux, J. R. Cooper, D. Jérome, P. Batail and C. Lenoir, *Physical Review Letters*, **62**, 2259 (1989). Copyright (1989) by the American Physical Society.

the orbit. An example is given in Fig. 1.36, which shows giant Shubnikov–de Haas oscillations in the magnetoresistance of β_H-(BEDT-TTF)$_2$I$_3$ with the field applied perpendicular to the *ab*-planes (Kang *et al.*, 1989). A single frequency of oscillations of 3730 T is observed, which corresponds to a value of A of about 50% of the area of the first Brillouin zone. This result is expected from the simple ideas of band structure described in this chapter. β-(BEDT-TTF)$_2$I$_3$ (β-(BEDT-TTF)$_2^+$ and I$_3^-$) corresponds to the case $N_{at} = 2$ with a three-quarters-band filling or equivalently to a half-band filling of the highest occupied band. Because of this half filling and the shape of the band the internal area covered by the FS corresponds to half of the total area of the first Brillouin zone. This can be clearly seen from Fig. 1.34(f).

Beyond the one-electron approximation

So far we have been discussing the case where the one-electron picture remains valid, that is when electronic correlations are weak. However, in many cases correlations can be strong and the one-electron scenario becomes no longer applicable. Ideal 1D systems are instrinsically electronically correlated, since a single electron cannot move freely: any movement implies the collective, and thus correlated, motion of all neighbouring electrons.[8] This forced collectivization might be relaxed at higher dimensionalities. Because of this the Fermi liquid model can no longer be applied in the case of strictly 1D interacting electron systems. Instead the Tomonaga–Luttinger liquid theory (Tomonaga, 1950; Luttinger, 1963) is more appropriate. This scenario is characterized by the separation of spin and charge and the absence of a sharp edge at E_F. Instead of the Fermi distribution function given by Eq. (1.13), the DOS decays as $|E - E_F|^\upsilon$, where $\upsilon = (K_\rho + K_\rho^{-1})/2 - 1$. The Luttinger liquid exponent K_ρ is a dimensionless parameter controlling the decay of all correlation functions ($K_\rho < 1$ and $K_\rho > 1$ represent repulsive and attractive interactions, respectively). We will discuss this point in Section 6.1. An amusing description of quantum physics in one dimension has been recently reviewed in Giamarchi, 2004.

As discussed earlier for linear chain systems, half-filled bands and three-quarter filled bands lead to metallic ground states because of incomplete filling of the highest occupied band, and some examples are TTF-TCNQ and BFS. However, when on-site and nearest-neighbour Coulomb interactions cannot be neglected, correlation gaps E_{corr} appear and materials exhibit semiconducting or insulating ground states. E_{corr} refers to the minimum energy required to make a charge excitation in a system

---

[8] Concerning the effects of dimensionality I strongly recommend the book entitled *Flatland* from E. A. Abbott, an Anglican clergyman (1838–1926), and in particular the chapter where the main character, a square in Flatland, meets the King of Lineland.

that would otherwise be gapless in the absence of correlation effects. In the limiting case of very strong electronic correlations ($U \gg W$), the maximum occupancy per orbital would be 1 ($N_{at} = 1$) since no spin degeneracy would be allowed. This means that for $N_{at} = 1$, $k'_F = \pm 2k_F$, where k_F refers to the non-correlated case, and thus structural instabilities would be associated with $4k_F$ modulations. Let us thus recall that $2k_F$ and $4k_F$ modulations are associated with $U < W$ and $U > W$, respectively.

2

Building molecules: molecular engineering

Toute la nature est à mes ordres, nous répondit la Durand, et elle sera toujours aux volontés de ceux qui l'étudieront: avec la chimie et la physique on parvient à tout.

All of nature is at my command, replied Madame Durand, and she will always be at the will of those who study her: with chemistry and physics one can achieve anything.

Marquis de Sade, *Histoire de Juliette*

Stating that molecules constitute the most important part of MOMs seems superfluous since they are indeed their fundamental building blocks. However, since most of the attention and glory has been devoted to the properties of MOMs, from both the scientific and technological points of view, the first and indispensable step in the chain that leads to MOMs, that is the synthesis of the constituent molecules without which MOMs cannot be built, has somehow been ignored. For a layman in the domain of synthesis of organic molecules (and this is my case) it is certainly easy to fall under the wrong impression that almost any molecule can be synthesized, given the seemingly countless number of available molecules that can be found in the literature. However, the creation of new molecules is, as expected, a complex matter and we will see a revealing example in the following section, where it will become evident that only a limited number of linear acenes have been prepared.

This chapter is essentially devoted to researchers unfamiliar with chemical synthesis and therefore it will be structured based on targeted examples of the most relevant molecules discussed in the book. Experts can skip this chapter if they wish but I hope they can still find some interesting issues. For the sake of generality we will proceed in line with the structural script given in subsection *Topology* of Section 1.2, first building molecules exclusively from C_6H_6 rings, then adding pentagons and finally including functional groups, in a LEGO-like way (in Spanish *lego* means layman!), consciously omitting the difficulties encountered in synthesis as a first approximation. Drawing two molecules in a particular geometry is easy, inexpensive and safe, but in the real world one has to induce such an assembly in sufficient quantities (some milligrams). We shall next explore the most relevant synthetic strategies. Details on the preparation of a host of molecules can be found in specialized handbooks and indexes of known chemical industries.

The usual and most rational route towards the synthesis of new complex molecules is by coupling of smaller molecules. However, we are not going to explore how all the required molecules are obtained, but proceed much in the way chemists would do in the laboratory, that is by starting from available products. A few examples will make this point clear.

2.1 Hexagons

We earlier recalled that one can build a huge number of molecules by simply assembling C_6H_6 rings, a scheme that was summarized in Table 1.4, and that a solid such as graphite can be obtained in this *gedanken* synthesis. Let us briefly discuss the case of linear acenes, which are linear polycyclic aromatic hydrocarbons (PAHs) composed of laterally fused C_6H_6 rings.

C_6H_6 is the smallest member of the acene family, and was discovered back in 1825 by M. Faraday who called it bi-carburet of hydrogen (Faraday, 1825). C_6H_6, naphthalene and anthracene can be extracted from coal, but higher homologues, such as pentacene or hexacene, can only be obtained by sequential or multistep synthesis. Longer acenes are expected to exhibit smaller HOMO–LUMO gaps, and hence their interest for molecular electronics. However, the preparation of such extended C_6H_6-based molecules is hindered by many undesirable factors such as easy photooxidation, insolubility, dimerization and polymerization. The longest reliably described isolated system with only moderate stability is hexacene, while the obtention of heptacene is still controversial. There thus seems to be a limit on the number of C_6H_6 rings that can be linearly assembled. The carcinogenic properties of PAHs are another serious drawback for their potential applications.

Pentacene can be efficiently prepared by reduction of pentacenequinone by an aluminium-cyclohexanol mixture (Goodings *et al.*, 1972). A scheme of this synthesis is shown in Fig. 2.1. Other synthetic routes with good overall yields have also been reported.

We also considered in Section 1.2 the possibility of preparing Möbius-band molecules from cycloacenes. In fact regular cycloacenes, that is molecules built from joining both ends of linear acenes without twisting, have never been obtained, despite several experimental attempts. They are interesting since they represent the basic cylindrical carbon units of zig-zag (n,0) nanotubes.

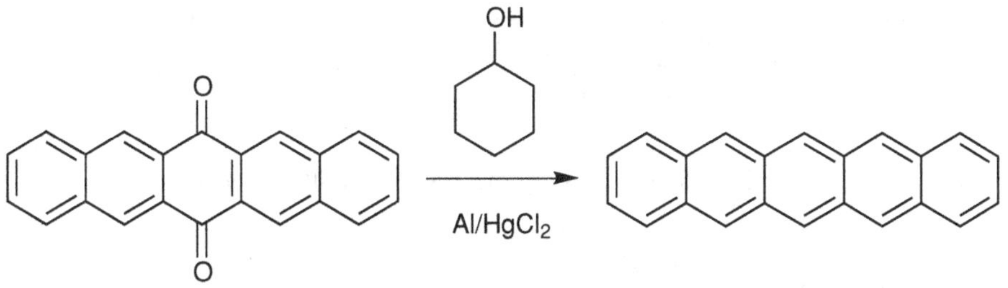

Figure 2.1. Synthesis of pentacene by reduction of pentacenequinone.

When 2D or 3D PAHs are considered then the choice becomes much larger. An exhaustive review on the advances in the synthesis of polycyclic aromatic compounds can be found in Harvey, 2004. The most used strategies are flash vacuum pyrolysis, cross-coupling, oxidative photocyclization, Diels–Adler cycloaddition, etc.

Linear and non-linear PAHs and their derivatives are produced not only in the Earth and in the terrestrial environment, e.g., as air pollutants resulting from incomplete combustion, and artificially in the laboratory, but also in space. In fact they may be the most abundant organic molecules, believed to compose up to 20% of the total cosmic carbon. Aromatic molecules have been identified in extraterrestrial matter such as meteorites, interplanetary dust particles and comets by measurements performed with the Infrared Space Observatory (ISO) satellite of the European Space Agency (ESA) during its 1995–8 operation period (Van Kerckhoven *et al.*, 2000). Although PAHs are not believed to be the key molecules forming the basis of life, they may have been important intermediates in the chemical pathways that led from space to the origin of life on Earth (Bernstein *et al.*, 1999). Irradiation of PAHs in water ice with UV light results in the formation of a diverse mixture of organic molecules, including ethers, quinones and alcohols. These may then further react to form amino acids and other biochemical molecules.

The experimental conditions found in space can be partially reproduced in a ground laboratory, i.e., at cryogenic temperatures in UHV conditions. Irradiation of CH_4, C_2H_4 and C_2H_2 ices, produced by condensation of these gases onto substrates held at very low temperatures (~ 10 K), with 7.3 MeV protons as well as 9.0 MeV He^{+2} nuclei (α-particles), aiming to simulate the interaction of galactic cosmic-ray particles with extraterrestrial organic ices, generates several aromatic species (Kaiser & Roessler, 1997). Under such conditions, from CH_4 ice typical FTIR absorptions of aromatic vibration modes are observed, including the C–H stretching, in-plane as well as out-of-plane ring vibrations and deformations. Irradiation of C_2H_2 ice with 9 MeV α-particles produces, among other molecules, chrysene, perylene, pentacene, coronene, etc. Thus physicists can also do chemical

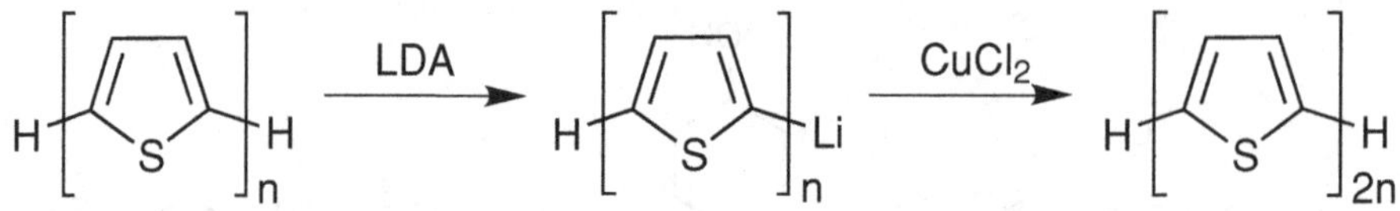

Figure 2.2. Scheme of oligomerization of α-thiophenes.

synthesis, although it would mean the ruin of the chemical industry. For instance, perylene can be more efficiently obtained from phenanthrene using anhydrous HF as a condensing agent (Calcott *et al.*, 1939).

An additional very expensive synthetic route, certainly not indicated for industrial production although scientifically sublime, is the generation of a single molecule by chemical reaction of two precursor molecules induced with a scanning tunnelling microscope (STM). We will see in Section 4.2 the example of the synthesis of biphenyl from C_6H_5I on a Cu(111) surface, thus reproducing Ullmann's reaction at the single molecule level (Ullmann *et al.*, 1904).

Finally, just a few words dedicated to the synthesis of polyphenylenes, extremely important polymers, and in particular substituted polyphenylenes such as PPV, which exhibit superb thermal and chemical resilience, semiconducting properties upon doping and applications such as OLEDs. Contrary to their linear acenes counterparts, long polyphenylenes can be obtained e.g., by Bergman's method consisting in the thermal cycloaromatization of enediynes (Lockhart *et al.*, 1981).

2.2 Pentagons

Thiophene, resulting from the substitution of a carbon by a sulfur atom in an all-carbon pentagon is a prominent representative of the MOMs building blocks. Although found in coal, in the laboratory it can be prepared by heating sodium succinate with phosphorus trisulfide or by passing C_2H_2 or C_2H_4 into boiling sulfur. α-thiophene oligomers, α-nT, can be obtained in good yields using a simple approach based on lithiation and oxidative coupling of organolithium compounds, schematized in Fig. 2.2.

The experimental procedure is exemplified here with the synthesis of α-6T ($n = 6$) from α-4T ($n = 4$) (Kagan & Arora, 1983). The reaction of two equivalents of α-2T ($n = 2$) with one equivalent of LDA and then with one equivalent of $CuCl_2$ yields α-4T in high yields. α-2T is obtained from thiophene in a similar way. Then, the reaction of two equivalents of α-4T with one equivalent of LDA followed by one equivalent of $CuCl_2$ produces α-6T ($n = 6$). Stoichiometry is thus important. In more detail, α-6T is prepared in the following way. *n*-BuLi is added dropwise to a stirred solution of diisopropylamine in dry THF at low temperature (about 195 K) under N_2. Then α-4T in dry THF is added dropwise, the mixture is stirred and anhydrous $CuCl_2$ is added. After further stirring and addition of HCl, solid α-6T is formed. Synthesis and crystallization, discussed in the next chapter, go hand in hand when the obtained materials form solids.

2.3 Hexagons and pentagons

Molecules made exclusively of hexagons necessarily form open structures, unless a large torus-shaped molecule is obtained (genus $= 1$), but can become closed simply by adding pentagons (genus $= 0$). The buckminsterfullerene, C_{60}, is a very special and rather unique molecule and perhaps the best example of combinations of hexagons and pentagons (see Table 1.1). It exhibits the truncated icosahedral structure and was named after the American architect and engineer R. Buckminster Fuller (1895–1983) because of his celebrated geodesic domes. C_{60} is prepared by *physical* synthesis instead of following a chemical route (Kroto *et al.*, 1985). It is obtained by laser evaporation of graphite using a focused pulsed laser and a high density helium flow (see Section 3.2 for details of the laser ablation technique). A large number of carbon clusters are obtained and under certain conditions, for a range of helium pressure, C_{60} dominates. A higher mass cluster, C_{70}, is also generated. The most effective way of producing C_{60} is the arc discharge technique with graphite electrodes in a helium atmosphere (Krätschmer *et al.*, 1990). The resulting black soot is scraped from the collecting surfaces inside the evaporation chamber and dispersed in C_6H_6. The solution is separated from the remaining soot and dried. C_{60} is stable at temperatures well above RT, is chemically active but stable in air. In addition to C_{60} the soot consists of other materials such as amorphous carbon, C_{36}, a range of other fullerenes (e.g., C_{70}), multilayered carbon onions, single and MWNTs and nanotube bundles. Purification of the soot is thus a major part of the preparation.

Nanotubes can be thought of as arising from rolling graphene layers and capping with two halves of C_{60} molecules. Multiwalled nanotubes (MWNTs) are made of concentric cylinders of rolled-up graphene sheets (Iijima, 1991). The length of a tube is in the range of a few micrometres and the diameter of 10–20 nm. A major breakthrough was achieved when single-walled nanotubes (SWNTs) were synthesized using metallic catalyst particles (Bethune *et al.*, 1993). These structures can show 1D metallic or semiconducting properties, depending on how the graphene sheets are rolled up. Carbon onions were first observed in an electron microscope after electron irradiation of graphitic materials (Ugarte, 1992). They consist of concentric spherical layers stacked one inside the other, like Russian dolls.

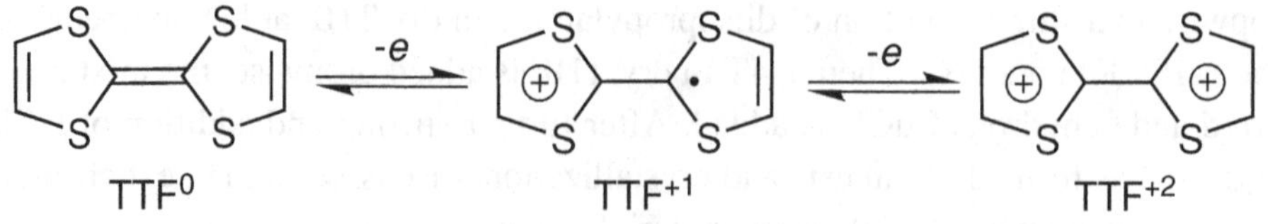

$$TTF^0 \quad\quad\quad TTF^{+1} \quad\quad\quad TTF^{+2}$$

Figure 2.3. Redox transformation of TTF.

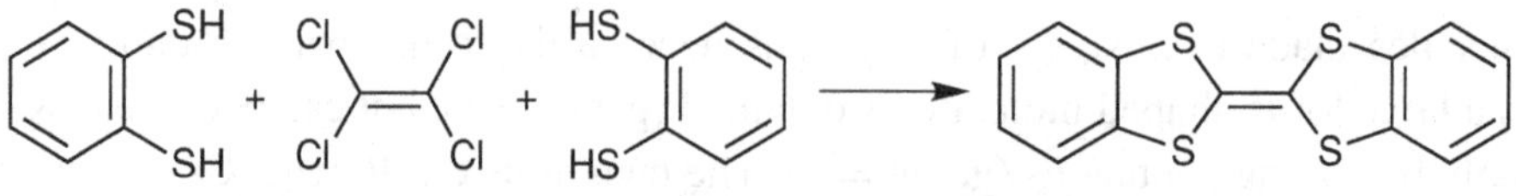

Figure 2.4. Scheme of the reaction leading to DBTTF (strategy S1).

2.4 Hexagons, pentagons and functions

In the wake of the discovery of TTF-Cl being a semiconductor (Wudl *et al.*, 1972) and TTF-TCNQ a metal (Ferraris *et al.*, 1973), TTF has become the core of an incredible number of heterocyclic systems in the search for not only new 2D metals and superconductors but also functional single-molecules. A plethora of TTF-derivatives have been synthesized following different strategies and in this section only a glimpse of the most common will be given. Among several reviews on TTF-chemistry I would recommend the recent Yamada and Sugimoto (2004) book and the November 2004 special issue of *Chemical Reviews* on Molecular Conductors (Volume 104, number 11). The activity in the synthesis of new TTF-based molecules is so intense that almost any attempt to report the state of the art soon becomes out of date.

TTF is an exceptional π-donor and has become so extensively studied because it can be oxidized to the cation radical (TTF^{+1}) and dication (TTF^{+2}), both thermodynamically stable species, sequentially and reversibly within a very accessible potential window ($E_1^{1/2} = +0.34$ and $E_2^{1/2} = +0.78$ V, vs. Ag/AgCl in CH_3CN), where $E_1^{1/2}$ and $E_2^{1/2}$ stand for the cyclic voltammetry first and second oxidation potentials, respectively. The redox transformation of TTF is shown in Fig. 2.3. In addition, such potentials can be tuned by the attachment of electron-donating and electron-withdrawing substituents and hence the interest in functionalizing the TTF core. The electronic band structure of neutral TTF as well as of partially ionized TTF will be discussed in Section 6.1.

Let us here explore different strategies developed for synthesizing derivatives centred on the TTF core. The resulting molecules are called functionalized TTF molecules. The first strategy (S1) is based on the condensation of two dichalcogenolate entities with C_2Cl_4. This strategy was used in 1926 for the preparation of

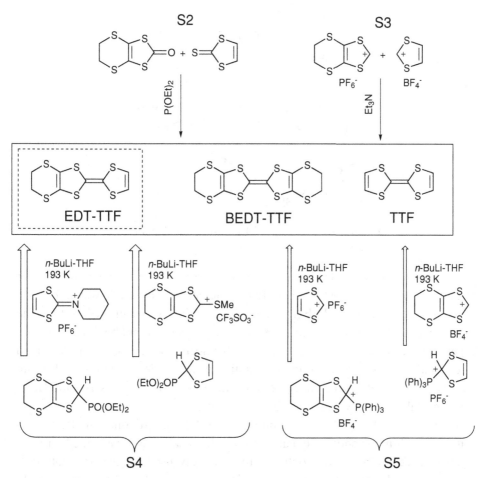

Figure 2.5. Scheme of the preparation of EDT-TTF by different synthetic routes (strategies S2, S3, S4 and S5). Increasing arrow width indicates higher selectivity (S4 is completely selective).

DBTTF by condensation of *o*-dithiolbenzene with tetrachloroethylene in a basic medium (Hurtley & Smiles, 1926). The scheme of the reaction is shown in Fig. 2.4. Although very intuitive this strategy is no longer used because the obtained yield is poor, the reaction is non-selective when using two different dichalcogenolates, and furthermore the 1,2-dichalcogenolates starting compounds (1,2-dithiolates in Fig. 2.4) are difficult to obtain (here we temporarily recover the numeric notation of the chemical formulation to avoid confusion).

The set of strategies leading to symmetrically and asymmetrically functionalized TTF-derivatives is discussed next (see Fig. 2.5). Strategies S2 and S3 are known as cross-coupling reactions and S4 and S5 are termed condensation reactions. S2 and

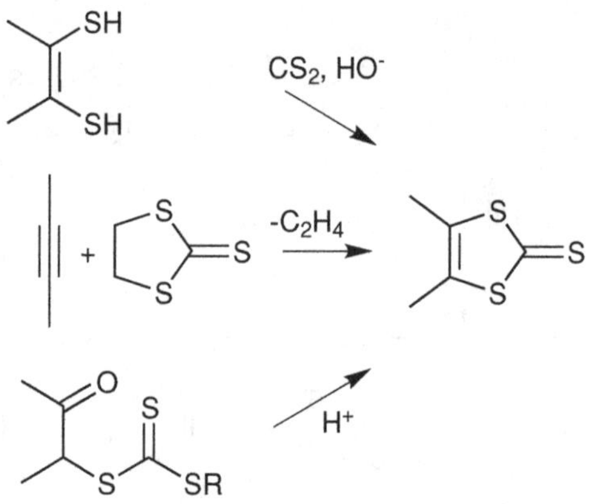

Figure 2.6. Synthesis of the 1,3-dithiole-2-thiones.

S3 are non-selective and lead to mixtures while S4 and S5 can be completely or partially selective. We illustrate these four strategies in a schematic way in Fig. 2.5 with the synthesis of EDT-TTF (Fourmigué *et al.*, 1993; Garreau *et al.*, 1995). The underlying mechanisms involving intermediate species will not be discussed here. In general such mechanisms are only partially understood. The methods described here can also be employed for the preparation of molecules containing selenium, such as TSF-derivatives, or molecules containing both sulfur and selenium. However, we shall only show examples involving sulfur.

S2 and S3 are based on the coupling of 1,3-dithiole-2-thiones (under the action of phosphine or phosphites) and of dithiolium salts (in a basic medium), respectively. S2 has been and continues to be largely used for the preparation of asymmetric TTFs as well as more complex structures when they can be easily separated from the formed mixtures. In general the yields are acceptable. The principal advantage of this strategy is the relatively easy access to a very large variety of 1,3-dithiole-2-thiones used as starting compounds in these reactions. Indeed, the synthesis of the 1,3-dithiole-2-thiones is very important and can be achieved in several ways (Schukat *et al.*, 1987). Figure 2.6 summarizes only a few of them.

When using substituents in the starting molecules the success of the cross-coupling strongly depends on such substituents. Many functions have thus been introduced on TTF and TSF, in particular ester groups, which can then be easily converted into carboxylic acid or amide or into the corresponding non-substituted derivative by a simple decarboxymethylation. Other functional groups incorporated include nitrile and halogens such as chlorine, bromine and iodine. The reaction yields obtained with S3 are lower than for S2 and hence this strategy is only occasionally used. This is in part due to the fact that the reactivities of the two dithiolium salts used can be quite different from each other.

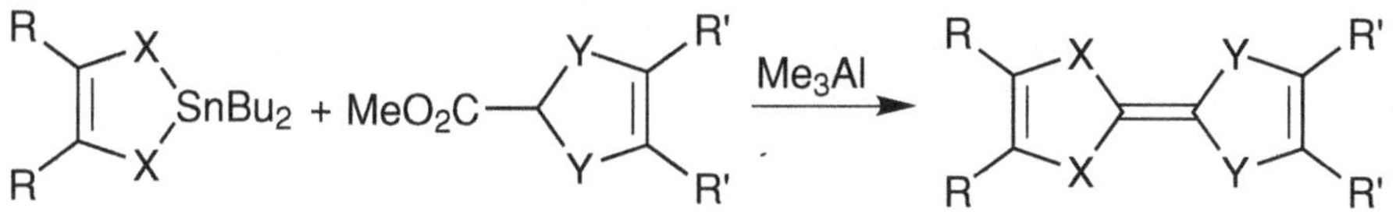

Figure 2.7. Scheme of the Me_3Al-promoted coupling reaction of organotin compounds with esters (strategy S6). X, Y = S, Se and R, R′ stand for substituents.

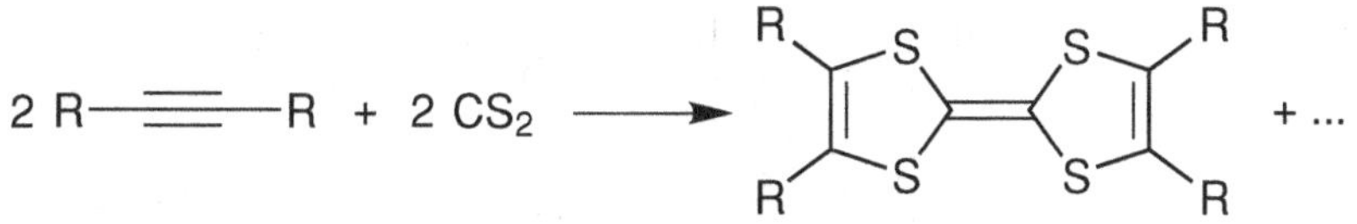

Figure 2.8. Scheme of the preparation of TTF-derivatives from C_2R_2 and CS_2 (strategy S7).

Strategies S4 (Wittig–Horner type) and S5 (Wittig type) enable a higher degree of selectivity in discriminating the symmetric products of the reaction. S4 can be considered as entirely selective while S5 is partially selective. In Fig. 2.5 only the initial and final products are shown. However, the reactions are much more involved. Note that two possible paths exist for each strategy. S4 has been employed in the synthesis of a great number of functionalized TTF derivatives and of TTF partially substituted with selenium atoms.

An additional entirely selective condensation synthesis for asymmetric TTF molecules is based on strategy S6, which consists in the reaction of an organotin dichalcogenolate with an ester in the presence of a Lewis acid such as Me_3Al (Yamada *et al.*, 2004). A scheme of the reaction is given in Fig. 2.7. The method applies to the preparation of varied asymmetric TTFs and can also be extended to the synthesis of selenium-containing analogues.

It is possible to add carbon disulfide to acetylenes having at least one electron-attracting substituent R (in particular with $R = -CF_3$) to give the product tetra-R-TTF (see Fig. 2.8) but in addition other compounds are obtained. This is our strategy S7. Their relative yields depend on the experimental conditions. However, in the presence of trifluoroacetic acid, the TTF derivative shown in the figure dominates. On the other hand, benzyne (C_6H_6 with a triple bond) is known to cycloadd rapidly to CS_2 giving DBTTF.

Functionalization of TTF can also be achieved by means of electrophilic substitution of metalated TTF. This strategy (S8) is based on the substitution of the hydrogen atoms of TTF by lithium in a metalation reaction with *n*-BuLi or LDA (see previous example of thiophene). Tetrathiafulvalenyllithium is highly reactive even at low temperature, and it reacts immediately with various electrophiles to

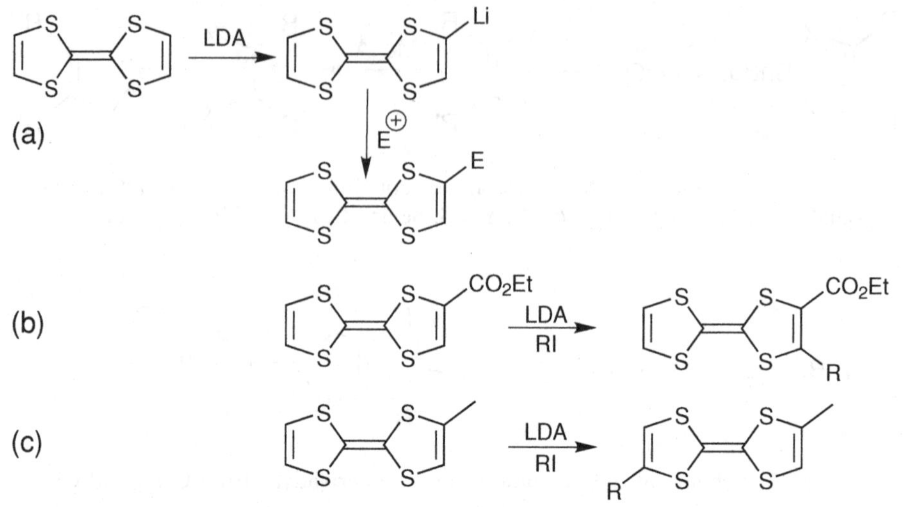

Figure 2.9. Electrophilic substitution of lithiated TTF (strategy S8). E stands for electrophilic substituent. (a) Generic reaction, (b) monosubstituted TTF with the electron withdrawing substituent CO_2Et and (c) monosubstituted TTF with the electron donor substituent CH_3. E = Br, I, CO_2H, CO_2Et, CH_2OH, CH_3, $CONR_2$, $CHNR_2$, CHO, etc.

give the corresponding functionalized or alkylated TTF by nucleophilic substitution (Fig. 2.9(a)). The main disadvantage of this strategy is the disproportionation of tetrathiafulvalenyllithium into multilithiated species and often results in a complex mixture of products along with the recovery of TTF.

In the case of further lithiation of monosubstituted TTF, the orientation of the lithiation depends on the nature of the substituent already present on the TTF heterocycle. An electron withdrawing substituent such as CO_2Et increases the acidity of the adjacent hydrogen atom and leads to a 4,5-disubstituted TTF (Fig. 2.9(b)) whereas an electron donor substituent (CH_3), which decreases this acidity, directs the lithiation to the unsubstituted 1,3-dithiole ring (Fig. 2.9(c)). One major advantage of this method is that it allows the preparation of a large variety of donors by direct functionalization of the TTF or TSF skeleton.

Let us end this section with the synthesis of three relevant molecules: TCNQ, α-nitronyl nitroxides and CuPc. The synthesis route that efficiently leads to the preparation of the strong acceptor TCNQ is depicted in Fig. 2.10. It is based on the ready condensation of malonitrile with 1,4-cyclohexadione. Further treatment with C_5H_5N and Br_2 gives TCNQ in excellent yields (Acker & Hertler, 1962). The electronic band structure of neutral TCNQ will be discussed in Section 6.1.

α-nitronyl nitroxides are a class of stable radicals with *p*-NPNN as the most prominent representative. The growth and physical properties of thin *p*-NPNN

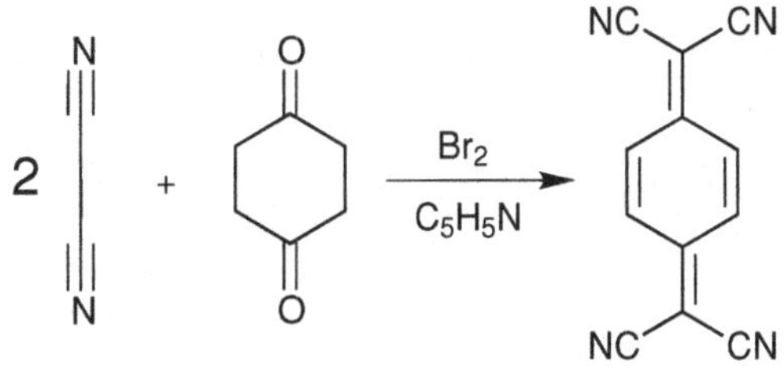

Figure 2.10. Synthesis of TCNQ.

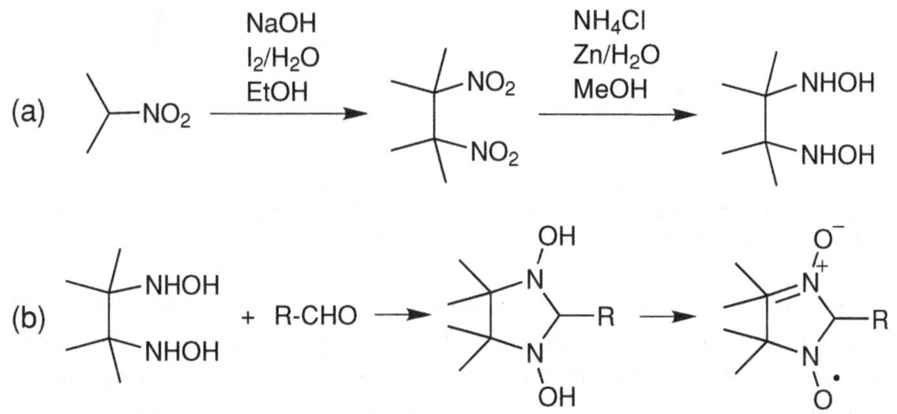

Figure 2.11. Scheme of the synthesis of α-nitronyl nitroxides.

films will be studied in Chapters 5 and 6, respectively. The synthesis route, shown in Fig. 2.11, can be divided into two steps (Osiecki & Ullman, 1968). The first is the synthesis of 2,3-dimethyl-2,3-bis(hydroxylamino)-butane from 2-nitropropane through dimerization of 2-nitropropane (Fig. 2.11(a)), followed by the reaction of the 2,3-dimethyl-2,3-bis(hydroxylamino)-butane species with aldehydes with sodium periodate or lead dioxide giving the nitronyl nitroxide (Fig. 2.11(b)) *p*-NPNN corresponds to the case R = nitrophenyl.

The synthetic pigment CuPc was obtained by serendipity in 1927 but not identified as such by the authors probably due to analytical limitations and/or because attention was focused on other compounds (de Diesbach & von der Weid, 1927). Upon reaction of *o*-$C_6H_4Br_2$ with cuprous cyanide and C_5H_5N a blue insoluble compound was obtained, which undoubtedly was CuPc. Basically there are two commercially important processes to produce CuPc. One is based on phthalonitrile and the other one uses phthalic anhydride. The phthalonitrile process often yields a product with fewer impurities and using metallic copper gives CuPc by cyclotetramerization.

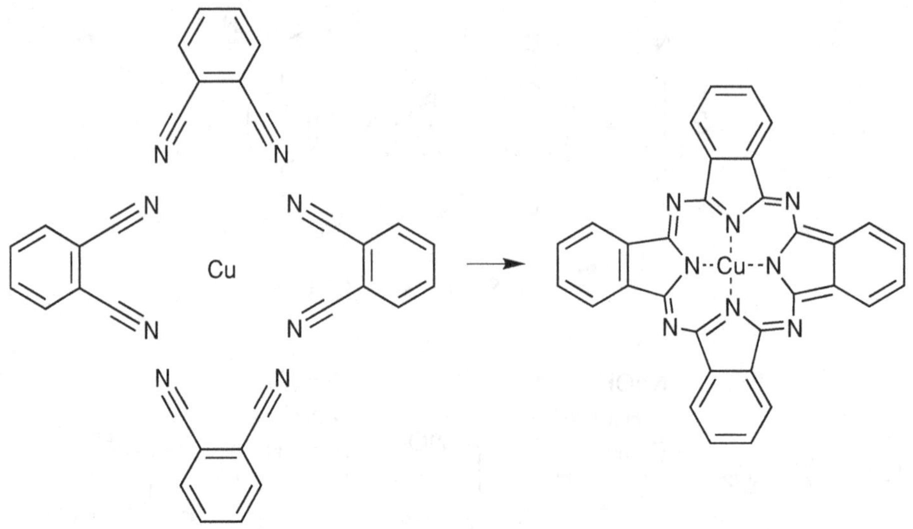

Figure 2.12. Scheme of the synthesis of CuPc by cyclotetramerization of phthalonitrile with metallic copper.

Figure 2.12 shows the case of the formation of CuPc from *o*-dicyanobenzene (Dahlen, 1939). CuPc is also produced by reacting phthalic anhydride, urea, a copper/copper salt and ammonium molybdate (acting as a catalyst) in a high-boiling solvent such as *o*-nitrotoluene, $C_6H_3Cl_3$ or alkyl benzenes. The choice of the solvent is very important as it plays a significant role in the generation of impurities. After the completion of the reaction, the solvent is removed by vacuum distillation and the crude product thus obtained is further purified by treatment with dilute acid and alkali solutions to remove the basic and acidic impurities. The product thus obtained is dried and pulverized. It is surprising that such a complex molecule is relatively easily obtained in the laboratory as compared to the simpler and closely related porphyrins. Porphyrins are tetrapyrrolic pigments widely occuring in nature and playing very important roles in various biological processes. They are present e.g., in hemoglobins and myoglobins, which are responsible for oxygen transport and storage in living tissues, and in the enzyme peroxidase.

2.5 Really big molecules

The chemistry of organic synthesis today has the tactics or means to increase the size of molecules by aggregating more and more parts resulting in complex supramolecular structures. TTF can be sequentially added forming bi-TTF, bis-TTF and so on, producing oligomers (Iyoda *et al.*, 2004). Bi-TTF can be obtained by making use of Ullmann's reaction by coupling two iodo-TTFs mediated by copper at about 410 K. At the single-molecule level this reaction may also be

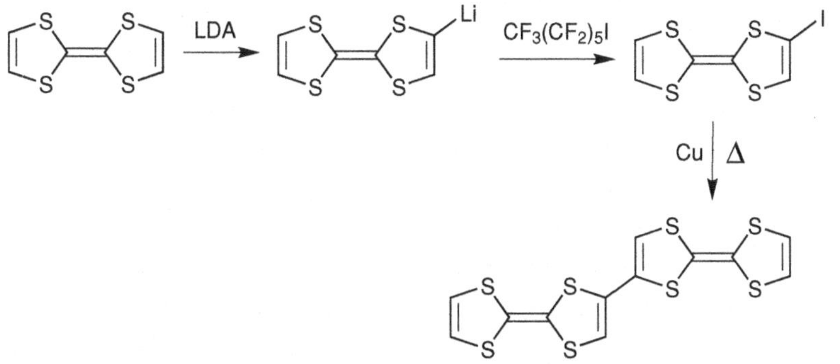

Figure 2.13. Scheme of the synthesis of bi-TTF.

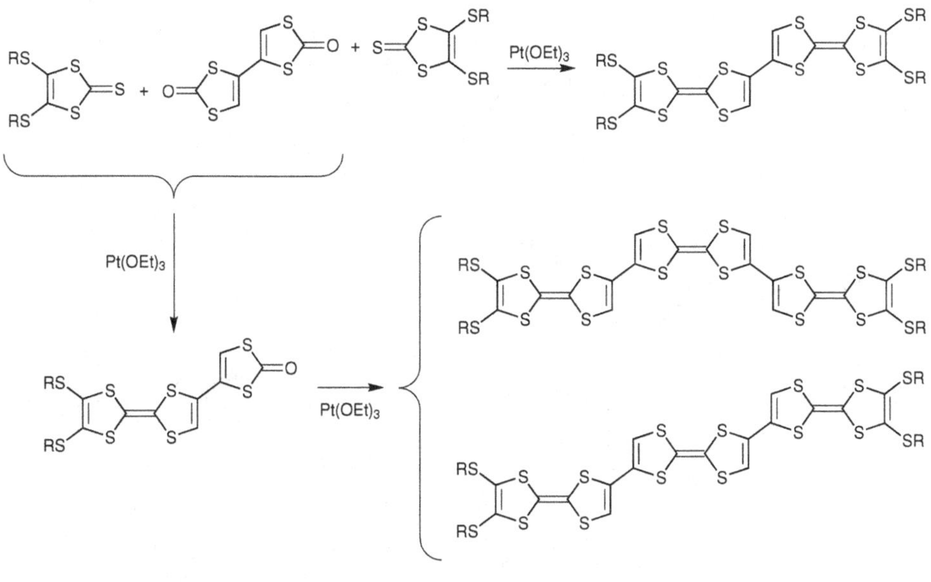

Figure 2.14. Scheme of the synthesis of tetraalkylthio-bi-TTF and tetraalkylthio-tri-TTF.

STM-induced, in analogy to the preparation of biphenyl (definitely something to be tried). Iodo-TTF can be obtained by lithiation of TTF with LDA. Figure 2.13 shows a scheme of this reaction.

From alkylthio-derivatives of bi-TTF tetraalkylthio-bi-TTF can be obtained using the phosphite-mediated cross-coupling method (strategy S2). Depending on the solvent, an intermediate molecule is obtained whose homocoupling yields the tetraalkylthio-tri-TTF (see Fig. 2.14).

Oligomerization leads to large molecules such as dendrimers, macrocycles, cyclophanes, etc. The ability to build growing molecules has also led to covalent

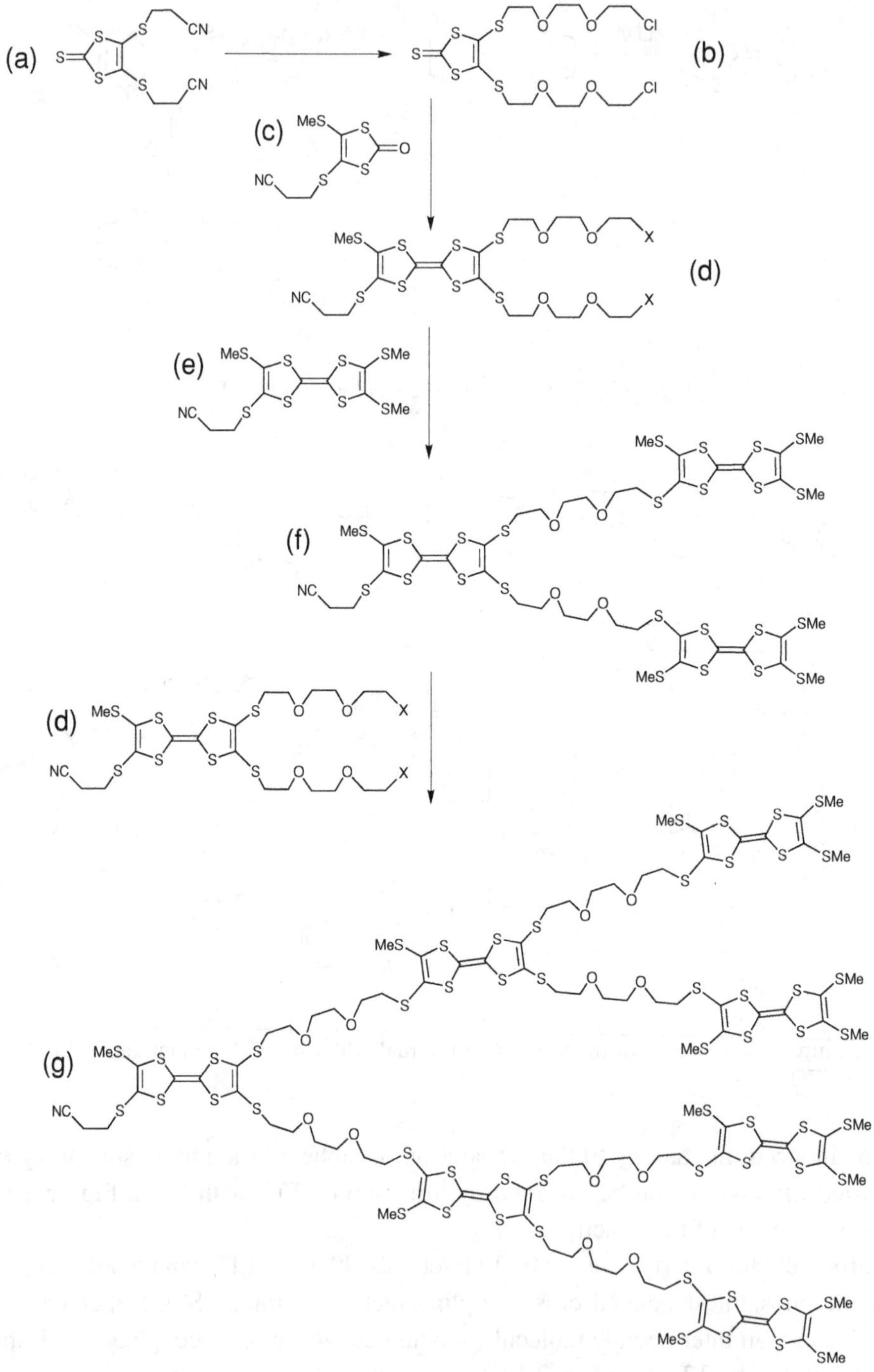

Figure 2.15. Scheme of the synthesis of a TTF-based dendrimer.

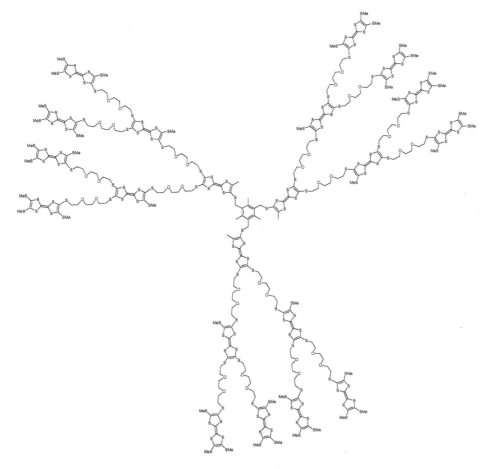

Figure 2.16. TTF$_{21}$-glycol dendrimer obtained with the method of deprotection/alkylation of cyanoethyl-protected TTF-thiolates described in Fig. 2.15 (Christensen *et al.*, 1998).

C$_{60}$-TTF molecules, interlocked molecules (catenanes, rotaxanes), donor–acceptor macrocycles, cage molecules, etc. (Jeppesen *et al.*, 2004). It is beyond the scope of this book to review such developments and I appeal to the curiosity of researchers really not familiar with such macromolecules to see how big the molecules can become.

One astute way to obtain macrocyclic systems with TTF is the stepwise method of deprotection/alkylation of cyanoethyl-protected TTF-thiolates. With this method molecular units can be built but with the precaution of preserving one cyanoethyl group in order to be able to iteratively proceed with the oligomerization. Combining such units, larger units can be produced. An example of a TTF dendrimer containing 21 TTFs is shown in Fig. 2.15 (Christensen *et al.*, 1998). Here only the main philosophy of the synthesis is discussed.

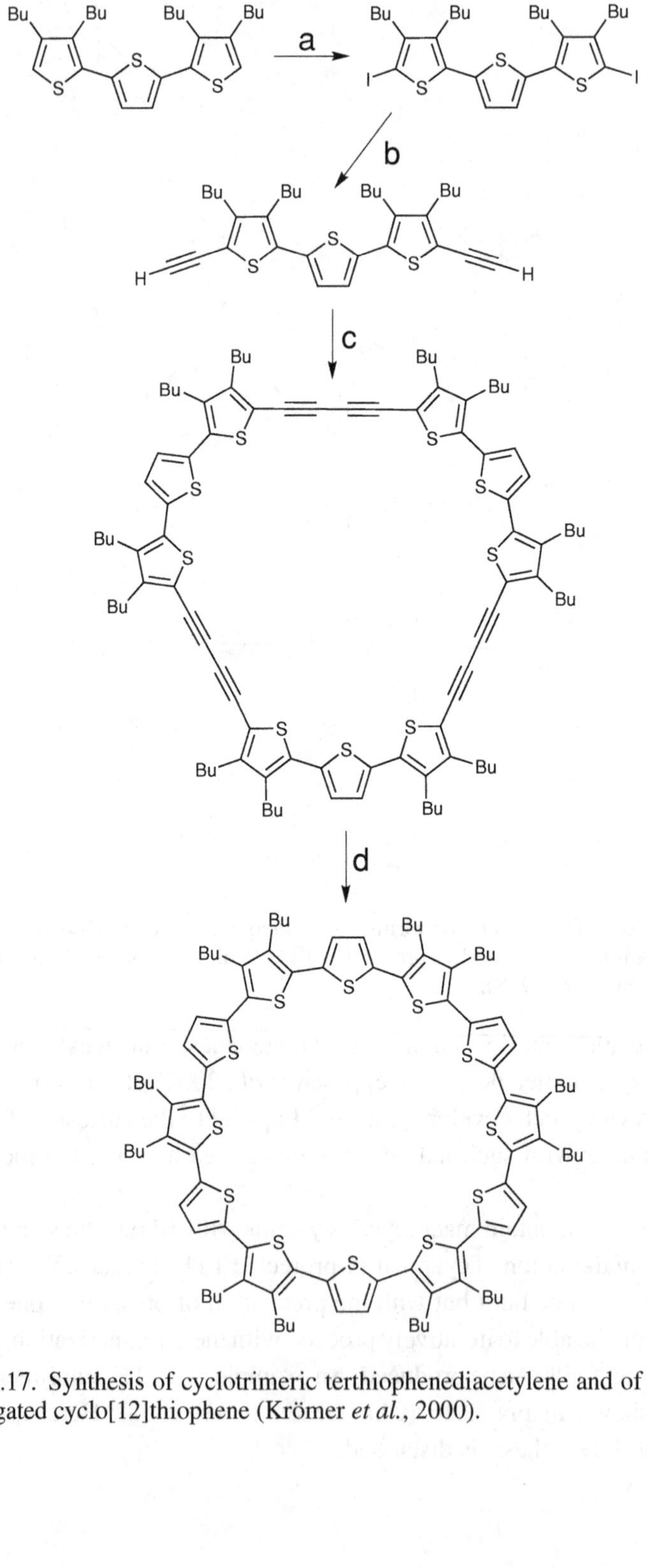

Figure 2.17. Synthesis of cyclotrimeric terthiophenediacetylene and of the fully α-conjugated cyclo[12]thiophene (Krömer *et al.*, 2000).

The starting cyanothioethyl thiolate molecule (a) is converted to the thione (b) and coupling of (b) with (c) yields the fundamental TTF-derivative brick (d). Reaction with (e) renders (f) and the iterative reaction of (f) with (d) gives larger molecules such as (g). Indeed the correct experimental conditions have to be met for each step. As pointed out above, it is mandatory in order to proceed with the dendrimerization to preserve the cyanoethyl *hook*.

Radial dendrimeric structures can be achieved by choosing a core, e.g., benzylic bromides, because (g) will attach to the core at the bromine location, as schematized in Fig. 2.16.

To finish this section and chapter we will briefly discuss the synthetic routes pursued in order to obtain cyclothiophenes, fully α-conjugated macrocyclic oligothiophenes, i.e., joining boths ends of linear α-thiophenes (Krömer *et al.*, 2000). One advantage of these macromolecules is that the mean diameter of the cavity can be tuned by selecting the initial building blocks. In Section 4.2 we will explore how such molecules, in particular cyclotrimeric terthiophenediacetylene, self-assemble on a surface.

Figure 2.17 shows the simplified scheme of the cyclization reaction. The chosen starting building block is butylated terthiophene. Symmetrical substitution of thiophene units with butyl groups has proven to be ideal for maintaining sufficient solubility and avoiding problems of regioisomerization. Thiophenes become iodated at the α-positions by elemental iodine and $Hg(CH_3COO)_2$ in $CHCl_3$ to yield diiodothiophenes (step a in Fig. 2.17). Diiodothiophenes become thiophenediynes by palladium-catalyzed coupling with TMSA (step b in Fig. 2.17). Macrocyclization is obtained by controlled addition of pyridine solutions of the resulting oligothiophenes to a slurry of anhydrous CuCl and $CuCl_2$ in pyridine at RT (step c in Fig. 2.17). Separation of the mixtures is achieved by preparative HPLC. The fully α-conjugated cyclo[n]thiophenes, after elimination of the triple bonds, are obtained from the reaction of the cyclooligothiophenediacetylenes with Na_2S (step d in Fig. 2.17). The full synthesis has been oversimplified here. Such experiments are in reality quite complex.

The preparation of large clusters is certainly not exclusive to organic molecules. Inorganic polyoxometallate clusters containing 154 molybdenum atoms per cluster or more have been obtained (Müller *et al.*, 1998).

3

Building materials: crystal engineering

Qué ley, justicia o razón
negar a los hombres sabe
privilegio tan suave,
excepción tan principal,
que Dios le ha dado a un cristal,
a un pez, a un bruto y a un ave?

For what law can so depart
From all right, as to deny
One lone man that liberty
That sweet gift which God bestows
On the crystal stream that flows,
Birds and fish that float or fly?

Pedro Calderón de la Barca, *La vida es sueño*
(Translation by D. F. MacCarthy)

The characterization of the intrinsic physical properties of materials strongly relies on the availability of high-quality single crystals of suitable dimensions. The high quality guarantees that the density of defects (impurities, dislocations, etc.) is low. The needed crystal dimensions and shape depend on the required specific characterization technique. Spectroscopic techniques with micrometre resolution can work with really small crystals (typically a few hundreths of micrometres) but those techniques where orientation is important and where external preparation is needed (e.g., evaporation of gold contacts for transport measurements) demand larger dimensions. The preparation of such ideally perfect crystals is in general a difficult task and it is not impossible that some exotic published phenomena arise from defective crystals of insufficient quality. Hence the importance of the synthesis of single crystals of highest quality. Most of the physics studied for MOMs corresponds to TTF-TCNQ, BEDT-TTF salts, BFS, oligomers such as α-6T and related compounds such as pentacene, because the synthesis of single crystals of extreme perfection has been mastered. In Section 1.7 a beautiful example of the observation of giant Shubnikov–de Haas oscillations in the magnetoresistance of β_H-(BEDT-TTF)$_2$I$_3$ was shown; only possible because of the quality of the single crystals.

When the synthesis of suitable single crystals implies extreme difficulty or when the sought crystallographic phases are metastable, their preparation in the form of thin films becomes the logical alternative. Nano or micrograms of MOMs can

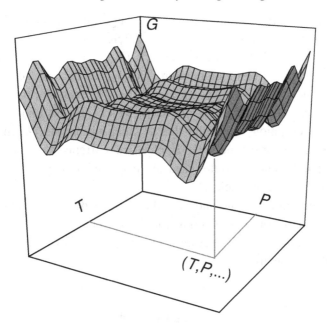

Figure 3.1. Free energy hypersurface $G(T, P, \ldots)$ in the parameter hyperspace.

be distributed with different degrees of homogeneity onto well-selected substrates covering sufficiently large areas. However, one has to be aware in this case of all the artifacts associated with the sample morphology, such as grain boundaries, degree of crystallization and orientation, wetting, etc. that can mask the intrinsic physical properties. We shall study some examples in Chapter 6.

In general, many externally accessible variables are involved in the crystallization process (temperature, pressure, molecular flux, distance, time, concentration, solvent, substrate, etc.). These variables, which may be continuous (e.g., temperature) or discrete (e.g., type of solvent), define the so-called *parameter hyperspace* and the crystallization process of a given crystallographic phase is achieved for a given combination of such variables, $(x_1, x_2, \ldots) = (T, P, \ldots)$ defined as a point in parameter hyperspace. Mathematically this can be regarded as a free energy G hypersurface $G(x)$ in N-dimensional space, defined by the external variables. Every material has its own free energy hypersurface, which will be characterized by several local minima with different energy barriers. Every minimum in G does not necessarily correspond to a polymorph. Distinct crystallographic phases should correspond to *absolute* minima, where $\partial^2 G / \partial x_i^2 > 0$ for all external variables x_i, with barrier heights of typically less than 0.2 eV. Figure 3.1 schematizes such an approach. It is important to note that for a given point in parameter hyperspace a single energy value exists.

The number of external variables can be very large and depends on the particular growth technique. Concerning temperature, different qualifiers can be assigned, e.g., substrate (T_{sub}), bath (T_{bath}), evaporation (T_{evap}), annealing (T_{ann}), etc. The spatial and time distribution of temperature, $T(x, t)$, can become a critical variable and as a consequence its gradients $\partial T/\partial x$ and $\partial T/\partial t$ become important factors. Analogous degrees of variety are found for the rest of the external variables such as pressure, concentration, molecular flux Φ_m, deposition rate D_t, chopper cycle time, external electric or magnetic fields, electrical currents, etc. The geometry of the specific apparatus and substrates used also plays an important role, in particluar when dynamic processes are relevant.

Let us next be more specific and describe in some detail the fundamentals of the most common preparation methods, for the production of both single crystals and thin films, illustrated with relevant case examples discussed in the book. Key references will also be given.

3.1 Single crystals

Wet preparation methods

Crystallization from solution

Perhaps the most simple way of obtaining small single crystals, albeit in an uncontrolled way, is the drop casting technique. This method consists in the deposition of a drop of a saturated solution of a soluble material on a clean substrate prior to the evaporation of the solvent in air, usually at RT. Some interesting examples are shown next.

Figure 3.2 shows optical microscope images of EDT-TTF-(CONHMe)$_2$ crystals ($P2_1/n$, $a = 0.763$ nm, $b = 1.612$ nm, $c = 1.352$ nm, $\beta = 99.61°$) grown by drop casting on a microfabricated electronic circuit prepared for transport measurements (Colin *et al.*, 2004). The solution was prepared by dissolving EDT-TTF-(CONHMe)$_2$ microcrystals in CH$_3$CN. As evident from the figure, straight and long needles are obtained (Fig. 3.2(a)). The width of the needles always remains smaller than 15 μm, while their typical length is *c.* 200 μm but can easily reach up to 1 mm. However, thinner needles of width below 1–2 μm are not necessarily straight and form long micro-nanowires (Fig. 3.2(b)). Similar results have been obtained for DTTTF (Mas-Torrent *et al.*, 2004) and (TMTSF)$_2$ClO$_4$ (Colin *et al.*, 2004).

When the solvent evaporation is performed in a well-controlled way, sufficiently large single crystals can be obtained. This is the case for pentacene, where single crystals are grown from solution in TCB by slowly evaporating the solvent over a period of four weeks at 450 K, under a stream of ultrapure N$_2$ gas.

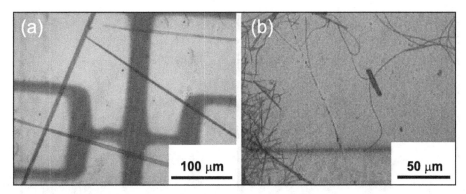

Figure 3.2. (a) Single crystals and (b) wires of EDT-TTF-(CONHMe)$_2$ observed with an optical microscope. The substrate is highly doped silicon with a 500 nm thermally grown oxide on top and titanium-gold contacts. Reprinted from *Synthetic Metals*, Vol. 146, C. Colin, C. R. Pasquier, P. Auban-Senzier, F. Restagno, S. Baudron, P. Batail and J. Fraxedas, *Transport properties of monocrystalline microwires of EDT-TTF(CONHMe)$_2$ and (TMTSF)$_2$ClO$_4$*, 273-277, Copyright (2004), with permission from Elsevier.

However, when the evaporation of the solvent occurs very fast (few minutes) then the resulting surface of the grown microcrystals is usually quite imperfect and sometimes really rough and peculiar, as for the case of the EDT-TTF-(CONHMe)$_2$ microcrystals discussed a few lines above. Figure 3.3 shows nanovolcanoes originating at the surface due to the eruption of the solvent. The amazing nanoworld that keeps captivating us!

Some binary MOMs form very stable phases in solution, which translates into rapid synthetic processes. TTF-TCNQ belongs to this class of materials and can be prepared quite simply as a black powder. In fact when highly purified TTF and TCNQ are combined in CH$_3$CN, the 1:1 complex precipitates from the solution.

A second example is the synthesis of the TTF[M(dmit)$_2$]$_x$ compounds, where M = Ni, Pd or Pt. Metathesis reaction under inert atmosphere of CH$_3$CN solutions of (TTF)$_3$(BF$_4$)$_2$ and the appropriate metal salt TBA[M(dmit)$_2$] immediately yields insoluble, black, shiny microcrystalline powders. Because TTF[M(dmit)$_2$]$_x$ is virtually insoluble in common solvents, its recrystallization from solution is precluded. Therefore, when single crystals are needed some alternative synthesis routes have to be undertaken.

In the case of TTF-TCNQ, single crystals with typical dimensions of $2 \times 0.2 \times 0.02$ mm^3 can be grown from solutions of multiply distilled CH$_3$CN using a U-tube diffusion technique in a glove box filled with argon. High quality crystals are usually obtained after approximately 3 days.

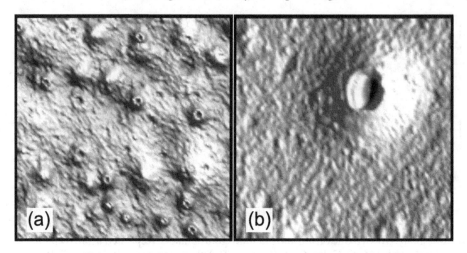

Figure 3.3. TMAFM images of nanovolcanoes formed at the surface of an EDT-TTF-(CONHMe)$_2$ microcrystal after eruption of the CH$_3$CN solvent to the atmosphere. The microcrystal was formed after the deposition of a drop of a saturated solution of EDT-TTF-(CONHMe)$_2$ in CH$_3$CN on a glass coverslip and exposed to ambient conditions. (a) 10 μm × 10 μm and (b) 2 μm × 2 μm. The illumination is set in such a way that the feature appears like a volcano seen by a satellite.

Single crystals of TTF[M(dmit)$_2$]$_x$ can be obtained by slow interdiffusion of saturated solutions of (TTF)$_3$(BF$_4$)$_2$ and TBA[M(dmit)$_2$]. These experiments must also be carried out under an inert atmosphere in e.g., a three-compartment H-tube with a central solvent chamber and porous glass frits between compartments. The concentration of the solution should be kept close to saturation during the entire diffusion process. This is accomplished by means of additional containers filled with an excess of solid starting reagent, placed in the appropriate compartment and communicating with it through a high porosity glass frit. The cell needs to be thermostated at T_{bath} = 300–320 K in a dark chamber for 10–30 days. Crystals with a maximum of $1 \times 0.1 \times 0.02$ mm^3 have been obtained under these conditions (Bousseau *et al.*, 1986).

Electrocrystallization

The EC technique is a general and versatile synthesis method for the preparation of high-quality single crystals involving molecular ions (Batail *et al.*, 1998). The method requires electroactive species, neutral or charged, leading their electrooxidation (or reduction) to stable radicals. If soluble, the generated radical species may diffuse into solution, but under suitable conditions of concentration, solvent, temperature and current density, they will crystallize on the electrode. The choice

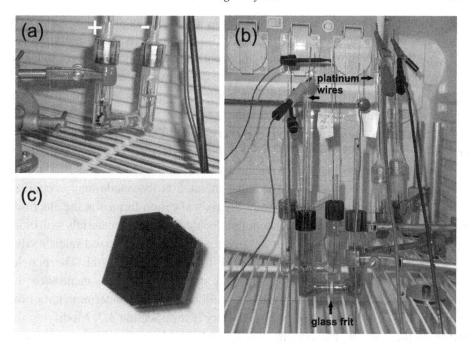

Figure 3.4. Picture of running EC cells: (a) detail showing the crystals at the anode (+) and (b) general view of the installation. (c) Single crystal of (BEDT-TTF)$_2$I$_3$ grown by EC. Courtesy of Dr P. Batail.

of the electrolysis solvent is limited by its ability to dissolve both the organic donor and the electrolyte. For this purpose, a mixture of two or more solvents may be required. It becomes evident that the salt of the electrooxidized species must not be soluble in those solvents.

The use of a stable and constant direct current source, the so-called galvanostatic mode, allows control of the local concentration of electroactive species and thus the rate of crystal growth. The other parameters involved (concentration, solvent and temperature) essentially control the solubility of the crystalline phase. The appropriate combination of these factors, together with the use of high-purity starting materials, ultimately determines the success of the EC experiment. A picture of a typical U-shaped EC cell is given in Figs. 3.4(a) and (b). Platinum wires, typically with diameters of 1 mm, are used as electrodes and both compartments are separated by a fine porosity glass frit.

EC allows high-purity materials to be reproducibly obtained only if all materials and chemicals involved are properly purified. This includes the glassware, which should be thoroughly cleaned. Solvents of high-purity grade should be dried on activated alumina or distilled just prior to use. Finally, the electrolyte, as well as the donors, should be recrystallized several times, and the latter should be ultimately purified by sublimation whenever possible. The electrolyte is typically introduced as a tetraalkylammonium or tetraphenylphosphonium salt to ensure its solubility in the organic solvents necessary to dissolve the donor molecules. CH_2Cl_2, CH_3CN, C_6H_5Cl, C_6H_5CN, TCE and THF are commonly used solvents. Also, the electrolyte counteranion needs to be stable at the working oxidation potential.

Let us consider the example of the synthesis of $(BEDT\text{-}TTF)_2I_3$ by oxidation of the neutral donor molecule BEDT-TTF. The electrolyte, e.g., $TBA\cdot I_3$, is introduced in both compartments together with the solvent, e.g., CH_3CN, and BEDT-TTF is introduced in the anode (+). The salt generated at the anode may crystallize with a 1:1 stoichiometry, but frequently the radical cation formed at the electrode associates with one or more neutral donor molecules producing materials with other stoichiometries, for example, 1:2 or 2:3, and thus giving rise to mixed valence salts. Figure 3.4(c) shows a high-quality $(BEDT\text{-}TTF)_2I_3$ single crystal. The platelet shape is intimately related to the 2D electronic structure. As already mentioned, the extremely high quality of these crystals has allowed e.g., the determination of the giant Shubnikov–de Haas oscillations as described in Section 1.7. Modifying the nature of the solvent, the current density and the electrolyte leads to different polymorphs, particularly important for BEDT-TTF-based RISs. The kinetics of crystallization is also dominated by the state of the electrode surface, thus becoming an interfacial problem.

Needle $TTF\text{-}Br_{0.76}$ crystals exhibiting widths ranging from 0.2 to 4 μm, lengths of 10–15 μm and heights of 0.1–0.75 μm have been obtained by EC on freshly cleaved HOPG electrodes (Hillier & Ward, 1994). The EC growth was performed in an AFM electrochemical cell using an electrolyte solution of 5 mM TTF in a 0.1 M solution of $TBA\cdot Br$ in ethanol in order to *in situ* characterize the growth by means of AFM images. This point will be discussed in Chapter 4.

In addition to the diffusion method discussed above, crystals of $TTF[Ni(dmit)_2]_2$ can also be obtained by galvanostatic EC of a CH_3CN solution containing $TBA[Ni(dmit)_2]$ and a large excess of neutral TTF. The procedure to obtain the crystals is the following. A nitrogen-flushed U-type cell where the anode and cathode are separated by a medium-porosity glass frit is used with platinum electrodes. With the cell thermostated at $T_{bath} = 293$ K and the electrical current held constant at $I = 5$ μA (current density $= 5 \times 10^{-5}$ A cm^2), black needle-like crystals form on the anode. Finally they are collected on a glass frit by filtration, washed with CH_3CN and vacuum-dried.

A last example is $(TMTSF)_2ClO_4$. High-quality crystals can be obtained by oxidation of TMTSF in TCE containing $TBA\cdot ClO_4$. Typical dimensions are $4 \times 0.2 \times 0.2$ mm^3 (Bechgaard *et al.*, 1981) but centimetre long crystals have been achieved. Again, the 1D shape is related to its electronic structure.

(a) (b)

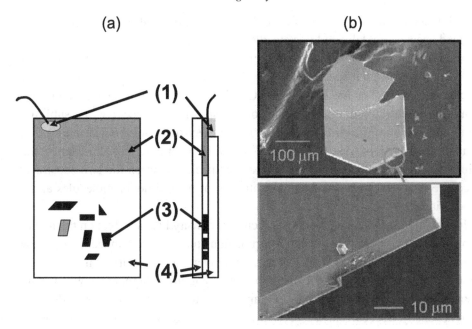

Figure 3.5. (a) Schematic drawing of a confined anode. (1) represents the connection with an external electric circuit, (2) the metal deposit, (3) the thin single crystals and (4) the insulating substrates. (b) SEM images of a thin single crystal of $(BEDT-TTF)Cu_2(NCS)_3$. The maximum thickness of this crystallite is c. 10 μm. Reprinted with permission from Deluzet *et al.*, 2002b.

Confined electrocrystallization

When the EC process is forced to occur within two opposite insulating flat substrates the technique is termed CEC and leads to thin single crystals, their thickness ultimately defined by the separation between both substrates (Thakur *et al.*, 1990). This technique can be universally applied, as the parent standard EC technique, to the synthesis of a variety of thin single crystals of conducting and insulating molecular materials.

As illustrated in Fig. 3.5(a), the spatial confinement is achieved by opposing two flat insulating substrates, e.g., glass, mica, a thick silicon dioxide layer grown on a silicon substrate, etc. On one of the substrates a metal film is deposited, e.g., gold, which acts as the working electrode during the EC experiment. The two substrates are mechanically held together and immersed in the electrolyte solution. The geometry of the confined anode appears to have little effect on the size and quality of the grown crystals.

In a CEC experiment, there are several salient differences as compared to standard EC, a prominent one being the duration of the experiment. While a standard EC experiment can last a few days, it takes typically several weeks for CEC experiments. Along the course of an EC experiment in solution, where neutral or charged species are delivered in a steady-state mode to the working electrode, the limiting kinetic step is essentially the electron transfer rate. In a confined environment, the molecular species will progress very much more slowly to the active electrode vicinity. For the growth of RCSs it is therefore indicated to deposit a thin film of neutral π-donor molecules on the substrate surface, e.g., by sublimation under vacuum, prior to immersion in the electrolyte solution, resulting in a combined dry–wet process (Miura *et al.*, 1996; Deluzet *et al.*, 2002b). Hence, the effect of slow movement of the electroactive species between the two substrates is counterbalanced by a sizeable concentration of neutral donor molecules at the anode vicinity.

In CEC, the set value of the constant current density has no effect on the nucleation of the RCSs because of the high current densities achieved. Indeed, with a few μm of gold deposits acting as the anode, for a given constant current intensity, the actual current density is typically three orders of magnitude higher than when using a standard platinum wire of 1 mm in diameter. Thus, in order to achieve identical current density, a set current of 1 μA in the confined experiment configuration ought to be increased to 1 mA in the classical EC cell. For example, EC of EDT-TTF in the presence of the polyoxometallate anion $[PW_{12}O_{40}]^{-3}$ has been achieved in a CEC experiment with a set constant current of 1 μA. A similar experiment conducted in standard EC conditions in solution also yielded single crystals, this time using a constant current set at 100 μA in order to ensure that the current density be kept as constant as possible between the two experiments. In both cases, the same phase, formulated $(EDT\text{-}TTF)_3[PW_{12}O_{40}](CH_3CN)(CH_2Cl\text{-}CHCl_2)$ was obtained (Deluzet *et al.*, 2002b).

In order to increase the thin crystal dimensions, a CEC experiment should be run at temperatures typically higher than those for the parent EC in solution. This, however, has its limit when one observes that the superconducting κ-phase, κ-$(BEDT\text{-}TTF)_2Cu(NCS)_2$ is selectively grown at 298 K, while the insulating phase, $(BEDT\text{-}TTF)Cu_2(NCS)_3$, is obtained instead at 333 K, all other parameters being identical. *Elevated temperatures favour fully oxidized over mixed valence formulations.* Figure 3.5(b) shows a scanning electron microscopy (SEM) image of a thin $(BEDT\text{-}TTF)Cu_2(NCS)_3$ single crystal.

Thin crystals of κ-$(BEDT\text{-}TTF)_2Cu[(N(CN)_2Br]$, a 2D metal with $T_c = 11.6$ K at ambient pressure, which forms bulk crystals with a parallelepiped morphology in standard EC experiments, have also been obtained with CEC displaying the same transition (Deluzet *et al.*, 2002b).

In addition, in Section 6.4 we shall see how CEC has allowed the preparation of polymorphs of BFS, a significant example because only one crystallographic phase was known, after countless EC syntheses in many laboratories around the world

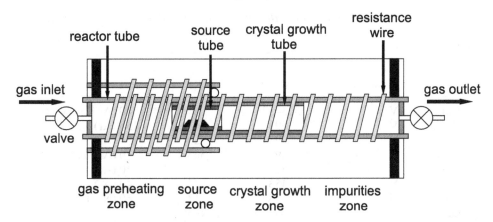

Figure 3.6. Schematic representation of an apparatus for production of sublimation crystals in a horizontal arrangement. The total length is typically 400 mm. Adapted from Laudise *et al.*, 1998.

obtained only the well-known triclinic phase. This implies that with CEC different regions of the parameter hyperspace can be explored.

Dry preparation methods

Physical vapour growth

Physical vapour growth in horizontal and vertical systems has been successfully used to grow mm–cm sized single crystals of single-component compounds, e.g., α-4T, α-6T, pentacene, anthracene and CuPc (Kloc *et al.*, 1997; Laudise *et al.*, 1998). A scheme of the apparatus used in a horizontal arrangement for producing sublimation crystals is shown in Fig. 3.6.

The source and the deposition regions are in separate tubes within the outer reactor tube. The starting material, usually in the form of purified powder, can be easily inserted with this geometry and both the final resulting residual material as well as the crystals can be removed without difficulty after growth. In this scheme, heating is applied by a resistance wire wound around the tubes designed in such a way that reproducible temperature gradients are achieved. Gas inlet and outlet tubes allow the introduction of inert gases such as argon, helium and nitrogen, as well as hydrogen. With the valves the growth can be performed either in a closed configuration (valves closed) enhancing convection or in an open configuration (valves open), precisely regulating the gas flow.

Table 3.1 displays experimental growth conditions for organic semiconductors, leading to high-quality single crystals. As for most growth methods the working conditions are critical and this is best illustrated here for pentacene (Mattheus

Table 3.1. *Typical experimental conditions for the growth of mm or larger single crystals of organic semiconductors*

Material	Arrangement	T_{source} (K)	T_{dep} (K)	Φ_{gas} (ml/min)
α-6T LT	V	553–573		40 (Ar,He,N_2+H_2)
α-6T HT	V	593	563	150 (Ar)
α-6T LT	V	593	523	150 (Ar)
α-6T HT	H	593	563	40 (Ar,He,N_2+H_2)
α-4T LT	V	413		40 (Ar)
α-4T HT	V	453		40 (Ar)
α-4T HT	H	458	413	60 (N_2+H_2)
anthracene	H	433	423	50 (He)
pentacene	H	558	493...558	30 (N_2+H_2)
CuPc	V	743	723	75 (He)

Note: This table is adapted from Table 1 of Laudise *et al.*, 1998. H and V represent horizontal and vertical. T_{source} and T_{dep} stand for source and deposition temperatures and Φ_{gas} corresponds to the gas flow rate.

et al., 2001). The source material was sublimed at 550 K and crystallized at the other end of the tube at approximately 490 K. The growth was performed under a stream of N_2 (99.999% purity) and H_2 gases (99.995% purity), with a volume percentage of 5.1% H_2. This yielded almost cm-sized violet crystals in the form of platelets and needles. Also, at a different part of the tube, a small amount of hydrogenated pentacene crystals (red needles) was found. Lowering the flow rate yielded a larger fraction of hydrogenated pentacene. At lower hydrogen content or if no ultrapure inert transport gas is used, the pentacene oxidizes, forming pentacenequinone, and small brown needles crystallize (see Section 2.1 for the synthesis of pentacene from pentacenequinone). Thin pentacene crystals in the form of lamellae up to 10 mm long and 1–3 mm wide have been obtained working in similar conditions (Siegrist *et al.*, 2001).

Crystallization of self-confined solutions

Exploring the surface morphology at the nanometre scale of MOMs leads to surprises as shown before with the formation of nanovolcanoes. Freshly prepared single crystals of EDT-TTF-(CONHMe)$_2$ exhibit the morphology depicted in Fig. 3.7. Dark needle-shaped single crystals are synthesized by reacting a CH$_3$CN solution of EDT-TTF-(CO$_2$Me)$_2$ with an excess of CH$_3$NH$_2$. After stirring at RT for 24 hours, the resulting precipitate is filtered, recrystallized twice from CH$_3$CN and dried in air. The tapping mode AFM (TMAFM) images shown here were taken at

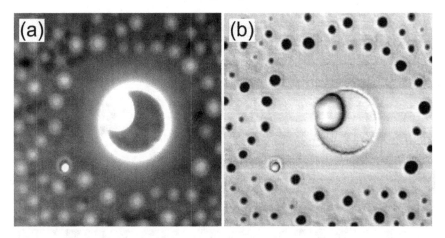

Figure 3.7. TMAFM images of a (001) surface of an as-received EDT-TTF-(CONHMe)$_2$ single crystal measured under ambient conditions (2.5 μm × 2.5 μm): (a) topography and (b) phase. The phase angle is defined as the phase shift observed between the cantilever oscillation and the signal sent to the piezo-scanner driving the cantilever.

ambient conditions using microfabricated silicon cantilevers with ultrasharp silicon tips (tip radius $R < 10$ nm).

Let us analyse the TMAFM images in some detail. The topography image of Fig. 3.7(a) depicts circular features with higher (brighter) central parts, which in turn show large phase contrasts of about 40 degrees (dark regions) in Fig. 3.7(b). The exception is the larger feature at the centre of the image inside the large circle. The large phase contrast along with the amplitude images strongly suggest that the structures exhibit clearly differentiated adhesion in comparison with the regions outside these features, thus suggesting a liquid or liquid-like nature in the form of droplets. Those features exhibiting low phase contrast show relatively high protrusions (*c.* 40 nm in height) indicating that for some reason spontaneous crystallization has occurred within the initial droplets. Confinement is achieved with the more external and lower circles surrounding the features, with heights of *c.* 7 nm, which we will term nanobeakers and which act as nanocontainers. The large circle in Fig. 3.7 is likely to have grown upon coalescence of smaller droplets.

Nanobeakers originate from droplets of a saturated solution of EDT-TTF-(CONHMe)$_2$ in CH$_3$CN, the crystallization solvent, which remain in air at the crystal surface after the synthesis process and undergo an outward capillary flow in which pinning of the drop-surface contact line of the drying drop ensures that solvent evaporating from the edge is replenished by solution from the interior, thus originating an EDT-TTF-(CONHMe)$_2$-rich edge (Fraxedas *et al.*, 2005). The microscopic picture of this process is exemplified by the formation of ring stains

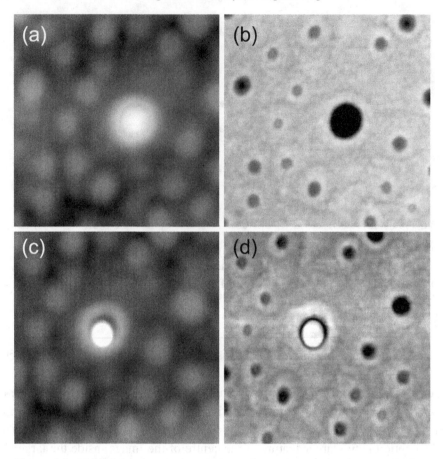

Figure 3.8. TMAFM images of a (001) surface of an as-received EDT-TTF-(CONHMe)$_2$ single crystal measured under ambient conditions (750 nm × 750 nm): (a) topography and (b) phase before perturbation and (c) topography and (d) phase after perturbation.

when a drop of coffee dries on a solid surface (Deegan *et al.*, 1997). The as-received droplets that are finally observed in the AFM images thus arise from the incomplete capillary flow process, which has been stopped at some stage probably in association with the large contact line curvature, and they thus retain a less concentrated solution inside the edges.

What is more amazing is the possibility of intentionally growing nanocrystals, confined in their nanobeakers, by controlled perturbation of the droplets with the AFM tip. Let us take again the as-received scenario and concentrate on a given nanodroplet. This is illustrated in the topography and phase TMAFM images from Figs. 3.8(a) and (b), respectively. The height of the central most prominent droplet is *c.* 15 nm. When the cantilever tip is brought into contact with the droplets they crystallize, resulting in *c.* 30 nm high nanocrystals, shown in Figs. 3.8(c) and (d). The effect of the tip is to release the solvent and the crystallization is enhanced by the ease with which EDT-TTF-(CONHMe)$_2$ grows along its *a*-axis.

3.2 Thin films

Wet preparation methods

Langmuir–Blodgett

The widely used Langmuir–Blodgett (LB) technique, named after I. Langmuir and K. Blodgett, is a way of preparing ultrathin organic films with a controlled layered structure and is based on the assembly of condensed monomolecular films on the surface of water followed by transfer to solid surfaces (Langmuir, 1917; Blodgett, 1935). LB films constitute the earliest example of man-made supramolecular assembly, providing a relatively high level of control over the orientation and placement of molecules in monolayer (ML) and multilayer assemblies that otherwise is hardly available. Furthermore, large uniform substrate areas can be covered by this method. A diagrammatic representation of the sequential process of depositing the layers is given in Fig. 3.9.

A short explanation of the fundamentals of this method is given next. A detailed description of the technique can be found in Petty, 1996. A trough is filled with pure water, the so-called subphase, and moveable barriers are used to skim the surface of the subphase permitting control of the surface area available to the floating ML. To form a Langmuir ML, the molecule of interest is dissolved in a volatile organic solvent, e.g., CHCl$_3$ or hexane, that will not react with or dissolve into the subphase. Surface active molecules are normally amphiphilic, with separate hydrophilic (–OH, –COOH, –NH$_2$, etc.) and hydrophobic (–CH$_3$) groups. A quantity of this solution is placed on the surface of the subphase, and as the solvent evaporates, the surfactant molecules spread. The result is that the hydrophobic tail points towards the air and the hydrophilic head points towards the water surface. Hence the molecules at the interface are anchored, strongly oriented and with no tendency to form a layer more than one molecule thick.

In fact a liquid surface in the absence of external perturbations such as mechanical vibrations is perhaps the most smooth (disordered) surface that can be achieved by the action of gravity. Clearly, LB films could not be grown in the absence of gravity, e.g., in a Space Shuttle. In the case of water, the flat water/air interface has a roughness of 0.3 nm, a value determined by X-ray reflectivity measurements (Braslau *et al.*, 1985). A further advantage of water is its relatively high surface tension γ_0, when compared to other liquids, which amounts to 72.8 mN m^{-1} for pure water at RT. This value originates in the formation of a network of weak

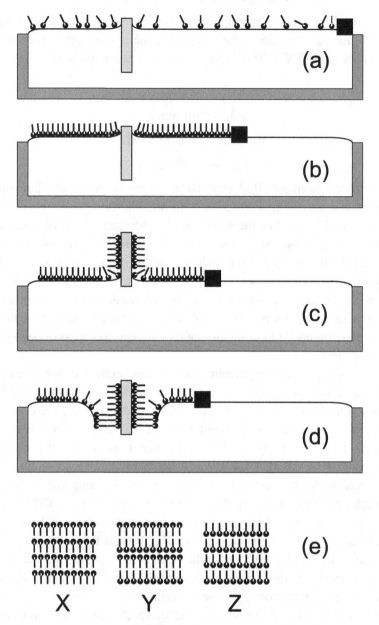

Figure 3.9. Scheme of the LB technique: (a) deposition of the amphiphiles on the water subphase with the solid substrate submerged; (b) by action of the barrier the ML is compressed; (c) by pulling out the substrate vertically a ML is transferred to both sides of the substrate; (d) a multilayered structure is built up by repeated up and down strokes of the substrate. (e) Distinct arrangements of LB films.

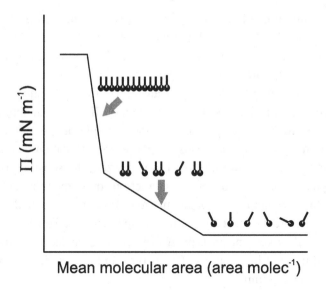

Figure 3.10. Idealized pressure vs. area isotherm indicating phase transitions of the ML.

O–H $\cdots$ O hydrogen bonds at the water/air interface. In general, the state of the ML on the water surface is monitored by measuring the surface pressure Π, defined as the difference between γ_0 and that of the film-covered surface γ. Π is of the order of a few mN m^{-1}.

As the ML is compressed on the water surface by the moveable barrier or piston, it will undergo several phase transitions, which can be readily identified by monitoring Π as a function of the area occupied by the film. An idealized pressure vs. area isotherm is shown in Fig. 3.10. This is the 2D equivalent of the pressure vs. volume isotherm for a gas–liquid–solid system. If the area per molecule is sufficiently high, the floating film will behave as a 2D gas phase (disordered) where the molecules are far enough apart, resulting in negligible intermolecular interactions. As the ML is compressed the pressure rises, signalling a change in phase to a 2D liquid state, called the expanded ML phase. Upon further compression, the pressure begins to rise more steeply as the liquid expanded phase transforms to a condensed phase, in fact to a series of condensed phases. The emergence of each condensed phase can be accompanied by constant pressure regions of the isotherm (plateaus), associated with enthalphy changes in the ML. MLs can be compressed to pressures considerably higher than the equilibrium spreading pressure. The surface pressure continues to increase with decreasing surface area until a point is reached where it is not possible to increase the pressure any further. This is referred to as collapse. The forces acting on a ML at this point are quite high. When collapse occurs, molecules are forced out of the ML.

The most common method of LB transfer is vertical deposition (see Fig. 3.9). However, horizontal lifting of Langmuir MLs onto solid supports, called Langmuir–Schaeffer deposition, is also possible. Either highly hydrophilic or highly hydrophobic substrates are desired. In the conventional LB mode the substrate is dipped vertically through the ML with transfer via the hydrophobic interactions between the alkyl chains and the surface, once the ML has been spread and compressed to the desired transfer pressure. In the example of Fig. 3.9 a hydrophilic substrate is submerged in the aqueous subphase prior to the spreading and compressing of the ML film. After the ML has stabilized, the substrate is withdrawn from the subphase, and the hydrophilic interactions drive the transfer.

Different film architectures can result upon deposition as depicted in Fig. 3.9(e). Y-type multilayers are the most common and can be prepared on either hydrophilic or hydrophobic substrates, and are typically the most stable due to the strength of the head–head and tail–tail interactions. The X-type and Z-type films, with head–tail interactions, are less common.

Several examples of LB films of molecule-based conductors and magnets have been reported (Bryce & Petty, 1995; Talham, 2004). Here we outline just some of them. By grafting alkyl side chains to either electron donating or accepting head groups, conducting LB films of e.g., TTF derivatives have been fabricated and subsequently doped by iodine vapour. EDT-TTF-$(SC_{18})_2$ with behenic acid as the matrix molecule (Dourthe *et al.*, 1992) and TTF-C(S)-O-$C_{16}H_{33}$ (Dhindsa *et al.*, 1992) both doped with iodine give $\sigma_{RT} \simeq 1 \ \Omega^{-1} \ cm^{-1}$. Behenic, also known as docosanoic acid, is a fatty acid with molecular formula $C_{21}H_{43}COOH$. Such conductivities are high for a LB film but low compared to the related solid-state salts. Examples of LB films with amphiphilic TCNQ, $C_{10}TCNQ$, exhibit $\sigma_{RT} \simeq$ 10 $\Omega^{-1} \ cm^{-1}$ when combined with BEDO-TTF using icosanoic acid as a matrix (Nakamura *et al.*, 1994). Icosanoic acid, also known as arachidic acid, is a fatty acid with the formula $C_{19}H_{39}COOH$. The highest σ_{RT} values, up to 25 $\Omega^{-1} \ cm^{-1}$, have been reported for a series of tridecylmethylammonium salts of $Au(dmit)_2$, again stabilized with icosanoic acid and electrooxidized (Nakamura *et al.*, 1989). A complete list of electrically conductive LB films can be found in Nalwa (1997: Vol. 1, Ch. 14).

Magnetism in LB systems is extremely interesting because of their intrinsic 2D structure. The earliest extensive investigation of magnetism in LB films was performed on manganese stearate (Pomerantz *et al.*, 1978). In these systems the Mn^{+2} ions are confined to a single layer between two stearate layers deposited head-to-head. Stearic acid is the common name for the fatty acid $C_{17}H_{35}COOH$. Weak ferromagnetic ordering below 2 K was hypothesized for such MLs (Pomerantz, 1978). The most extensive magnetic studies on metal salts of alkylphosphonic acids correspond to manganese phosphonate LB films (Seip *et al.*, 1997). Magnetic

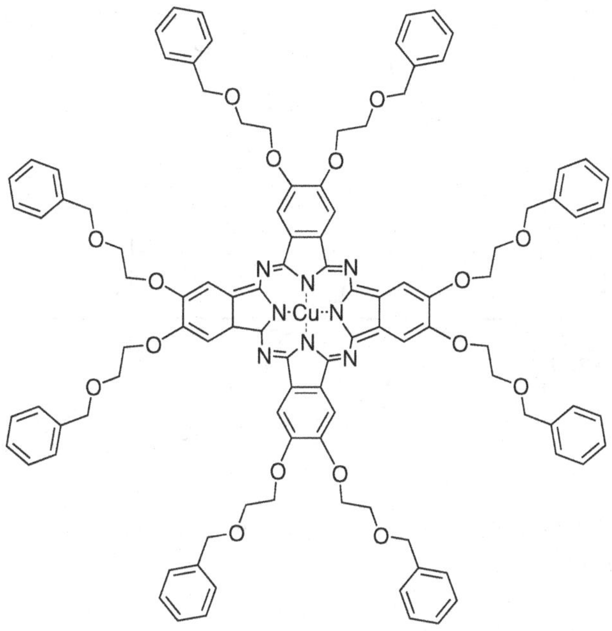

Figure 3.11. Scheme of the CuPc(OC$_2$OBz)$_8$ molecule.

exchange within the layers is antiferromagnetic and a magnetic ordering transition is observed at 13.5 K.

The LB technique provides some degree of control over intermolecular magnetic interactions. This has been proven for the radical C$_n$PNN. Magnetic studies performed on a LB ML were compared to the same molecule in a cast film. Superconducting quantum interference device (SQUID) magnetometry revealed weak antiferromagnetic interactions in the cast film, with $\theta_W \simeq -0.9$ K. In contrast, a ML deposited onto silica displays ferromagnetic interactions with $\theta_W \simeq 2.5$ K (Gallani *et al.*, 2001).

From the mechanical point of view unusually robust fibre bundles can be achieved by compression of large planar molecules. This is the case for Pcs modified at eight positions with benzyloxyethoxy groups, CuPc(OC$_2$OBz)$_8$ and H$_2$Pc(OC$_2$OBz)$_8$, which form closed packed arrays (see Fig. 3.11). When these self-organizing molecular systems are compressed on the surface of a LB trough, the Pc columns align parallel to the compression barriers and form rigid bilayer films that can be mechanically removed from the trough, preserving their integrity. This behaviour arises from the multiple π–π interactions between adjacent Pcs, resulting in strong non-covalent interactions (Smolenyak *et al.*, 1999).

Finally, although this book has been intentionally not focused on polymers, there is an example worth being discussed, since it illustrates the importance of some

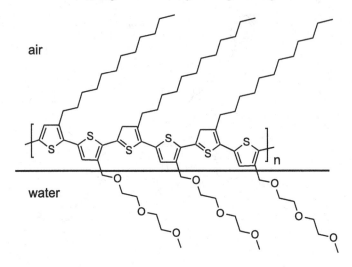

Figure 3.12. Molecular structure of the amphiphilic polythiophene derivative discussed in the text.

crystal engineering concepts that are also relevant to the fabrication of conducting molecule-based films. A series of amphiphilic regioregular polythiophene derivatives have been prepared with alternating hydrophobic and hydrophilic side chains as illustrated in Fig. 3.12 (Bjørnholm *et al.*, 1999). At the air/water interface, the hydrophilic groups are directed to allow interaction with the water surface while the hydrophobic tails can then orient away from the water surface as in a traditional amphiphile. Langmuir films readily form under isothermic compression leading to densely packed MLs in which the polythiophene backbones are π-stacked parallel to the water surface. The stacking of the polymers is highly ordered. When compressed beyond the collapse point on the trough the MLs spontaneously fold into micrometre-long wire-like structures. Undoped bundles of wires exhibit $\sigma_{RT} \sim 10^{-5} \, \Omega^{-1} \, cm^{-1}$ but when the bundles are exposed to iodine or AuCl$_3$ vapours, σ_{RT} increases up to 40 $\Omega^{-1} \, cm^{-1}$.

Self-assembled monolayers

SAMs are ordered molecular assemblies formed by the adsorption of an active surfactant on a solid inorganic surface. As for the LB films case the molecules exhibit two well-differentiated end groups: head and tail. The adsorbate interacts with the surface through its head group, forming *strong covalent bonds* (sulfur–gold, carbon–silicon), thus defining robust ML/solid interfaces. The tail chemical function can be selected (methyl, carboxylic acid, amides, etc.) and thus the ML/environment interfaces can be chemically controlled. In SAMs the packing and ordering are essentially determined by the contribution of the chemisorptive interaction with the surface and of both intra- and interchain interactions (van der Waals, steric, electrostatic, etc.). A detailed discussion of the technique can be found in Ulman, 1996 and Schreiber, 2004. The preparation of MLs on gold substrates is rather simple (Nuzzo & Allara, 1983; Bain *et al.*, 1989). It consists in immersing thin evaporated gold films in dilute solutions. Gold is used because of its stability in air. The two external parameters that are varied are the solution concentration (typically few mM) and the immersion time (from few minutes to several hours) in order to control the degree of assembly.

Examples of surface-active sulfur-based organic compounds that form MLs on gold are alkanethiols, dialkyl sulfides, thiophenols, thiophenes, mercaptopyridines, cysteines, etc. The most extensively studied are alkanethiols, with the general formula C_nSH. Their kinetics of adsorption on Au(111) shows two clearly differentiated steps: fast and slow. The initial fast step follows the diffusion-controlled Langmuir adsorption and strongly depends on thiol concentration. Also, the kinetics is faster for longer alkyl chains, intimately related to the van der Waals interactions. The slow step is essentially a crystallization process where alkyl chains change from the disordered state to the formation of an ordered 2D phase, analogous to the condensation process encountered in LB films (Barrena *et al.*, 2004; Schreiber, 2004).

It is also possible to produce covalently bonded alkyl MLs on Si(111) surfaces using a variety of chemical reactions with passivated H-terminated Si(111), but the preparation methods are more complex than the immersion strategy in part due to the higher reactivity of silicon. This is a major achievement because it allows direct coupling between organic and bio-organic materials and silicon-based semiconductors. Both pyrolysis of diacyl peroxides (Linford & Chidsey, 1993) and Lewis acid-catalyzed hydrosilylation of alkenes and direct reaction of alkylmagnesium bromide (Boukherroub *et al.*, 1999) on freshly prepared Si(111)-H produce surfaces with similar characteristics. These surfaces are chemically stable and can be stored for several weeks without measurable deterioration. Thienyl MLs covalently bonded to Si(111) surfaces have also been obtained, in which a Si(111)-H surface becomes brominated, Si(111)-Br, and is further reacted with lithiated thiophenes (He *et al.*, 1998).

Electrocrystallization

We saw in Section 3.1 that CEC consists in substituting the anode platinum wire by a gold film mechanically squeezed between two flat insulating substrates and that the physical confinement favours a quasi-2D growth. We can here raise the question, what would happen if we directly use free large surface anodes as substrates? We next see two examples, using gold plates and silicon wafers as electrodes.

The first selected example consists in electrodeposition carried out with gold plates, with typical dimensions $\sim 10 \times 5 \times 0.17$ mm^3, as anodes and platinum wires as cathodes (Wang *et al.*, 2002). The gold electrode was cleaned with piranha solution before use and connected to an external constant current power supply with no glass frit between cathode and anode. The target coatings, $(BEDT\text{-}TTF)_2PF_6$, have been prepared with BEDT-TTF, TBA·PF_6 supporting electrolyte and purified TCE. Low current densities (≤ 5 µA cm^{-2}) lead to conventional EC and after 48 h tiny black needle crystals scattered on the gold electrode can be observed. However, for higher current densities of about 10 µA cm^{-2}, densely covered black microcrystalline coatings are produced after 22 h identified as $(BEDT\text{-}TTF)PF_6$ salts. Increasing the current to 20 µA cm^{-2} results in $(BEDT\text{-}TTF)(PF_6)_2$ and soaking the films in a BEDT-TTF solution overnight leads to the desired $(BEDT\text{-}TTF)_2PF_6$ salt.

The second example uses large surface silicon electrodes as anodes (Pilia *et al.*, 2004). Intrinsic-type silicon wafers have a sufficiently high conductivity ($\sim 10^{-3}$ Ω^{-1} cm^{-1}) to be used as electrodes. As far as large surface electrodes are concerned, the use of silicon affords evident benefits versus platinum or gold in terms of maintenance and cost. We illustrate here the case of the preparation of θ-$(BEDT\text{-}TSF)_4[Fe(CN)_5NO]$, where $[Fe(CN)_5NO]^{-2}$ is the photochromic nitroprusside anion. The experimental electrodeposition procedure uses an intrinsic Si(100) one-face-polished wafer (5 cm diameter, 275 mm thickness) as anode and a platinum wire as cathode. The silicon wafer is stripped in an NH_4HF solution and then washed in distilled water before use. Solutions of BEDT-TSF and $(PPh_4)_2[Fe(CN)_5NO]$ in TCE and ethanol are introduced into a Schlenk-type cylindrical one-compartment electrochemical cell. The galvanostatic oxidation of the donor is performed at 298 K and at a current density of 0.1 mA cm^{-2}. Within 2 days, microcrystals are observable, which grow homogeneously on the silicon electrode. A ~ 1 µm thin film is obtained after typically 7 days of electrolysis. When the electrolysis is conducted for 4 weeks typically 60 mg of product can be collected by scratching the substrate using a glass slide. This is a clear advantage because it can be regarded as an alternative synthesis method. Figure 3.13 shows a picture of the experimental set-up.

Following this procedure thin films of $TTF[Ni(dmit)_2]_2$ have been prepared using the following experimental conditions and their metallic character will be discussed in Section 6.4. A solution of TTF and TBA·$[Ni(dmit)_2]$ in freshly distilled and degassed CH_3CN is introduced into the anodic compartment of an H-shaped electrochemical cell. TBA·$[Ni(dmit)_2]$ as supporting electrolyte in CH_3CN is introduced into the cathodic compartment. Galvanostatic electrolysis is conducted at constant current density of 1.5 µA cm^{-2} (de Caro *et al.*, 2004).

Dipping

In the case of TTF-TCNQ, sequential immersion at RT of suitable substrates, inside a glove box, on separate CH_3CN solutions of TTF and TCNQ leads to surprising

Figure 3.13. Experimental set-up for θ-(BEDT-TSF)$_4$[Fe(CN)$_5$NO] deposition on a silicon wafer. Courtesy of Dr L. Valade.

results (de Caro *et al.*, 2000a). When the substrates are stainless steel sheets covered with conversion layers, TTF-TCNQ nanowires are obtained. Such substrates are prepared from austenitic stainless steel sheets dipped in a sulfuric acid bath containing sodium thiosulfate at 318 K. After an electrolytic activation, the conversion coating grows by chemical treatment in the same bath. The sample is then washed with water and dried before use. Adsorption of TTF is done first. Then, immersion of the TTF-containing surface in the TCNQ solution results in the formation of nanowires and a few platelets. As shown in Fig. 3.14(a) the nanowires are anchored on the surface, some of them bridging the grains of the conversion coating (boundary separation $\sim$1–2 μm). The nanowires are organized in bundles having an average thickness of 20 nm and a width between 20 and 200 nm (Fig. 3.14(b)). They can be up to 20 μm long and, occasionally, form closed loops (Fig.3.14(c)).

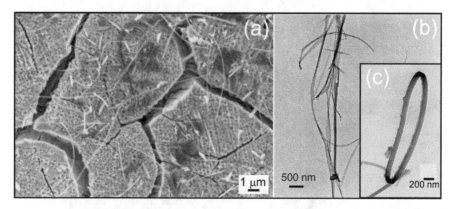

Figure 3.14. (a) SEM image of a stainless steel surface with a conversion coating covered by TTF-TCNQ nanowires. TEM images of (b) a bundle of nanowires and (c) a loop. Reprinted from *Comptes Rendues de l'Académie des Sciences Paris, Série IIC, Chimie*, Vol. 3, D. de Caro, J. Sakah, M. Basso-Bert, C. Faulmann, J.- P. Legros, T. Ondarçuhu, C. Joachim, L. Ariés, L. Valade and P. Cassoux, 675–680, Copyright (2000), with permission from Elsevier.

From the mathematical point of view the existence of torus-shaped nanotubes should not be a surprise since a rectangle (graphene sheet), a cylinder (nanowire) and a torus have the same topology, albeit, with different curvatures. A rectangle can be curved and transformed into a cylinder by gluing two edges together and the cylinder can easily transform into a compact torus by gluing both ends together. The key point is whether we want to think in terms of 2D or 3D representations.

Wires or strips at the nanometre scale are surprisingly flexible (surprising if we insist on thinking in terms of our macroworld!) and can even be twisted in order to form Möbius strips, as recently achieved with $NbSe_3$ single crystals (Tanda *et al.*, 2002).

An interesting, relatively easy but hardly universal method of preparing thin films (thickness ~ 10 μm) consists in anion replacement by immersing a single crystal template into an electrolyte solution containing the new anion. This has been achieved for thin $(TMTSF)_2ClO_4$ films grown on top of particular crystal faces of $(TMTSF)_2PF_6$ (Angelova *et al.*, 2000). In the immersion experiments, the template $(TMTSF)_2PF_6$ crystals (typical sizes: $10 \times 1 \times 0.5$ mm^3) are introduced into an anhydrous TCE solution, containing TMTSF and a large excess of TBA·ClO_4. The tightly closed vessels are maintained at RT in the absence of oxygen. The immersion process results in both morphological and chemical modifications of the various crystal faces of the template. The $(01\bar{1})$ faces exhibit the formation of continuous $(TMTSF)_2ClO_4$ films. The temperature dependence of the zero-field resistance of the $(TMTSF)_2ClO_4$ film below 16 K exhibits the metal to SDW transition at 12 K specific to the $(TMTSF)_2PF_6$ template. For temperatures below 3.5 K, the

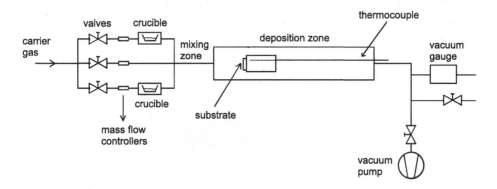

Figure 3.15. Schematic representation of a horizontal hot-wall CVD apparatus.

film exhibits the properties of (TMTSF)$_2$ClO$_4$ alone, since its resistivity decreases towards a sharp superconducting transition with an onset critical temperature of $T_c \simeq 1.4$ K. This method is thus restricted to the use of single crystals as templates and to the replacement of anions, but has proved its validity. The overlayer films can also be prepared electrochemically.

Dry preparation methods

Chemical vapour deposition

The chemical vapour deposition (CVD) technique can be regarded as an extension of the sublimation method used for the preparation of single crystals discussed before, but adapted to thin film growth implying deposition on substrates. A schematic diagram of a horizontal hot-wall CVD apparatus is shown in Fig. 3.15.

After introducing the substrates into the reactor, usually made out of glass, and charging the vaporization crucibles with the selected precursor materials, the entire experimental set-up is pumped down for several hours (typically overnight) while the mixing zone and the reactor vessel are baked out at the final chosen temperatures. The experiment starts by vaporizing the precursors at low pressures and transporting them by a carrier gas (typically helium or argon) through heated lines to the mixing zone and then to the deposition zone. The mixing zone is heated to a higher temperature than the vaporization zone in order to avoid condensation of the starting compounds. Obviously, all temperatures have to be set below the thermal decomposition temperatures of the precursors.

What differentiates CVD from other evaporation techniques is that growth units, the fundamental bricks essential for growth, are chemically generated at the mixing zone. The formation of growth units implies either chemical reaction of the

Table 3.2. *Typical experimental CVD conditions for the formation of*
$[Fe^{+3}(Cp^*)_2][TCNE]$ and $V(TCNE)_x$ thin films

Precursor	Mass (mg)	T_{vap} (K)	He flow rate (sccm)	T_{mix} (K)	T_{sub} (K)	P (Torr)
$Fe^{+3}(Cp^*)_2$	134	383	33			
				433	348	0.5
TCNE	105	348	17			
$V(C_6H_6)$	94	418	17			
				448	348	0.5
TCNE	119	348	10			

Note: This table corresponds to Table 1 from de Caro *et al.*, 2000b. T_{vap}, T_{mix}, T_{sub} stand for temperatures of vaporization, in the mixing zone and of the substrate, respectively, and P corresponds to the measured total pressure.

Figure 3.16. SEM image of (a) $[Fe^{+3}(Cp^*)_2][TCNE]$ and (b) $V(TCNE)_x$ thin films deposited on KBr pellets. Reprinted with permission from de Caro *et al.*, 2000b. Copyright (2000) American Chemical Society.

precursors, where the resulting final products differ from the precursors (type I), or collective assembling via intermolecular interactions, where the final products and the precursors are identical (type II).

We illustrate the CVD preparation of thin films with some examples. The selected materials are ferromagnetic $[Fe^{+3}(Cp^*)_2][TCNE]$ and the solvent-free $V(TCNE)_x$ phase. $[Fe^{+3}(Cp^*)_2][TCNE]$ is a ferromagnet with $T_C = 4.8$ K (Miller *et al.*, 1988) and $V[TCNE]_x$ orders magnetically above 350 K (Manriquez *et al.*, 1991), as previously discussed in Section 1.5. Table 3.2 summarizes the CVD conditions for growth of the thin films. $V(TCNE)_x$ and $[Fe^{+3}(Cp^*)_2][TCNE]$ correspond to type I and II, respectively.

As shown in Fig. 3.16, the films are polycrystalline formed of micrometre-sized crystals. Magnetic measurements performed on the $[Fe^{+3}(Cp^*)_2][TCNE]$ films indicate that $T_C \simeq 3.7$ K (de Caro *et al.*, 2000b). When purely organic precursors are involved (organometallic precursors are excluded), the CVD technique is termed OCVD, which stands for organic CVD, and corresponds to type II growth. However, although the growth units are expected to be built in the mixing zone, there is no experimental evidence. Thin film TTF-TCNQ grown from its precursors TTF and TCNQ is a good OCVD example (Figueras *et al.*, 1999). In this case TTF and TCNQ, previously recrystallized from CH_3CN, are used as source materials. Co-evaporated from stainless steel crucibles and transported through heated lines to the mixing zone, they reach the deposition zone carried by an argon gas flow. Stoichiometric films are obtained at evaporation temperatures of 330 K and 393 K for TTF and TCNQ, respectively, with T_{mix} maintained at 443 K and T_{sub} ranged between 333 and 355 K. The argon flow is typically maintained at 12.5 sccm. Within these conditions typical deposition rates of about 1.5 μm h^{-1} are obtained.

Physical vapour deposition

ML control over the growth of organic thin films with extremely high chemical purity and structural precision can be achieved in UHV by means of organic molecular beam deposition (OMBD), also referred to as organic molecular beam epitaxy (OMBE). We will preferentially use OMBD instead of OMBE because the utilization of the term epitaxy is rarely justified for organic materials, as will be discussed in Chapter 4. This technique is extremely reliable but complex and expensive. Growth in an UHV environment assures that a minimal density of impurities will be incorporated into the film. As shown above, ML control is also possible using the LB and SAM deposition techniques. However, UHV growth has the advantage of providing an atomically clean environment. When combined with the ability to perform *in situ* high-resolution structural and electronic diagnostics of the films as they are being deposited (see Chapter 4), OMBD provides the ideal scenario for understanding many of the fundamental structural and physical properties of ultrathin organic film systems.

The study of vapour-deposited organic films has been applied to a large number of molecular systems. However, most of the work has concentrated on the study of the growth and optoelectronic characteristics of planar stacking molecules such as the Pc and polycyclic aromatic compounds based on naphthalene, and perylene. In particular, PTCDA has become extensively studied.

Let us explore the essentials of the OMBD technique. A detailed description can be found in Forrest, 1997. The use of UHV systems is mandatory for OMBD growth. Typically, growth occurs by the sublimation in a background vacuum below 10^{-8} mbar of a highly purified powder. The evaporant is collimated by passing through a series of orifices, after which it is deposited on a substrate held perpendicular to the beam located 10–20 cm away from the source. Φ_m is controlled by a combination of oven temperature, which is maintained somewhat above the sublimation temperature of the source and well below its chemical decomposition temperature, with a mechanical shutter, which allows for an accurate control of the deposition time. By sequentially shuttering the beam flux from more than one Knudsen cell, multilayer structures consisting of alternating layers of different compounds can be grown. Typical growth rates range from 0.0001 to 10 nm s^{-1}, as monitored by a quartz microbalance. At very low deposition rates there exists the danger of adsorbing contaminants onto the surface at a higher rate than the organic molecules, whereas at the higher rates, growth is extremely difficult to control, leading to undesired 3D growth (small grains) and polymorphs. For these reasons, the most reliable growth data are frequently obtained at rates from 0.01 to 0.5 nm s^{-1}. Between growth runs, the Knudsen cell is often maintained slightly below the material sublimation temperature in order to continuously outgas the source. In this manner, very high purity source materials can be achieved and maintained over long periods of time. UHV storage of the source material at elevated temperatures is believed to be a key factor in eliminating impurities and moisture which lead to film defects during growth. This vacuum storage leads to extreme material purity over the long term, far exceeding that which is at first obtained in the thermal gradient purification process.

There are several techniques for purification, including gradient sublimation, zone refining from the melt, chromatography, etc. Gradient sublimation is the most useful means for purification of the powders employed in OMBD since most of these compounds do not have a liquid phase at atmospheric pressure or below. Gradient sublimation purification is based on the sublimation method previously described. T_{sub} is also an important variable in determining the film structure. A typical range of growth temperatures is from 80 to 400 K. Note that at lower temperatures impurities tend to quickly adsorb onto the substrate. This is particularly important when growth occurs at only modest background pressures.

As mentioned before, one significant advantage of using UHV is that numerous *in situ* film characterization techniques can be employed both during and immediately after growth. Many growth chambers are equipped with reflection high-energy electron diffraction (RHEED) to allow for real-time determination of the thin-film crystalline structure. In Section 4.4 we shall see the difference between *in situ* and real time. The surface-sensitive RHEED technique consists in irradiating a surface with a high energy beam (20–30 kV) at grazing incidence, resulting in a suitable geometry for *in situ* studies. However, the films can be damaged by irradiation and contaminated by cracking of hydrocarbons present in the residual base pressure. This technique is preferentially used for inorganic films grown by MBE. In addition, a residual gas analyser (RGA) can be employed to determine the molecular species within the chamber during growth, as well as to determine the energy of film–substrate adhesion using thermal desorption spectroscopy (TDS). Other diagnostics include low-energy electron diffraction (LEED). In this case, low-energy electrons ($< 100\,\text{eV}$) impinge perpendicularly onto the sample surface. This method reduces radiation-induced damage but the working geometry precludes real-time growth studies and the *in situ* experiments have to be performed in an additional chamber attached to the growth chamber. Finally, many other vacuum analysis techniques such as scanning tunnelling microscopy, XPS, Auger electron spectroscopy, ellipsometry, etc. can also be employed in conjunction with OMBD simply by installing the appropriate equipment in a chamber connected to the growth chamber. Concerning the type of pumping used to achieve the UHV environment necessary for OMBD systems, there is considerable variation. Due to the volatility of some organic materials, high throughput pumps are generally employed. These include turbomolecular pumps, ion pumps and cryopumps.

Several examples of OMBD-grown films will be discussed in Chapter 4 and here we briefly study the case of highly ordered DIP films (Dürr *et al.*, 2002a). DIP was first purified by temperature gradient sublimation. All samples were prepared on oxidized Si(100) substrates. DIP layers with varying thickness between 7 and 110 nm were grown at $T_{\text{sub}} \simeq 418\,\text{K}$ and at a rate of $1.2\,\text{nm min}^{-1}$ under UHV conditions ($\leq 2 \times 10^{-9}$ mbar during deposition). The spatial coherence of the crystalline order has been quantitatively determined by high-resolution XRD measurements. In the so-called specular diffraction mode the momentum transfer q is perpendicular to the surface normal, thus probing the electron density profile along the surface normal. Figure 3.17(a) displays a typical scan along the specular rod for a DIP thin film (thickness $\simeq 20.6$ nm) prepared under the conditions described above. For all samples, Bragg reflections associated with the DIP film up to at least the seventh order are visible, from which a lattice constant of 1.660 ± 0.005 nm is derived. The large number of Bragg reflections reveals the high structural order achieved in the DIP films. Figure 3.17(b) shows a magnification of a small-q_z region of Fig. 3.17(a). The periodicity of the interference fringes, the so-called Kiesing fringes, equals 2π divided by the film thickness. The almost undamped Laue oscillations around the Bragg reflections are evidence for a laterally homogeneous coherent thickness, e.g., a well-defined number of ordered MLs.

A recently developed alternative to the OMBD technique is the hyperthermal molecular beam deposition (HMBD) method, which has been used to grow e.g., highly ordered pentacene films on metal surfaces (Casalis *et al.*, 2003). The technique is based on the acceleration of the impinging molecules to a few eV using seeded supersonic free jet expansion from a molecular beam source (typical diameter of 0.1 mm) in which different inert gases (He, Ar and Kr) can be used as a carrier gas. The fine control of the kinetic energy leads to high-quality films. While

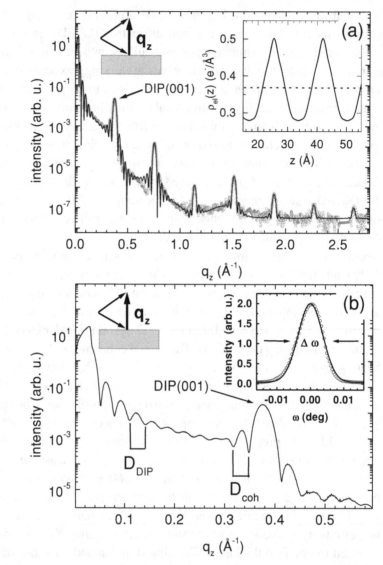

Figure 3.17. (a) Specular X-ray intensity of a DIP film with 20.6 nm thickness. Bragg reflections up to the seventh order are seen. (b) Magnification of the specular scan from the region of total external reflection across the first specular DIP Bragg reflection. The inset displays a rocking scan across the DIP(001) reflection which exhibits FWHM as small as 0.0087°. Reprinted with permission from A. C. Dürr, F. Schreiber, M. Münch, N. Karl, B. Krause, V. Kruppa and H. Dosch. *Applied Physics Letters*, **81**, 2276 (2002). Copyright 2002, American Institute of Physics.

working with $T_{sub} \leq$ RT OMBD results in the formation of amorphous films, highly ordered pentacene films have been grown at low substrate temperatures (200 K) when pentacene molecules with high kinetic energies of a few eV were deposited on a silver substrate.

By varying the nature and/or the pressure of the carrier gas and/or the temperature of the organic material, the translational kinetic energy of the molecules can be controlled. In particular, given a carrier gas pressure of about 300 Torr and the vapour pressure of pentacene being 1.2×10^{-3} Torr at 493 K, the translational kinetic energy is estimated to vary between 5 and 0.4 eV by changing the carrier gas from He to Kr. UHV is still needed in order to keep substrate surfaces atomically clean prior to deposition. The growth of good quality pentacene thin films at relatively low substrate temperature points to the fact that the local annealing induced by the impact of the hyperthermal molecules is a more efficient route to order than using either a global annealing of the film or higher substrate temperatures.

If we allow for a relaxation of the stringent experimental conditions inherent to OMBD by working in a HV environment, high-quality thin films can also be obtained. As case studies we consider TTF-TCNQ and *p*-NPNN, and the characterization of their physical properties will be discussed in detail in Chapter 6.

High-quality thin TTF-TCNQ films can be obtained by thermal sublimation in HV ($\sim 10^{-6}$ mbar) of recrystallized TTF-TCNQ powder. *Ex situ* cleaved alkali halide substrates such as NaCl(100), KCl(100) and KBr(100) are held at RT and the evaporation temperature set to 433 K. TMAFM images of such films are depicted in Figs. 3.18(a) and (b). The films consist of highly oriented and strongly textured rectangular-shaped microcrystals, which are oriented with their *a*- and *b*-axis parallel to the [110] and [$\bar{1}$10] substrate directions, respectively, due to the cubic symmetry of the substrates. FTIR spectra of the films, grown on KBr(100), confirm the existence of CT. This is shown by the observed shift of the CN stretching mode shown in Fig. 3.18(c), which is associated with ϱ from the π-donor molecule TTF to the acceptor TCNQ (Chappell *et al.*, 1981).

The second example corresponds to thin films of *p*-NPNN, grown on glass, on *ex situ* cleaved alkali halides and on mica (muscovite) substrates, obtained by sublimation of the recrystallized powder *p*-NPNN precursor at *c.* 10^{-6} mbar, sublimation temperature 413 K and $T_{sub} =$ RT, which crystallize in the monoclinic α-phase (see Fig. 1.23(a)). The films are (002)-oriented (*ab*-planes parallel to the substrate surface) as illustrated in Fig. 3.19 for a glass substrate with a high degree of orientation, indicated by a small FWHM of the rocking curve: $\Delta\Omega \simeq 1°$. In the case of highly oriented pyrolytic graphite (HOPG) substrates, the films crystallize in the monoclinic δ-phase (see Fig. 1.23(d)) and are (020)-oriented (*ac*-planes parallel to the substrate surface). Note that the HV-grown films never crystallize in the

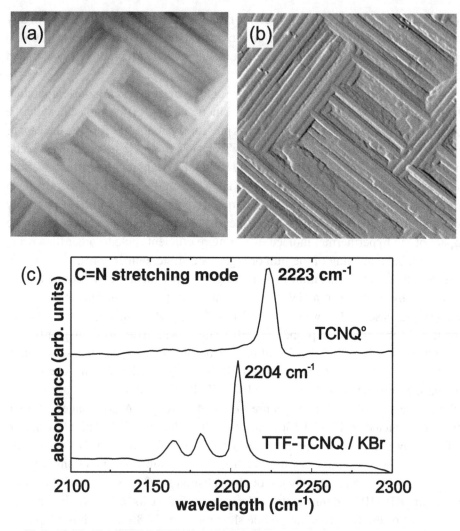

Figure 3.18. Thin TTF-TCNQ film (thickness $\simeq 1$ μm) HV-grown on a KCl(100) substrate. (a) Topography and (b) amplitude TMAFM images. The scale is 5 μm × 5 μm. (c) FTIR spectra of the CN stretching mode in neutral TCNQ (powder) and in a TTF-TCNQ thin film (thickness $\simeq 1$ μm) HV-grown on KBr(100). Reprinted from *Surface Science*, Vol. 482–485, C. Rojas, J. Caro, M. Grioni and J. Fraxedas, *Surface characterization of metallic molecular organic thin films: tetrathiaful-valene tetracyanoquinodimethane*, 546–551, Copyright (2001), with permission from Elsevier.

thermodynamically most stable β-phase, a matter that will be discussed in detail in Section 5.5 together with their exotic surface morphology. Under well-defined experimental conditions the films exhibit 2D spherulitic morphology.

Finally, let us conclude this section by highlighting how critical the growth parameters can be concerning the structure–property relationship. This is best

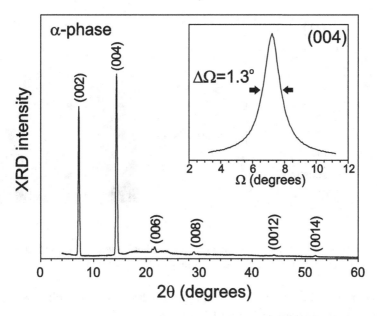

Figure 3.19. XRD pattern of a thin film (thickness ~ 1 µm) of the α-phase of *p*-NPNN HV-grown on a glass substrate. The inset shows the rocking curve of the (004) reflection. Reprinted with permission from J. Fraxedas, J. Caro, J. Santiso, A. Figueras, P. Gorostiza and F. Sanz, *Europhysics Letters* 1999, Vol. 48, 461–467.

exemplified for pentacene OFETs (Gundlach *et al.*, 1997). When deposited by flash evaporation onto substrates held at temperatures between 293 and 433 K, only very low mobilities are obtained (~ 0.0001 cm^2 V^{-1} s^{-1}). Flash evaporation is often used for organic semiconductor depositions to avoid problems with thermal decomposition of the material being evaporated. Using rapid thermal annealing to rapidly heat the fabricated OFETs to 623 K for a few seconds increases the field-effect mobility to about 0.002 cm^2 V^{-1} s^{-1}. The initial poor mobility is clearly due to poor molecular ordering in the flash evaporated film, and the mobility improvement comes from an improvement in ordering with the anneal. However, ordered films are obtained for depositions at low to moderate rates (0.1–0.5 nm s^{-1}) onto substrates held at temperatures from 293 to 393 K. Relatively high mobility devices are obtained for depositions over this entire range: typically 0.3 cm^2 V^{-1} s^{-1}, with a maximum observed value of 0.7 cm^2 V^{-1} s^{-1}. Mobilities measured in thin films have reached values as high as 1.5 cm^2 V^{-1} s^{-1} or more, a point that will be discussed in Section 6.3 (see Table 6.1).

Pulsed laser deposition

The pulsed laser deposition (PLD) technique is widely used for inorganic materials but is becoming increasingly employed for the preparation of thin films of polymers

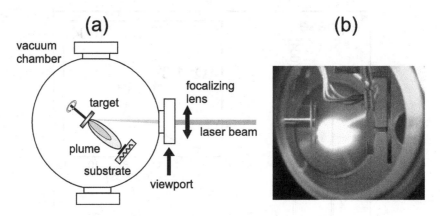

Figure 3.20. (a) Schematic diagram showing the basic elements of a PLD deposition system. (b) Picture of a plume. Courtesy of Dr J. Santiso.

and biomaterials (Chrisey *et al.*, 2003). Figure 3.20(a) shows a schematic diagram of the basic elements comprising a PLD system.

A focused laser pulse passes through an optical window (viewport) into a vacuum chamber and interacts with a solid surface, the target. The laser pulse is absorbed, and above a given power density significant material removal occurs in the form of an ejected forward-directed plume (see Fig. 3.20(b)). The threshold power density needed to generate such a plume depends on the target material, its morphology, the laser pulse wavelength and the laser pulse duration, parameters to be added to the parameter hyperspace. The organic material targets are usually formed by casting the powder under high mechanical pressure.

The use of PLD to deposit these materials is complicated by the fact that their structures cannot be subjected to a laser-induced pyrolytic decomposition and re-polymerization process that works for addition polymers. In fact, the perfect reconstruction upon deposition of the organic molecular structures for these materials is never observed once ablated at laser fluences higher than their ablation threshold. However, PLD is possible at laser fluences near the ablation thresholds of these materials and some examples are given next.

Growth of amorphous thin films of CuPc and Alq$_3$ using PLD has been reported (Matsumoto *et al.*, 1997). In this case the thin-film deposition was carried out in vacuum at $\sim 10^{-5}$ Torr using a KrF (248 nm) laser beam with a pulse length of ~ 25 ns. The films were grown on Si(100) and glass substrates kept at RT located 3 cm from the target. Crystallinity can be achieved by substrate heating, irradiating the film with a second laser and/or applying a dc electric field between the target and the substrate.

Pentacene thin films grown with PLD under vacuum ($\sim 10^{-5}$ Torr) and at substrate temperatures ranging from 300 to 473 K have been obtained using again a KrF excimer laser (248 nm), operated at 50 Hz and with a power of 6–8 W. Relatively large field-effect mobilities ($\sim 3 \times 10^{-2}$ cm^2 V^{-1} s^{-1}) have been reported for such films (Salih *et al.*, 1996).

In order to minimize photochemical damage that results from the interaction of the laser light with the organic target, a novel deposition technique, known as matrix-assisted pulsed laser evaporation (MAPLE), has been developed. MAPLE is essentially a variation of the conventional PLD technique and has been successfully used to deposit thin and uniform layers of MOMs, polymers and biomolecules (Chrisey *et al.*, 2003). In MAPLE, a frozen matrix consisting of a dilute solution of an organic compound in a relatively volatile solvent is used as the laser target. The solvent and concentration are selected so that (i) the material of interest can dissolve to form a dilute, particulate free solution, (ii) the majority of the laser energy is initially absorbed by the solvent molecules and not by the solute molecules, and (iii) there is no photochemical reaction between solvent and solute.

The light–material interaction in MAPLE can be described as a photothermal process. The photon energy absorbed by the solvent is converted to thermal energy that causes the organic molecules to be heated and the solvent to vaporize. As the surface solvent molecules are evaporated into the gas phase, the organic molecules attain sufficient kinetic energy through collective collisions with the evaporating solvent molecules to be transferred into the gas phase.

By careful optimization of the MAPLE deposition conditions (laser wavelength, repetition rate, solvent type, concentration, temperature, background gas and gas pressure), this process can occur without any significant chemical decomposition. When a substrate is positioned directly in the path of the plume, a coating starts to form from the evaporated organic molecules, while the volatile solvent molecules, which have very low sticking coefficients, are evacuated by the pump in the deposition chamber.

Even when the solute absorbs the laser light, thin films can be prepared by carefully adjusting the laser fluence. This has been shown for MAPLE-grown anthracene films (Gutierrez-Llorente *et al.*, 2003). Anthracene exhibits a noticeable absorption at 266 nm, the wavelength of a frequency-quadrupled Nd:YAG laser. Solutions of 2 wt% anthracene in orthoxylene or in CHCl$_3$ were in this case used as targets. Prior to deposition, the chamber was evacuated to a background pressure of 10^{-6} mbar. A pulsed Nd:YAG laser (266 nm, 5 ns, 10 Hz) was used to irradiate the solutions, which were frozen at around 203 K. The substrates were placed at a distance of 35–40 mm from the target. Anthracene thin films were grown either

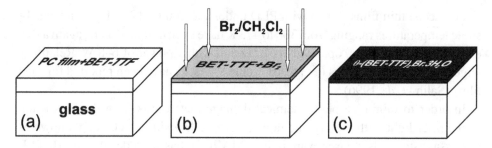

Figure 3.21. Scheme of the preparation by MRDT of a conducting surface layer of θ-(BET-TTF)$_2$Br·3H$_2$O nanocrystals on a polycarbonate (PC) substrate. Adapted from Mas-Torrent *et al.*, 2001.

on double-sided polished crystalline Si(100) or on quartz substrates kept at 243 K to avoid the incorporation of solvent molecules. Pure anthracene films are grown if the deposition is carried out by irradiating the target at fluences lower than a threshold value, while hydrogenated amorphous carbon films are formed at higher fluence values because of the fragmentation of the target molecules. This threshold value is around 0.6–0.7 J cm^{-2}.

The MAPLE technique is analogous to the analytical matrix-assisted laser desorption ionization (MALDI) technique, a method developed for studying large organic molecules and polymeric materials to accurately determine their molecular weight distributions. The main differences between the MAPLE and MALDI techniques lie in the treatment of the evaporated organic molecules and in the selection of the matrix. In the MAPLE process, the organic molecules are not deliberately ionized and are collected on a substrate to form a coating rather than being directed into a mass spectrometer.

Vapour-exposure

We have seen in previous sections that doping of insulating films can be achieved through exposure to e.g., iodine gas. This practical and universal method can also be employed for the formation of conducting layers on transparent and mechanically flexible polymeric substrates as discussed next. The example discussed here refers to conducting layers of θ-(BET-TTF)$_2$Br·3H$_2$O on a polycarbonate matrix (Mas-Torrent *et al.*, 2001).

Polycarbonate films with a typical thickness of 10–15 μm containing 2 wt% of molecularly dispersed BET-TTF are cast following the modified reticulate doping technique (MRDT) (Laukhina *et al.*, 1995) as illustrated in Fig. 3.21. Then the surfaces of the samples are treated with the vapour of a Br$_2$/CH$_2$Cl$_2$ solution, producing the BET-TTF + Br_2 reaction in the swollen film surface. The ratio of the reagents employed during the reaction is critical because it determines if either a completely ionic non-conducting salt or a mixed valence metallic salt is formed. An optimum mass of Br_2 absorbed per unit volume of a swollen film surface layer exists, at which a conducting layer formed by θ-(BET-TTF)$_2$Br·3H$_2$O nanocrystals is achieved. The films exhibit metal-like transport properties down to liquid He temperatures and an exceptionally high σ_{RT} value of $c.$ 100 Ω^{-1} cm^{-1} and are completely transparent because of their small thickness.

To end this section devoted to vapour deposition, just a few words concerning gravity. We saw previously that gravity is essential for the preparation of LB films, but does it exert any influence on the preparation of vapour-deposited thin films? To answer this question let us have a look at an illustrative and pedagogical example, with a component of exoticism for some readers.

Physical vapour transport experiments designed to prepare deposits of dimethyl-dicyanovinylaniline, a material of interest because of its large optical non-linearities, yield thin films when the experiments are conducted in microgravity but bulk crystals when carried out *apparently* under the same conditions in Earth-normal gravity. The microgravity experiments were performed in the Space Shuttle (Carswell *et al.*, 2000). Carrying out experiments under identical conditions is nearly impossible, as every experimentalist knows, hence the emphasis on the word apparently. The explanation of this paradox is really interesting and tells us how carefully all experiments have to be performed.

Background nitrogen pressure appears to be the key to transforming from the bulk crystal to thin-film growth. Vacuum in the ground-based laboratory is provided by a combination of mechanical pumps, while in orbit vacuum was provided by a valve that vented the experiment container volume directly to the vacuum of space. For safety reasons NASA requires the installation of both a 60 μm effluence filter and a small quantity of 0.5 nm molecular sieve material. For engineering reasons the evacuating gases pass first through the valve, then through the molecular sieve material, then through the 60 μm filter material before passing into the shuttle cargo bay. This is the origin of the discrepancy: the evacuation rate in the space experiment is significantly lower than in the ground control experiment, so that the nitrogen background is different. Furthermore, in the weightless environment the presence of powder can hinder the correct operation because it does not remain fixed.

The case of thin CuPc films is also intriguing. When grown by physical vapour transport under unit gravity conditions the films contain mixtures of both α- and β-polymorphs (see Fig. 1.26), while under microgravity, again in the Space Shuttle, the films contain predominantly a new polymorphic form designated as M-CuPc

(Debe & Kam, 1990). It remains to be seen whether gravity is ultimately responsible for such behaviour or if subtle differences in the experimental conditions are the cause.

3.3 Patterning

Micrometre and nanometre-sized artificial structures based on MOMs can be produced by patterning with the goal of devising technological applications. This is a necessary step to adapting such materials into real devices. Because of the molecular character of these materials the conventional lithographic procedures used for inorganic materials and polymers (electron beam, UV irradiation, chemical etching, etc.) can hardly be applied, hence several soft (non-aggresive) strategies are being developed. Here we briefly review some of these strategies and illustrate some selected examples.

Micropatterning

Micrometre-scale patterned thin films of small molecular weight organic semiconductors can be achieved by organic vapour phase deposition (Shtein *et al.*, 2003). This growth technique is essentially type II CVD, or OCVD, described before, but using only one type of molecule and cooling the substrate. In the literature it is quite common to find many different names to describe very similar processes. For researchers not familiar with such techniques it can be puzzling at the beginning. Patterned structures are achieved by means of masks placed close to the substrates. It is clear that the optimal experimental conditions (pressure of the carrier gas, distance mask–substrate, temperature, etc.) have to be met in order to produce efficient structures. For instance, the background pressure of the carrier gas has to be of the order of the mask dimensions (0.1–10 Torr approximately corresponds to a mean free path of 100–1 μm). This technique has been successfully applied e.g., to Alq_3.

A different way of preparing patterned films is by using patterned substrates. Such substrates can be prepared by well-controlled lithographic techniques or simply by mechanical preparation. The most common example is the case of PTFE films obtained by friction transfer onto smooth substrates such as silicon, glass, ITO, etc. PTFE films are easily prepared by sliding a PTFE rod at a constant pressure against a clean glass slide held at 573 K. The PTFE surfaces consist of a succession of atomically flat areas and oriented few nm high steps separated by typically 25 to 500 nm. PTFE films are chemically inert, thermally stable up to *c.* 600 K, optically transparent in the visible range and, very important, of low cost. The oriented PTFE films induce the preferential orientation of organic films grown on top of them,

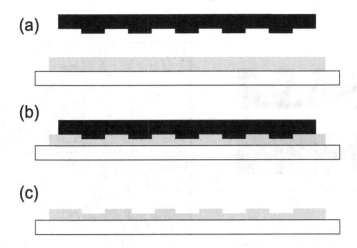

Figure 3.22. Schematic diagram of the process of the NIL technique. The master, the organic film and the substrate are represented by black, grey and white components, respectively.

e.g., by vapour deposition, essentially by the process known as graphoepitaxy (Moulin *et al.*, 2002).

Nanoimprint lithography

One of the most successful lithographic techniques is nanoimprint lithography (NIL), mainly because both its working principle and operation are quite simple (Chou *et al.*, 1996). A schematic diagram of the NIL technique is shown in Fig. 3.22.

NIL uses a hard mould or master containing nanoscale features defined on its surface ready to emboss into an organic material prepared on a substrate by e.g., vapour deposition, spin-coating, etc., thus creating a thickness contrast in the organic material. NIL has the capability of patterning down to *c.* 10 nm features. Hard moulds are usually made of silicon and the grating is fabricated by well-established lithographic techniques such as electron beam or interference lithography. Once the mould grating is available the processing is rather simple. Gratings are used as stamps and pressed against the organic film. After release of the mask the pattern is transferred to the film. This is analogous to a classical typewriter used without the ink tape. This is the reason why expectations are high for this deposition method. The NIL process can be performed either in air or under controlled conditions (vacuum, inert gas atmosphere), depending on the stability of the organic materials and on the relevance of contaminants, e.g., oxygen and water, which are detrimental e.g., for OLEDs. Examples of NIL-patterned light-emitting films preserving their

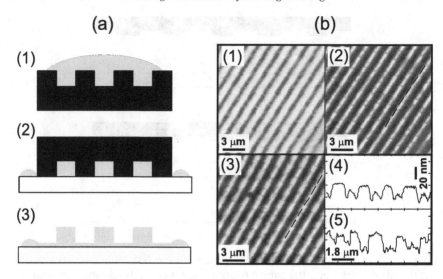

Figure 3.23. (a) Schematic illustration of the procedure for modified microtransfer moulding: (1) a drop of solution is applied to the patterned surface of the mould and without removing the excess of solution, (2) the mould is placed in contact with a substrate and (3) once the liquid has evaporated the mould is peeled away, leaving a continuous patterned film. The master, the organic film and the substrate are represented by black, grey and white components, respectively. (b)(1) AFM image of the mould, (2) image obtained by inverting the contrast in (1), (3) printed Alq$_3$ thin film. Line profiles of (4) the mould and (5) the printed film, along the dashed lines in (2) and (3), respectively. Reprinted with permission from Cavallini *et al.*, 2001. Copyright (2001) American Chemical Society.

optical properties are Alq$_3$ (Wang *et al.*, 1999) and derivatives of oligothiophenes (Pisignano *et al.*, 2004).

Modified microtransfer moulding

The microtransfer moulding process of systems with solvents is a well-known soft lithographic method where a drop of liquid containing a precursor of e.g., a polymer, is applied to the patterned surface of a mould and the excess of liquid is removed. The mould is then placed in contact with a substrate and either heated or irradiated. After the liquid has cured the mould is peeled away, leaving a polymeric replica on the substrate. This already simple procedure can be further simplified for soluble MOMs by not removing the excess liquid and not heating the sample (Cavallini *et al.*, 2001). The evident difference between microtransfer moulding and emboss-ing is the absence of solvent in the latter technique. The resolution of solvent-assisted micromoulding can be under 60 nm. A schematic illustration of the simplified modified microtransfer moulding procedure is shown in Fig. 3.23(a). This method is of general application to a large variety of soluble materials.

The chosen example describes patterned Alq_3 films (Cavallini *et al.*, 2001). The selected mould consists of a piece of a gold-coated recordable CD. Before gold coating a periodic sequence of 1 0 0 in binary language is recorded on the whole CD. A blank CD consists of a sequence of grooves with a typical intergroove distance of 1.4 μm, depths of $\simeq 230$ nm and widths of $\simeq 400$ nm. The writing of 1 0 0 sequences on the CD yields a regular sequence of holes $\simeq 20$ nm deep and with a pitch of $\simeq 750$ nm along a groove. The mould is then coated with 500 nm of gold by evaporation in HV. An AFM image of the mould is shown in Fig. 3.23(b1) and its negative, obtained by inverting the grey scale in the topography image, is displayed in Fig. 3.23(b2). The line profile in Fig. 3.23(b4) clearly shows the finer structure recorded within the grooves.

A solution of Alq_3 in CH_2Cl_2 is prepared for the moulding process. Alq_3 was previously purified by sublimation in a vacuum. The substrate is a silicon wafer with native oxide. Typically 10 μL of the solution is poured onto the patterned mould, and then the mould is placed on the clean surface. The solution to be transferred has to become more viscous, but should not solidify. It is important to prevent the stamp from floating over the mould, which would result in inhomogeneous printed features. After a few minutes the film solidifies and then the mould has to be carefully removed from the sample. Figure 3.23(b3) and the line profile in Fig. 3.23(b5) show the printed Alq_3 pattern. The motifs in the groove are perfectly reproduced along the molecular lines. From the comparison between the profiles of the inverted mould and the moulded Alq_3 film, it should be noted that both the pitch and the aspect ratio of the structures along the grooves are retained in the moulding and transfer process. In other words, the information recorded on the CD is effectively transferred to the organic material.

Lithographically controlled wetting

Lithographically controlled wetting (LCW) is a rather simple, fast and low-cost fabrication process suitable for large-area nanopatterning. The technique is based, like the previously described method, on stamp-assisted deposition of a soluble material from a solution (Cavallini & Biscarini, 2003). A scheme of the LCW working principle is shown in Fig. 3.24.

As the stamp is gently placed in contact with a solution spread on a substrate, capillary forces drive the solution to form menisci under the stamp protrusions. The solution remains pinned to the protrusions upon solvent evaporation. As the critical concentration is reached, the solute precipitates from solution onto the substrate, giving rise to a structured thin film replicating the protrusion of the stamp. By optimizing the conditions, it is possible to print isolated structures with a size comparable to or even smaller than the lateral width of the stamp protrusions,

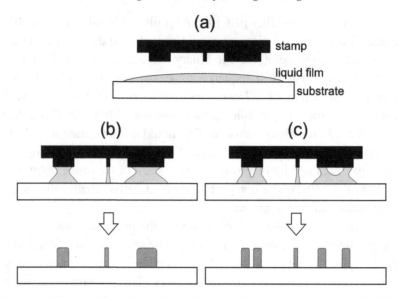

Figure 3.24. Schematic illustration of printing nanostructures by LCW. (a) A drop of solution is applied to the surface of the substrate and the stamp is placed on top of it. Process undertaken with (b) dilute solution and (c) with extremely dilute solution. The stamp, the organic film and the substrate are represented by black, grey and white components, respectively.

related to the meniscus diameter, yielding nanometre resolution. LCW is of general application to a large variety of soluble materials. The experimental parameters involved are the stamp specifications (material, distance from the substrate, profile, etc.), the solution (solute, solvent, viscosity, density, etc.), and the substrate (material, roughness, etc.). The choice of the concentration of solution and the stamp–surface distance is crucial to the effectiveness of the method.

Here we discuss the example of the preparation of arrays of nanometre-sized aggregates of single-molecule magnets derived from Mn_{12} complexes, depicted in Fig. 3.25 (Cavallini *et al.*, 2003). The complex used in this work is $[Mn_{12}O_{12}(CO_2C_{12}H_9)_{16}\text{-}(H_2O)_4]$, which consists of a Mn_{12} core stabilized by an outer shell of 16 biphenyl carboxylate ligands. The ligands make the outer shell hydrophobic, so it is expected that the deposition of the complex from a solution onto a hydrophilic surface will result in droplets. The result of the deposition of a solution of the Mn_{12} complex on native oxide on Si(100) substrates is shown in Fig. 3.25. The stamp used in the experiments reproduced here consists of a piece of polycarbonate coated with a ~ 100 nm thick gold film. The stamp exhibits parallel grooves with typical spacing of 1.4 µm, depth of $\simeq 230$ nm and width of $\simeq 550$ nm.

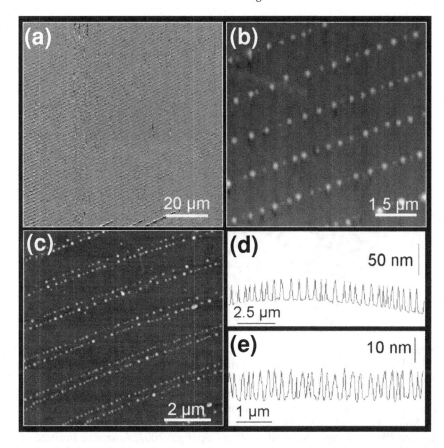

Figure 3.25. (a) AFM image (error signal) of patterns of the Mn_{12} complex fabricated on a large area (100 μm × 100 μm). (b) Image of printed features using a dilute solution. (c) Image of printed features using a very dilute solution. Typical line profile along a line obtained using (d) a dilute solution (corresponding to b) and (e) a very dilute solution (corresponding to c). Reprinted with permission from Cavallini *et al.*, 2003. Copyright (2003) American Chemical Society.

A spacer, made by evaporating three columns of gold 50 nm high through a contact mask serves to control the stamp–surface distance.

In Fig. 3.25(a), it is shown that patterning is effective across at least a 100 × 100 μm² area. The pattern is made of parallel lines that replicate the periodicity of the stamp. At greater magnification, Fig. 3.25(b), the lines appear to consist of droplets aligned along the stretching direction: their average size is $\simeq 270$ nm, and their distance is $\simeq 440$ nm, as shown in Fig. 3.25(d). The remarkable result is that the arrays of droplets are obtained by using a featureless stamp with a much larger size. The technological advantage lies in the ability to pattern on length scales much smaller than those of the features present in the stamp. By using a more dilute

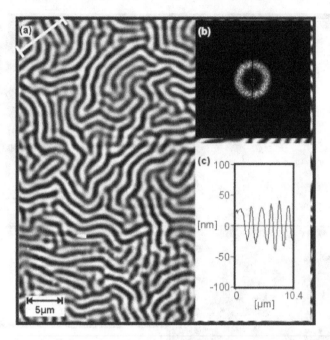

Figure 3.26. (a) AFM image of the self-organized pattern formed by annealing a thin layer system, consisting of 0.15 mm glass substrate, $\simeq 110$ nm Spiro-TAD and $\simeq 30$ nm SiN$_x$H$_y$, at 404 K for 60 s. (b) Fourier transform of the image. (c) Height profile of a cross section of the image in (a). The lateral wavelength is 1.5 μm in this case and the height modulation 70 nm. Reprinted with permission from M. Müller-Wiegand, G. Georgiev, E. Oesterschulze, T. Fuhrmann and J. Salbeck, *Applied Physics Letters*, **81**, 4940 (2002). Copyright 2002, American Institute of Physics.

solution (Fig. 3.25(c)), the droplets split, with a height of $\simeq 8$ nm and spaced by $\simeq 200$ nm.

Spinoidal patterning

The capability of organic multilayer devices to self-assemble in lateral patterns is an intrinsic property of such constructions. In these devices, the organic layers are sandwiched between a substrate and a cladding layer, typically consisting of oxidic materials or metals. In the example discussed here a reference system consisting of a glass substrate, a ~ 100 nm thick vacuum evaporated film of the molecular glass Spiro-TAD (see Table 1.1) and a cover layer of silicon nitride SiN$_x$H$_y$ ($\simeq 30$ nm), deposited in a low-temperature process (340 K) by plasma-enhanced CVD, has been used (Müller-Wiegand *et al.*, 2002). The samples were heated on a temperature-controlled microscopy hot-stage up to the glass transition temperature

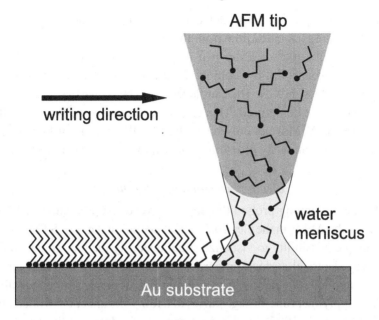

Figure 3.27. Schematic representation of the DPN technique. A water meniscus forms between the AFM tip coated with alkanethiols and the gold substrate. The size of the meniscus, which is controlled by relative humidity, affects the molecular transport rate, the effective tip–substrate contact area, and DPN resolution.

of the organic layer (404 K) at fast heating rates (100 K min^{-1}). The samples become opalescent after about 1 min at the final temperature, indicating defined light diffraction behaviour as known from photonic crystal structures. Atomic force microscopy measurements reveal a morphology with well-defined periodic structures of the order of 1.5 μm, which are shown in Fig. 3.26.

Scanning probe lithography

Dip-pen nanolithography

Controlled delivery of collections of molecules onto a substrate with nanometre resolution can be achieved with the tip of an AFM. This positive printing mode technique is called dip-pen nanolithography (DPN) and its working principle is illustrated in Fig. 3.27. DPN uses an AFM tip as a *nanopencil*, a substrate as the *paper* and molecules with a chemical affinity for the substrate as the *ink*. Capillary transport of molecules from the AFM tip to the solid substrate is used in DPN to directly write patterns consisting of a relatively small collection of molecules in submicrometre dimensions. The first example introducing the technique was the transfer of octadecanethiol onto gold surfaces (Piner *et al.*, 1999).

When an AFM tip coated with octadecanethiol is brought into contact with a sample surface, the molecules flow from the tip to the sample by capillary action, a process much like that of a dip pen (Fig. 3.27). Silicon nitride tips become coated with octadecanethiol by dipping of the cantilever into a saturated solution of octadecanethiol in CH_3CN. The cantilever has to be blown dry with compressed difluoroethane before being used. This elegant technique, which makes full use of the capabilities of AFMs, can only be applied for the writing of large areas if an efficient way of continuously providing solution to the tip is achieved.

Electrochemical nanopatterning

Electrochemical scanning probe lithography involves the operation of STMs and conducting probe AFMs in the electrochemical operational mode. STMs and AFMs offer the unique possibility to visualize simultaneously the topographical features and the local electrical surface properties of the substrate. This mode has been successfully applied for e.g., the local oxidation of inorganic semiconductors such as silicon by applying probe voltage pulses of a few volts under controlled humidity (Pérez-Murano et al., 1995). In this case nanometric dots and lines are routinely obtained. The confinement of these electrochemical reactions in the nanometre-scale range implies that the electrical current promoting these reactions flows only in very limited regions between the probe apex and the substrates.

Smooth nanometre-scale insulating barriers on the (001) face of $(TMTSF)_2PF_6$ single crystals have been obtained by means of an ambient-air operated conducting probe AFM working in the oxidative mode (Schneegans et al., 2001). At zero bias there are no surface modifications in contrast to the surface modifications achieved when the tip–substrate bias is applied. For positive substrate bias up to +4 V, smooth insulating lines are obtained, which increase in width for increasing voltage potential.

The capabilities of scanning probe microscopes (SPMs) have not yet been fully applied to MOMs!

4

Interfaces

I call our world Flatland, not because we call it so, but to make its nature clearer to you, my happy readers, who are privileged to live in Space.

Edwin A. Abbott, *Flatland*

This chapter is devoted to interfaces involving small organic molecules. We start with the ideal 0D case of a single molecule on an inorganic surface. The simple theoretical arguments from R. Hoffmann based on MOs will be discussed and complemented with experiments making use of STMs, demonstrating the incredible capabilities of this technique, which permits not only the chemical identification of such isolated molecules through the determination of their vibrational spectrum but the possibility of directly imaging the MOs in real-space. The 0D case will be followed by examples of the formation of small 2D aggregates, allowing molecule–molecule lateral interactions. Finally, the 2D case where molecules form ordered compact layers on top of the substrate surfaces within the ML regime will be analysed.

A ML can be simply defined as a one-molecule thick 2D film, but the molecular surface density has to be defined for each molecule–substrate system because it depends on the shape, size and relative orientation of the molecules. To clarify this point let us consider the examples of PTCDA and C_{60} on the Ag(111) surface. The surface density of the substrate is 1.4×10^{15} atoms cm^{-2}, which is usually defined as 1 ML as a reference limit. The surface density of the (102) plane of PTCDA, the cleavage plane, is 8.4×10^{13} and 8.3×10^{13} cm^{-2} (molecules cm^{-2}) for the monoclinic α and β polymorphs, respectively. Therefore, full coverage corresponds to 0.02 ML according to this definition. On the other hand, the surface density of a full hexagonal layer of closed-packed C_{60} molecules corresponding to the (111) plane in the fcc-C_{60} crystal is 1.2×10^{14} cm^{-2}. Thus, C_{60} would fully cover the Ag(111) surface at a coverage of 0.09 ML. However, other authors define 1 ML as a full compact layer on a substrate surface. In the absence of a common consensus one has to look for the specific definition adopted by the authors. The reading of the original work of I. Langmuir on adsorption of molecules on surfaces is highly recommended (Langmuir, 1918) as well as Chapter 2 from Redhead, 1993.

The 2D case heterostructures and multilayers (also called superlattices) will then be studied. Heterostructures consist of thin films with typical thickness below *c.* 5 nm and multilayers are built by the stacking (not necessarily periodic) of several organic and/or inorganic layers. Thicker (3D) structures, with thickness above 100 nm, will be discussed in Chapter 5. Although a thick film can indeed be considered as a heterostructure, we intentionally restrict the term heterostructure to ultrathin films in the spirit of associating the physical properties of the system to the interface. This point will be discussed at a later stage. The present chapter tries to highlight the relevance of interfaces for MOMs, which is in part due to the nearly planar geometry of most molecules.

4.1 Surfaces

Let us imagine that the infinite periodic 3D solid discussed in Section 1.7 is separated into two halves, leading to two semi-infinite 3D solids, preserving their 3D bulk periodicity but becoming aperiodic in the direction perpendicular to the generated surfaces. Because the translation symmetry is lost in this direction, the Born–von Karman boundary conditions can no longer be applied, hence the apparent paradox that a semi-infinite problem becomes more complex than the infinite case. This fact inspired W. Pauli to formulate his famous sentence: *God made solids, but surfaces were the work of the Devil.*

The free energy change, or reversible work done, to cleave or separate a material into two halves and to take the two halves to infinite distance from each other in vacuum is called the work of cohesion W_c, which is given by $W_c = 2A\gamma_s$, where A and γ_s stand for the generated area and the surface or material–vacuum interface tension (surface energy per unit area), respectively. Strong (weak) interatomic or intermolecular forces imply large (small) W_c, so that for MOMs we expect moderate or low γ_s values. For anisotropic materials, as is the case for the vast majority of MOMs, W_c and thus γ_s depends on orientation. As a result, the generated surfaces exhibiting the lowest γ_s values, corresponding to the molecular planes containing the strongest intermolecular interactions, should be quite stable in air because of the low degree of reactivity, except for materials containing reactive radicals, and they exhibit negligible reordering either in the form of reconstruction or relaxation. Surface reconstruction is known to be important for inorganic semiconductors like Si, Ge, GaAs, etc. leading to complex surface reorganization due to the strong covalent bonding. Well-known examples are Si(111)-(7 × 7) and Ge(111)-c(2 × 8) surface reconstructions. In the case of graphite no reconstruction or relaxation is observed because of its low γ_s value. This is the reason why graphite is very easily mechanically cleaved with e.g., sticky tape, exposing atomically flat surfaces with micrometre-range terraces. In addition the generated surfaces are hydrophobic. The stability of surfaces of many MOMs upon exposure to air is clearly evidenced by the quality of the measured STM and AFM images, which exhibit molecular or even submolecular resolution as thoroughly discussed in Magonov & Whangbo, 1996. This stability often allows *ex situ* analysis of such surfaces, avoiding the rather involved *in situ* UHV measurements, although investigations performed as a function of temperature impose the operation in UHV.

The STM was developed by G. Binnig and H. Rohrer (Binnig *et al.*, 1982) and its working principle is based on the quantum mechanical tunnelling effect. When two conducting samples are brought into close proximity (few nm), electrons from one sample may flow (tunnel) into the other. The probability of an electron to get through the tunnelling barrier decreases exponentially with the distance, hence the tunnelling current I_t at a constant applied bias V_t is an extremely sensitive measure of the distance between the two conducting samples. The STM makes use of this sensitivity. A sharp metal tip is brought mechanically into close proximity of a flat conducting surface and V_t is applied between the sample and the tip. In a STM electrons flow from the tip to the sample for $V_t > 0$, thus exploring the sample's unoccupied electronic states, and from the sample to the tip for $V_t < 0$, thus exploring the sample's occupied electronic states. This is closely related to inverse and direct photoemission, respectively. Representing the tip by an atom with a single s-orbital and assuming small bias voltages, J. Tersoff and D. R. Hamann (Tersoff & Hamann, 1985) showed that the spatial variation of I_t can be described by that of the partial local electron DOS associated with the energy levels lying in the vicinity of E_F. For metals the highest occupied and lowest unoccupied states (around E_F) are similar in nature, so that the characteristics of their STM images are independent of the bias polarity. Typically $|V_t| \leq 1$ V and I_t lies in the pA–nA range. Note that the applied electric field is rather large, ~ 1 V nm^{-1} = 10^9 V m^{-1}, which can strongly perturb the surface molecular arrangement. In fact this effect is used for the manipulation of single molecules.

Few examples of surface relaxation have been reported for MOMs. One of them corresponds to the *ac*-surface of β-(BEDT-TTF)$_2$PF$_6$ (Ishida *et al.*, 2001), an organic 1D conductor along the *c*-axis exhibiting a metal–insulator transition at ~ 297 K associated with the doubling of the *c*-parameter below the transition temperature (see discussion in Section 1.7 concerning the band gap opening at E_F). The maximum value of σ along the *c*-axis is ~ 10–20 Ω^{-1} cm^{-1} and below the transition temperature $E_a \simeq 0.2$ eV (Kobayashi *et al.*, 1983). In the crystal, BEDT-TTF and PF$_6$ layers are alternately stacked along the *b*-axis. In the case of PF$_6$-terminated

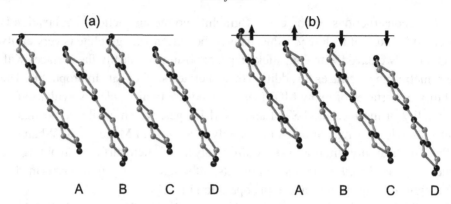

Figure 4.1. β-(BEDT-TTF)$_2$PF$_6$ (*Pnna*, $a = 1.496$ nm, $b = 3.264$ nm, $c = 0.666$ nm). Structural models for the (a) unrelaxed and (b) relaxed *ac*-surface layer. C and S atoms are represented by black and medium grey balls, respectively. Adapted from Ishida *et al.*, 2001.

ac-surfaces STM images taken in air show that PF$_6$ molecular rows are alternately missing along the crystal *a*-axis. For the BEDT-TTF-terminated *ac*-surfaces, the unit cell consists of four BEDT-TTF molecules, which form two dimers (A-B and C-D in Fig. 4.1(a)). If the surface retains the ideal crystal structure, molecules A and C should be observed with the same shape and brightness in the STM image plane because they have the same structure and are the same distance from the surface. However, alternately modulated molecular rows appear as schematized in Fig. 4.1(b) with a net shift between the A-B and C-D dimers.

As already discussed in Chapter 1, this kind of mixed valence salt becomes conductive due to the transfer of one electron from two BEDT-TTF molecules to the anion layers. However, at the surface, the charge can become unbalanced, resulting is an incomplete CT. This leads to differentiated surface vs. bulk nesting vectors and to the existence of surface CDWs (Ishida *et al.*, 1999). The Peierls transition has also been observed on the *ab*-planes of single crystals of TTF-TCNQ with a variable temperature STM (Wang *et al.*, 2003) and will be discussed in Section 6.1.

Detailed reviews on the characterization of surfaces of single crystals of small and large biomolecules such as catalase, insulin, canavalin, etc. can be found in McPherson, 1999 and Ward, 2001.

4.2 Organic/inorganic interfaces

We will now deal with interfaces formed between organic and inorganic materials. The convention that will be followed is that for a generic A/B interface, A and B will correspond to organic and inorganic counterparts, respectively. Thus, when considering e.g., semiconductor/semiconductor systems it should be clear that the left and right members correspond to organic and inorganic materials, respectively.

Single molecules on surfaces

When trying to understand the formation of organic/inorganic interfaces the first and most fundamental step consists in the study of the interaction of a single molecule with a surface. This requires, from the experimental point of view, working at extremely low coverages in order to avoid molecule–molecule lateral interactions. This situation is possible for sufficiently large molecule–substrate interactions (chemisorption) and/or at cryogenic temperatures (physisorption) where surface diffusion and thus the formation of 2D islands can be suppressed. A few remarkable recent examples of single individual molecules on metallic surfaces are found in the literature, thanks to the availability of low temperature STMs, and will be discussed here. As mentioned before, UHV is needed when experiments have to be performed at low temperatures.

In order to model the chemisorptive molecule–surface interactions let us perform a comparison of orbital interactions in discrete molecules, as discussed in Section 1.7, and of a single molecule with a surface following the conceptually simple approach of R. Hoffmann (Hoffmann, 1988). Figure 4.2(a) shows a typical molecule–molecule interaction diagram, while Fig. 4.2(b) illustrates a schematized molecule–metallic surface interaction diagram. In this simplified picture it is assumed that molecule–metal interactions can be explained in terms of the interaction of the frontier orbitals, HOMO and LUMO. The most relevant interactions are expected to be the two-orbital, two-electron stabilizing interactions **1** and **2**. For the molecule–metal case, the MOs combine to give a lower energy two-electron MO and a higher energy unoccupied MO (Fig. 4.2(c)).

Depending on the relative MO energies and on the degree of overlap, both interactions will involve CT from one system to the other. Interaction **3** is a two-orbital, four-electron one, which for the molecule–molecule system is destabilizing and repulsive. However, in the molecule–surface system (Fig. 4.2(d)) interaction **3** may become attractive if the antibonding component of this four-electron, two-level system is located above E_F. In this case electrons will transfer to the solid and the system becomes stabilized. Interaction **4** is expected to have no effect on molecule–molecule interactions, since both orbitals are empty, but in the molecule–surface case (Fig. 4.2(e)) it may contribute significantly if the bonding level lies below E_F because in this case CT from the solid to the molecule is expected. Again this would lead to an attractive molecule–surface interaction.

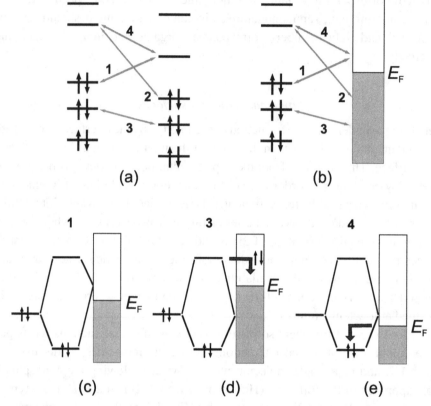

Figure 4.2. Schematic diagrams of (a) molecule–molecule orbital interactions and (b) molecule–metallic surface interactions. Cases (c), (d) and (e) represent the **1**, **3** and **4** molecule–metallic surface interactions, respectively. Adapted from Hoffmann, 1988.

Let us exemplify these ideas with the $C_2H_2/Pt(111)$ system. C_2H_2 brings to the adsorption process a degenerate set of occupied π and unoccupied π^* orbitals (top of Fig. 4.3(a)).

When adsorbed on Pt(111), C_2H_2 undergoes significant rehybridization and its linear symmetry is broken, removing the π and π^* degeneracy (bottom of Fig. 4.3(a)). The point group of the unperturbed molecule, $D_{\infty h}$, changes either to C_{2v} in the *cis*-planar configuration or C_2 in the non-planar configuration. The valence orbitals are shown in Fig. 4.3(b). Both π_σ and π_σ^*, which lie in the bending plane and point towards the surface, result from some *s* mixing and should interact with the Pt d_{z^2} surface more than the π and π^* orbitals. π_σ and π_σ^* interact preferentially with the bottom and the top of the d_{z^2} band, respectively, because of the symmetry-dependent mixing (Fig. 4.3(b)). As a result of the interaction with C_2H_2 the d_{z^2} band

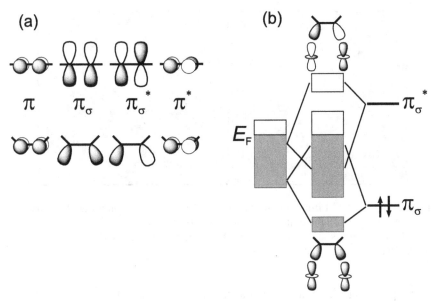

Figure 4.3. (a) $\pi - \pi^*$ C_2H_2 MOs: (top) free molecule and (bottom) molecule interacting with a surface. (b) Schematic interaction diagram between orbitals π_σ from C_2H_2 and d_{z^2} from platinum. Adapted from Hoffmann, 1988.

becomes modified, weakening the metal–metal bonding. Some of the metal–metal bonding levels that were at the bottom of the band are pushed up (lower binding energies) while some metal–metal antibonding levels are pushed down. Therefore, the molecule–surface bonding is accomplished at the expense of bonding within the surface and the chemisorbed molecule.

Hoffmann's simplified model explains the most important features but has to be considered, because of its simplicity, as an approximation. In fact, high-resolution X-ray emission spectroscopy results performed on the N_2/Ni(100) system, where the N_2 molecules exhibit an upright adsorption geometry, have shown that the frontier orbitals approximation is insufficient because the chemisorptive bonds affect all valence states, down to the inner 2σ states (see Fig. 1.1), which are located about 25 eV below E_F (Nilsson *et al.*, 1997).

Let us now review some relevant measurements performed on individual molecules adsorbed on metallic surfaces with a STM. The first chosen molecule is C_2H_2, thus connecting with the discussion given above. Figure 4.4(a) shows a STM image taken in UHV of a single C_2H_2 molecule on a Pd(111) surface held at 28 K. The molecule appears as a combination of a protrusion (brighter part) with two shallower depressions (darker parts), suggesting a famous cartoon character.

A cross section through the protrusion and depressions is shown in the lower part of Fig. 4.5(a). The maximum is 0.013 nm high, the minimum 0.007 nm deep, and

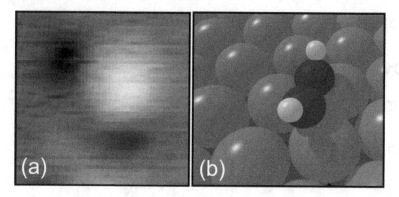

Figure 4.4. (a) STM image of an individual C_2H_2 molecule on a Pd(111) surface taken at 28 K. (b) Representation of this C_2H_2/Pd(111) system with a ball model. C, Pd and H atoms are represented by black, medium grey and light grey balls, respectively. Courtesy of Dr M. Salmeron.

they are separated by 0.35 nm. The total apparent corrugation of the molecule is thus 0.02 nm. The lower part of Fig. 4.5(b) shows calculated cross sections through the molecule as a function of the molecular tilt angle θ_t, defined as the angle between the molecular z-axis (see Section 1.2) and the surface normal (see Fig. 4.5(c)), using the electron scattering quantum chemistry (ESQC) technique (Sautet & Joachim, 1988; 1991).

   This technique is well suited to treating the challenging problem of electron tunnelling through adsorbates. The ESQC method offers a means of studying the transmission of elastic electrons through a localized defect inserted in an infi-nite periodic medium. The method has been further developed in order to study the tunnelling of electrons in a STM. In this case the apex of the STM tip, the adsorbate to be imaged and the surface atoms of the substrate are treated as the defect. The observed STM image shape can be explained qualitatively in a rather simple way as illustrated in Fig. 4.5(c). Following the approximation of frontier orbitals discussed in the previous section, the STM images are interpreted in terms of overlap between the molecular π orbital with a tip apex s orbital. The C_2H_2 π orbital (Fig. 4.3(a)) lies in the molecular xy-plane and hence parallel to the Pd(111) surface for $\theta_t = 0°$. The calculated image for an untilted molecule has a depression slightly offset from the hollow site with a much smaller protrusion on the other side. As the molecule tilts, the relative size of the protrusion increases as the depression decreases and shifts away from the centre. At $\theta_t = 13°$, the image shape is in good agreement with the experiment. When this angle increases, the positive and nega-tive lobes of the π orbital extend out on each side of the molecule and overlap with the tip s orbital. Because of the 13° tilt of the molecule, the overlap with one lobe makes a dominant contribution to the image (brighter part) and the depressions

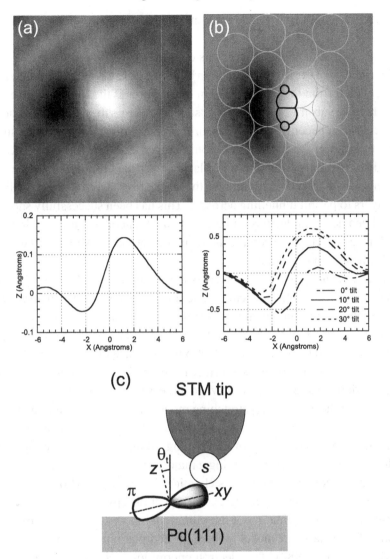

Figure 4.5. (a) 1.2 nm × 1.3 nm image of a single C_2H_2 molecule ($V_t = 40$ mV, $I_t = 400$ pA). The graph below shows a cross section along a path through the maximum and minimum of the image. (b) 1.2 nm × 1.3 nm ESQC calculated STM image for the minimum energy geometry. The calculated image is overlaid on a schematic of the adsorbed molecule. The graph shows the calculated molecular profile vs. molecular tilt angle θ_t. As θ_t increases, the protrusion increases as the depression decreases. Reprinted with permission from J. C. Dunphy, M. Rose, S. Behler, D. F. Ogletree, M. Salmeron and P. Sautet, *Physical Review B*, **57**, R12705 (1998). Copyright (1998) by the American Physical Society. (c) Schematic representation of the orbital interaction of the STM tip/C_2H_2/Pd(111) system.

in the image are the result of the interference between the remaining molecular lobes.

As mentioned at the beginning of this chapter, a STM is able to probe the real-space distribution of the MOs and this can be achieved if the molecules are electronically decoupled from the substrate, or in other words if the MOs are unperturbed. In fact, when the substrates are metals or semiconductors the electronic structure of the molecules can become strongly perturbed. Hence the influence of the substrate has to be reduced, if the inherent electronic properties of individual molecules are to be studied, by interposing a thin insulating layer between the molecule and the conducting substrate. The thickness of this insulating layer has to be only a few atomic layers to permit tunnelling across it. This has been successfully shown with individual pentacene molecules on layers of NaCl on copper surfaces. In this case the STM images perfectly resemble the structures of the HOMO and LUMO of the free molecule (Repp *et al.*, 2005).

In the case of individual C_{60} molecules adsorbed on clean Si(111)-(7 × 7) surfaces the experimental intramolecular structure compares well to *ab initio* calculations based on density functional theory (DFT) in spite of the covalent character of the substrate (Pascual *et al.*, 2000). In this case a certain degree of uniaxial strain has to be considered to simulate the electronic influence of the substrate.

For C_2H_2 adsorbed on a Cu(100) surface the tunnelling conductance as measured with a STM, defined as dI_t/dV_t, is shown to increase at $V_t = 358$ mV, resulting from the excitation of the internal ν_{CH} stretching mode. The inelastic scanning tunnelling spectroscopy (ISTS) technique makes such determination possible and has opened up the extraordinary possibility of performing vibrational spectroscopy with single molecules on surfaces, representing a major achievement, since the vibrational fingerprint of an adsorbate allows its chemical identification (Stipe *et al.*, 1998). Earlier high-resolution electron energy loss spectroscopy (HREELS) experiments performed for the determination of the vibrational modes of chemisorbed MLs of C_2H_2 on (111) surfaces of nickel, palladium and platinum gave $\nu_{CH} \simeq 370$ meV (Gates & Kesmodel, 1982). The electron energy loss spectroscopy (EELS) technique consists of irradiating a sample with monochromatic electrons with energies lower than typically 500 eV and measuring the energy of the inelastically back-scattered electrons. Details in the meV regime can only be achieved in the high-resolution (HR) mode.

The example of ISTS of a single C_6H_6 molecule chemisorbed on a Ag(110) surface is illustrated in Fig. 4.6(a). The isolated C_6H_6 molecules exhibit inelastic peaks at ±4 and ±19 mV, while fully C_6H_6 covered Ag(110) (Fig. 4.6(b)) exhibits peaks at ±7 and ±44 mV, where C_6H_6 molecules are in a very weakly adsorbed state. These differences in the spectra between isolated molecules and MLs, where lateral molecule–molecule interactions are present in addition to the

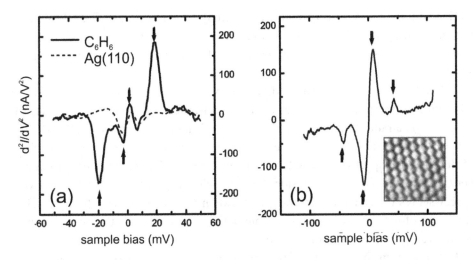

Figure 4.6. (a) $d^2 I_t/dV_t^2$ spectra acquired above a single C_6H_6 molecule (continuous line) and on the bare Ag(110) surface (dashed line). The peaks at 4 and 19 meV represent a change of the junction conductance of about 1 and 8%, respectively. (b) $d^2 I_t/dV_t^2$ spectra acquired on a C_6H_6 ML grown at 4 K (STM image shown in the inset: 2.5×2.5 nm^2, $I_t = 0.1$ nA, $V_t = 100$ meV). The peaks at ± 7 and ± 44 meV represent a change of the junction conductance of about 6 and 1%, respectively. The authors define 1 ML as a fully covered Ag(110) surface. Reprinted with permission from J. I. Pascual, J. J. Jackiw, Z. Song, P. S. Weiss, H. Conrad and H.-P. Rust, *Physical Review Letters*, **86**, 1050 (2001). Copyright (2001) by the American Physical Society.

molecule–substrate interactions, verify the extraordinary sensitivity of ISTS to the chemical environment of an adsorbate (Pascual *et al.*, 2001). The energies of these excited modes are well below the expected values for internal vibrations of C_6H_6 adsorbed on *sp*-metal surfaces and are attributed to the excitation of vibrations of the C_6H_6 molecule with respect to the silver surface. In the case of single C_{60} molecules adsorbed on Ag(110) the internal $H_g(\omega_2)$ mode is identified at 54 mV with ISTS (Pascual *et al.*, 2002).

Inelastic tunnelling electrons can also be used to selectively induce either the translation over a metallic surface or desorption from the metallic surface of individual molecules, as has been shown for NH_3 on Cu(100) surfaces (Pascual *et al.*, 2003). Activation of either the stretching vibration of ammonia ($\simeq 408$ meV) leading to lateral translation on the surface, or the inversion of its pyramidal structure (umbrella mode $\delta_s(NH_3) \simeq 139$ meV) leading to desorption, can be achieved by adjusting I_t and V_t.

A STM-induced Ullmann's reaction between two molecules has also become possible as demonstrated for C_6H_5I on Cu(111), producing a single biphenyl molecule (Hla *et al.*, 2000). Here I_t is used to dissociate the C_6H_5I molecule into

C_6H_5 and I, the STM tip is then able to manipulate and bring two C_6H_5 fragments together and again induce a chemical reaction by application of I_t. Such manipulations are successful when the molecules or fragments are immobilized, e.g., at surface steps.

Tunnelling electrons from a STM have also been used to excite photon emission from individual molecules, as has been demonstrated for Zn(II)-etioporphyrin I, adsorbed on an ultrathin alumina film (about 0.5 nm thick) grown on a NiAl(110) surface (Qiu *et al.*, 2003). Such experiments have demonstrated the feasibility of fluorescence spectroscopy with submolecular precision, since light emission is very sensitive to tip position inside the molecule. As mentioned before the oxide spacer serves to reduce the interaction between the molecule and the metal. The weakness of the molecule–substrate interaction is essential for the observation of STM-excited molecular fluorescence.

Depending on the strength of the molecule–surface interaction and on the experimental conditions, molecular adsorption may result in surface relaxation or even in surface-atom bond breaking and thus mass transport. This was in part discussed before for the case of C_2H_2 on Pt(111) surfaces, where it was shown that the molecule–metal bonding implies a weakening of the metal–metal bonds. Two further remarkable examples are discussed next: C_{60}/Ge(111) (Wirth & Zegenhagen, 1997) and $C_{90}H_{98}$, known as the *Lander*, on Cu(110) (Rosei *et al.*, 2002). In both cases the structures were prepared by OMBD in UHV.

We start with C_{60}/Ge(111) because the perturbation caused by the incoming molecules is readily evident from the STM images (Fig. 4.7(a)). When prepared in UHV under certain experimental conditions, the clean Ge(111) surface reconstructs with a $c(2 \times 8)$ structure. The top bilayer is decorated with 0.25 ML of adatoms in T_4 positions that are arranged in double rows along the $[1\bar{1}0]$ direction, as shown in Fig. 4.7(a). The distances between neighbouring adatoms are 0.80 nm and 0.69 nm. Each adatom bonds to three atoms in the first layer. When depositing less than 0.005 ML of C_{60} on the Ge(111)-$c(2 \times 8)$ surface in UHV with the substrate held at RT, the C_{60} molecules remain isolated at the surface, due to limited diffusion. Here 1 ML is defined as the surface density of the substrate, 7.2×10^{14} cm^{-2}, and for C_{60} the full coverage as that corresponding to the (111) plane in the fcc-C_{60} crystal, 1.2×10^{14} cm^{-2}, as discussed above. Hence, C_{60} will fully cover the Ge(111) surface at a coverage of 0.17 ML. The C_{60} molecules perturb the surface reconstruction by inducing shifts of the adatom rows. The most common experimentally observed case is highlighted in the rectangle on Fig. 4.7(a) and is schematized in Fig. 4.7(b). The shift is found in the row indicated by the arrow. The row has moved upwards by one surface lattice constant. As a result one triple row and one single row arise below the C_{60} molecule, whereas two regular double rows are retained above. The row shift encompasses the existence of dangling bonds that induce the

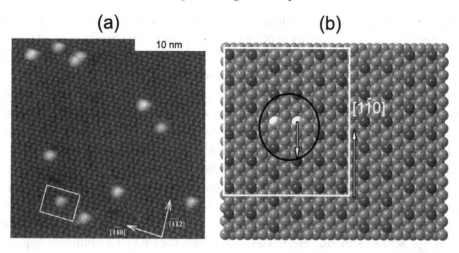

Figure 4.7. (a) STM image showing the Ge(111)-$c(2 \times 8)$ surface with isolated C_{60} adsorbates ($V_t = 1$ V, $I_t = 0.05$ nA). The section enclosed in the rectangle represents the most common molecule-induced defect. Reprinted with permission from K. R. Wirth and J. Zegenhagen, *Physical Review B*, **56**, 9864 (1997). Copyright (1997) by the American Physical Society. (b) Representation of the Ge(111)-$c(2 \times 8)$ surface with the shift on one row induced by an individual C_{60} molecule. The large, distorted circle is the reproduction of the C_{60} imaged by the STM.

chemisorption of the molecule with the surface, hence the low diffusion. In addition to this molecule-induced defect other perturbations have been also observed.

Our second example deals with the adsorption of a $C_{90}H_{98}$ molecule on a clean Cu(110) surface. The Lander molecule can be regarded as a conducting board (π-system) and four spacer legs that elevate the board from the substrate, with the aim of electronically isolating it from the surface. By manipulating individual molecules at low temperature, it is found that a single Lander molecule can act as a template, self-fabricating metallic nanostructures at step edges. This restructuring process can be described in terms of the structure of the molecule, which reshapes a portion of the step edge, leading to the formation of a nanostructure two Cu atoms wide and seven Cu atoms long below the molecular board.

In principle a STM should be adequate to measure the electrical resistance of a single molecule since it suffices to measure I–V curves of the metal (tip)–molecule–metal (substrate) system. However, published results in the literature concerning this subject have to be considered cautiously because of the generally unknown nature of the molecule–metal contacts. An illustrative experiment demonstrates the relevance of the interface (Kushmerick *et al.*, 2002). This experimental work studies charge transport using the cross-wire tunnel junction technique, where two metal wires, one of them modified with a SAM of the molecule of interest, are mounted in a crossed geometry with one of the wires perpendicular to an external magnetic field. The junction separation is controlled by deflecting this wire with the Lorentz force generated by a small dc current. The selected SAM was based on linear oligophenyleneethylene with either one or two thioacetyl functional groups on the ends of the molecule. Upon metal–SAM–metal contact, molecules with one or two functional groups behave as molecular diodes or wires, respectively, due to the different molecule–metal contacts. Here the importance of chemisorption in order to improve the electrical contact becomes evident.

Another way to improve the metal–molecule contact consists of applying a force, e.g., with a conducting AFM tip, while measuring the electrical conductance. For the case of SWNTs connected to a gold electrode and explored with metallized AFM tips acting as mobile electrodes, it has been shown that a minimum applied force of ~ 30 nN is required to obtain an optimum electrical contact and that above 50 nN the radial deformation of the molecule induces a semiconducting-like behaviour due to the opening of an electrical gap (Gómez-Navarro *et al.*, 2004).

In conclusion, the development of nanometre-scale techniques such as scanning tunnelling microscopy, which permit imaging, manipulation and the performance of electronic, vibrational and optical spectroscopy at submolecular level offers a myriad of possibilities in the coming years to understand the phenomenology of interactions between single molecules and conductive substrates. Spin-polarized STMs have also emerged in recent years (Bode *et al.*, 1998), adding magnetic characterization of single molecules at the nanometre scale.

Self-assembly at submonolayer coverages

In the previous section we have explored the 0D case of individual molecules on inorganic surfaces, which is experimentally achieved for very low coverages. If the density of molecules at the surface is increased or T_{sub} increased then molecules may interact laterally and form aggregates or assemblies. We thus have the interplay of molecule–molecule in addition to the molecule–substrate interactions. Indeed the geometry and chemical nature of the molecules are at the origin of the potential assemblies. Here one can realize the great possibilities that chemical synthesis can offer, since molecules can be chemically modified by e.g., adding functional groups and thus engineering the assemblies.

The first chosen example refers to the control of surface-supported supramolec-ular assembly by judicious distribution of substituents in porphyrin molecules (Yokoyama *et al.*, 2001). Substituted porphyrin molecules OMBD-deposited on a Au(111) surface form monomers, trimers, tetramers or extended wire-like structures depending on the number and distribution of substituents. The experimentally found structures can easily be predicted from the geometric and chemical nature of the porphyrin substituents, which mediate the interactions between individual adsorbed molecules. The porphyrins used in this study are based on H_2-TBPP, which has a free-base porphyrin core and four ditertiarybutylphenyl substituents (Fig. 4.8(a)). Isolated H_2-TBPP single molecules are found on the reconstructed gold surfaces. CTBPP is formed by replacing a ditertiarybutylphenyl group of H_2-TBPP by a cyanophenyl substituent (Fig. 4.8(b)). At low coverage, most CTBPP molecules assemble into triangular clusters on Au(111) (Fig. 4.8(e)). Substituting one more cyano group in CTBPP gives rise to BCTBPP, which forms two types of isomers (*cis* and *trans*) with respect to the configuration of two cyanophenyl substituents (Fig. 4.8(c) and (d), respectively). The *cis*-BCTBPP molecules are aggregated into a supramolecular tetramer (Fig. 4.8(f)). In this structure, the antiparallel intermolecular connections of all the cyanophenyl substituents lead to a macrocyclic arrangement of the porphyrin molecules, forming a molecular ring. Sequential aggregation is achieved for the *trans*-BCTBPP molecules (Fig. 4.8(g)). In this structure, the antiparallel configuration between the cyanophenyl substituents results in a linear arrangement of the *trans*-BCTBPP molecules, forming supramolecular wires. The maximum observed length of the straight wires is above 100 nm.

Linear chains have also been obtained on Ag(111) surfaces by OMBD-growth of *trans*-PVBA (Barth *et al.*, 2000). In this case hydrogen bonding is the main factor responsible for this 1D configuration. PVBA is a planar molecule composed of a pyridyl group as the head and a carboxylic acid group as the tail. The molecular stripes are oriented along the $[11\bar{2}]$ direction of the silver lattice. These 1D superstructures consist of two chains of PVBA. In each chain the molecular axis is oriented along the chain direction, with a head-to-tail distribution. The orientation of the molecular chains reflects a good match between the *trans*-PVBA subunits and the high-symmetry lattice positions.

Hydrogen bonding is again at the origin of the striking 2D supramolecular clusters and chains observed upon submonolayer deposition of NN onto reconstructed Au(111) (Böhringer *et al.*, 1999). NN molecules, prepared from naphthalene and a mixture of nitric and sulfuric acids at 320 K, adsorb planarly on Au(111) and it is interesting to note that the 2D confinement of NN introduces a chirality not present in the gas phase: the molecule and its mirror image cannot be superimposed by translation and rotation within the surface plane. This is a case of surface-induced chirality. In a wide range of low coverages (0.05–0.2 ML) NN molecules self-assemble mainly in decamers and to a lesser extent in tetramers and undecamers. 1 ML is defined here as a close-packed molecular layer. The decamers consist of an eight-molecule ring surrounding a two-molecule core. Manipulation experiments at decreased tunnelling resistances show that the decamers are surprisingly robust

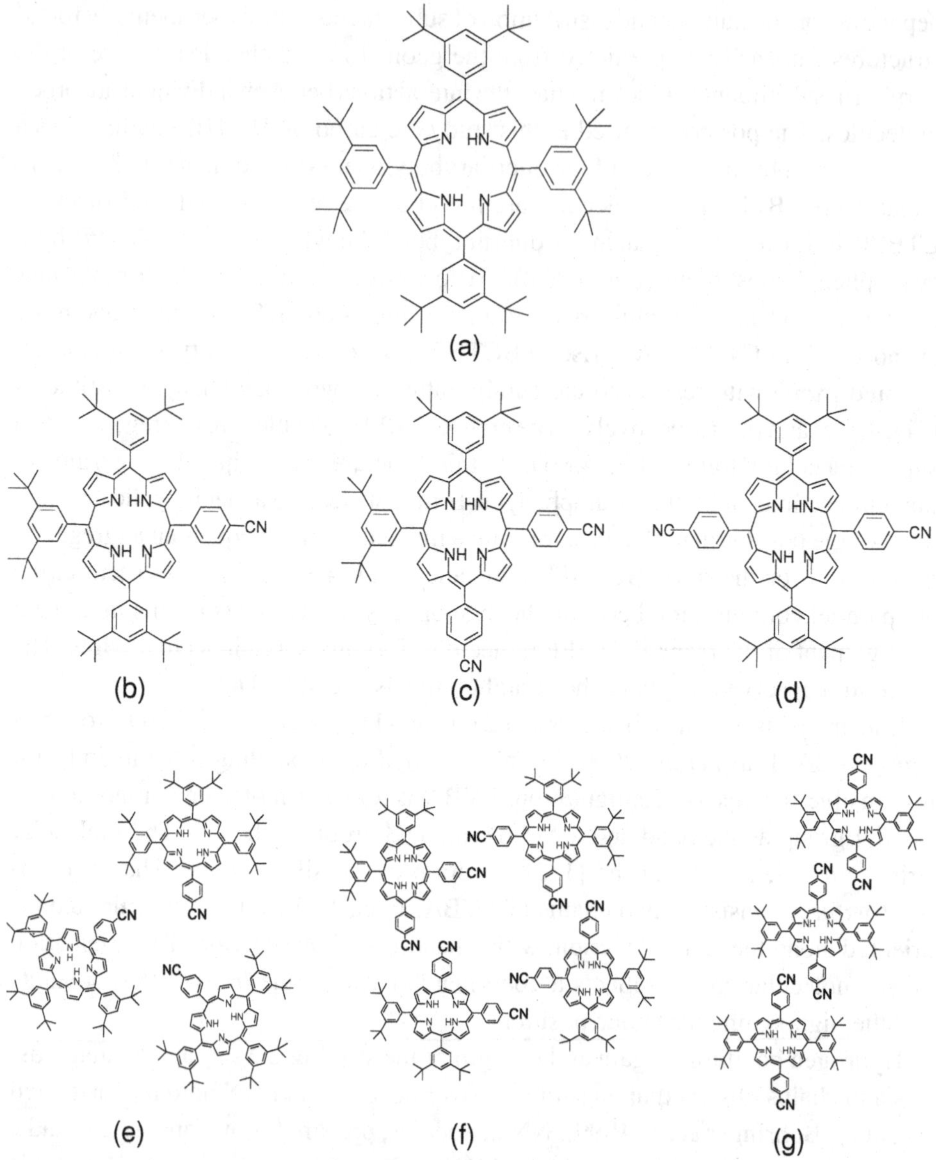

(a)

(b)   (c)   (d)

(e)   (f)   (g)

Figure 4.8. Schemes of the porphyrins discussed in the text. (a) H_2-TBPP, which is composed of a free-base porphyrin and four ditertiarybutylphenyl substituents. (b) CTBPP, where a ditertiarybutylphenyl substituent of H_2-TBPP is replaced with a cyanophenyl substituent. (c) *cis*-BCTBPP, where two cyanophenyl groups have been substituted at the *cis* position. (d) *trans*-BCTBPP, where two cyanophenyl groups have been substituted at the *trans* position. (e) CTBPP trimer, (f) *cis*-BCTBPP tetramer and (g) *trans*-BCTBPP wire.

supramolecules: they can be rotated around the surface normal or displaced laterally. When two decamers are pushed towards each other they separate spontaneously to a distance of 0.9 nm. The interaction between NN molecules on Au(111) is related to the asymmetry of the molecular charge distribution induced by the NO_2 group. Below a critical number of molecules, open linear chains with unsaturated bonds at their ends energetically prefer to close into rings as observed for the decamers. The C_2 symmetry of the decamers in the experimental images requires the number of molecules of each type of chirality, n_1 and n_2, in the cluster to be even. Thus, the decamers have a well-defined overall chirality and each cluster has a mirror-symmetric isomer with the same energy in which n_1 and n_2 are interchanged. Its structure allows the hydrogen bonding of all oxygen atoms in the cluster.

We end this section with two examples of the substrate's surface reorganization induced by the incoming molecules: CuPc/Ag(110) (Böhringer *et al.*, 1997) and HtBDC/Cu(110) (Schunack *et al.*, 2001). In the CuPc/Ag(110) case, CuPc molecules interact with steps in a rather surprising way, inducing their reorganization as a function of coverage. After adsorption of 0.6 ML and annealing, a drastic change in the morphology of the clean Ag(110) surface occurs. The authors define 1 ML as one molecule per 15.8 substrate surface atoms, corresponding to a dense molecular superstructure. Extended areas of substrate (110) terraces are separated by bunches of steps aligned along the substrate $[1\bar{1}0]$ direction. The surface thus becomes 3D faceted. However, after deposition of 1 ML CuPc 2D step faceting occurs.

At low coverages HtBDC molecules deposited on Cu(110) surfaces decorate the substrate steps, when held at RT, indicating that the diffusion barrier for an individual molecule on the flat Cu(110) surface is low enough to allow the molecules to be mobile. However, at low T_{sub} STM images show individual immobile molecules away from steps with a planar geometry with six lobes arranged in a distorted hexagon with threefold rotational symmetry (see Fig. 4.9(A)). Each lobe can be assigned to one of the tertbutyl appendages. At $T_{sub} > 160$ K, a 1D diffusion of the single molecules along the $[1\bar{1}0]$ substrate direction sets in.

When a controlled manipulation of the molecules is performed by reducing the tunnelling resistance by changing I_t or V_t or both, the HtBDC molecules are pushed outside the scanned area. The resulting *cleaned* surface area reveals the existence of local disruption of the topmost copper surface layer. About 14 copper atoms are expelled from the surface in two adjacent $[1\bar{1}0]$ rows, forming a trench-like base for anchoring of the molecules as shown in Fig. 4.9(B). The spontaneous surface disruption formed underneath the molecules during the adsorption process is a generic way to reduce the mobility of the molecules and bind them to the surface at even low coverages.

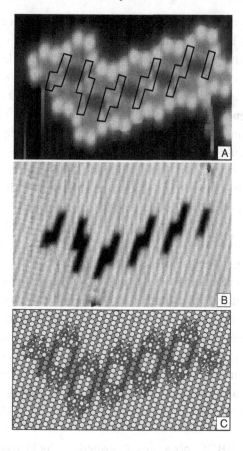

Figure 4.9. Constant current images (10.5×6.9 nm^2) at 41 K. (A) HtBDC double row structure ($V_t = 1.070$ V, $I_t = 0.45$ nA). The trenches in the underlying surface are sketched. (B) The trenches in the surface layers are disclosed after manipulating the molecules aside ($V_t = 7$ mV, $I_t = 1.82$ nA). Atomic resolution along the close-packed direction was obtained in the left part of the image (vertical fast scanning direction), whereas it was lost when the tip scanned the restructured area. (C) Ball model of the double row structure. The substrate atoms are shaded darker the deeper the layers lie, while the molecules are shown on top. Reprinted with permission from M. Schunack, L. Petersen, A. Kühnle, E. Laegsgaard, I. Stensgaard, I. Johannsen and F. Besenbacher, *Physical Review Letters* **86**, 456 (2001). Copyright (2001) by the American Physical Society.

Monolayer regime: heterostructures

We now deal with the structures that molecules build on substrate surfaces at full coverage, that is in the ML regime. Such hybrid systems are known as heterostructures. H. Kroemer defined heterostructures as heterogeneous semiconductor structures built from two or more different semiconductors, in such a way that the transition region or interface between the different materials plays an essential role in any device action (Kroemer, 2001). The term heterostructure can be generalized to any material, not necessarily restricted to semiconductors, and the regular stacking of several heterostructures builds superlattices, also termed multilayers. By defining the chemical nature of the materials involved and their thickness, the electronic and optical properties of the artificial structures can be engineered to some extent. As already mentioned in the definition, interfaces represent the key part of such artificial solids.

A large number of molecule–substrate systems have been explored. We are not interested here in reviewing all such heterostructures but instead in highlighting some relevant issues. Recommended dedicated reviews are e.g., Forrest, 1997 and Witte & Wöll, 2004.

Commensurate, coincident and incommensurate interfaces

In order to classify the relative ordering or registry that an ordered overlayer can achieve on an underlying crystalline substrate surface we follow the geometrical model proposed by D. E. Hooks, T. Fritz and M. D. Ward (HFW) (Hooks *et al.*, 2001), which generalizes the matrix formulation previously introduced (Wood, 1964; Park & Madden, 1968). A detailed description of the nomenclature recommended by the IUPAC for surfaces can be found in Bradshaw & Richardson, 1996. The reference substrate surface will be described by the lattice parameters a_s, b_s and the angle α_s ($\alpha_s > 0$) formed between the corresponding vectors a_s and b_s ($a_s^2 = a_s \cdot a_s$, $b_s^2 = b_s \cdot b_s$ and $a_s \cdot b_s = a_s b_s \cos \alpha_s$). Analogously, the lattice parameters of the overlayer will be represented by a_o, b_o and the angle α_o ($\alpha_o > 0$) between both vectors ($a_o^2 = a_o \cdot a_o$, $b_o^2 = b_o \cdot b_o$ and $a_o \cdot b_o = a_o b_o \cos \alpha_o$). Finally, both lattices are related by the azimuthal angle θ between the lattice vectors a_s and a_o ($a_s \cdot a_o = a_s a_o \cos \theta$). The substrate and overlayer lattice vectors for a given azimuthal orientation θ are related through the general expression:

$$\begin{pmatrix} R_{11} & R_{12} \\ R_{21} & R_{22} \end{pmatrix} \begin{bmatrix} a_s \\ b_s \end{bmatrix} = \begin{bmatrix} a_o \\ b_o \end{bmatrix}, \tag{4.1}$$

where the matrix coefficients R_{ij} (from registry) are defined as:

$$R_{11} = \frac{a_o \sin(\alpha_s - \theta)}{a_s \sin \alpha_s}, \tag{4.2a}$$

$$R_{12} = \frac{a_o \sin \theta}{b_s \sin \alpha_s}, \tag{4.2b}$$

$$R_{21} = \frac{b_o \sin(\alpha_s - \alpha_o - \theta)}{a_s \sin \alpha_s}, \tag{4.2c}$$

$$R_{22} = \frac{b_o \sin(\alpha_o + \theta)}{b_s \sin \alpha_s}. \tag{4.2d}$$

Registry is classified according to the R_{ij} values as commensurate, coincident and incommensurate as discussed next. This classification relies on the construction of a primitive overlayer unit cell based on translationally equivalent molecules. Indeed, the values of the R-matrix elements depend on the selected primitive lattice vectors describing the primitive unit cell but the equivalent R-matrices must have equal determinants.

(i) *Commensurism*   All $R_{ij} \in \mathbb{Z}$. All the overlayer lattice points coincide with symmetry equivalent substrate lattice points. This can also be described as *point-on-point* coincidence and an example is given in Fig. 4.10(a). In this figure the $\begin{pmatrix} 2 & 3 \\ -2 & 3 \end{pmatrix}$ experimentally determined registry of PTCDA on Ag(110) has been selected (Umbach *et al.*, 1998).

(ii) *Coincidence-I*   Among R_{ij} there are at least two integers confined to a single column of the matrix. Certain lines of the overlayer coincide with one set of primitive substrate lines. Every lattice point of the overlayer lies on at least one primitive lattice line of the substrate, a condition that has been described as *point-on-line* coincidence. This kind of coincidence can be subdivided into two classes depending on the rational ($\in \mathbb{Q}$) or irrational nature of the non-integer matrix elements:

(a) *Coincidence-IA* $R_{ij} \in \mathbb{Q}$. Figure 4.10(b) illustrates the hypothetical case $\begin{pmatrix} 2 & 7/2 \\ -2 & 7/2 \end{pmatrix}$.

(b) *Coincidence-IB*   At least one of the non-integer elements in the matrix is an irrational number. The irrational element(s) produce(s) an incommensurate relation between the overlayer and substrate along the coinciding primitive lattice vector. See the hypothetical example represented by the matrix $\begin{pmatrix} 2 & 7/2 \\ -2 & 3.14 \end{pmatrix}$ in Fig. 4.10(c).

(iii) *Coincidence-II* $R_{ij} \in \mathbb{Q}$, $R_{ij} \notin \mathbb{Z}$. Figure 4.10(d) shows the hypothetical case $\begin{pmatrix} 3/2 & 7/2 \\ -3/2 & 7/2 \end{pmatrix}$.

(iv) *Incommensurism*   At least one R_{ij} is irrational and neither column of the translation matrix consists of integers. Under this condition, no distinctive registry between the substrate lattice and the deposit lattice exists. See the example $\begin{pmatrix} 3/2 & 7/2 \\ -3/2 & 3.14 \end{pmatrix}$ in Fig. 4.10(e).

Table 4.1 lists some commensurate heterostructures formed by different molecules and substrates. Some of them will be discussed in detail throughout this chapter. Examples of *coincident-IA* are EC3T/Ag(111) (Soukopp *et al.*, 1998) with a transformation matrix $\begin{pmatrix} -4 & -8 \\ 8.5 & 0 \end{pmatrix}$, PTCDA on In-terminated InAs(111)A-(2 × 2) (Cox & Jones, 2000) with $\begin{pmatrix} 14/3 & -4 \\ 4/3 & 2 \end{pmatrix}$ and OMTTF/Ag(100) (Poppensieker *et al.*,

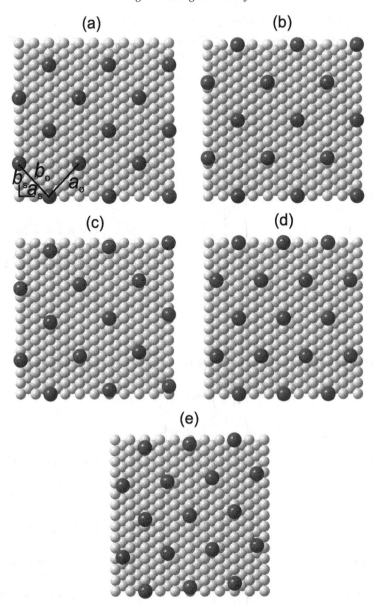

Figure 4.10. Schematic representation of the possible modes of registry. Overlayer and substrate lattice points are depicted as dark grey and light grey balls, respectively, and the primitive cell vectors a_s, b_s and a_o, b_o are also indicated. Ag(110) has been selected as substrate ($Fm\bar{3}m$, $a = 0.409$ nm). Examples of: (a) commensurate registry, (b) coincidence-IA registry, (c) coincidence-IB registry, (d) coincidence-II registry and (e) incommensurate registry.

Table 4.1. *Examples of commensurate heterostructures*

Heterostructure	Transformation matrix	References
coronene/Ag(111)	$\begin{pmatrix} 4 & 0 \\ 0 & 4 \end{pmatrix}$	Lackinger *et al.*, 2002
CuPc/Cu(100)	$\begin{pmatrix} 5 & -2 \\ 2 & 5 \end{pmatrix}$	Buchholz & Somorjai, 1977
CuPc/InAs(100)	$\begin{pmatrix} 3 & -1 \\ 1 & 3 \end{pmatrix}$	Cox *et al.*, 1999
CuPc/InSb(100)	$\begin{pmatrix} 3 & 0 \\ 0 & 3 \end{pmatrix}$	Cox *et al.*, 1999
CuPc/InSb(111)	$\begin{pmatrix} 2 & 2 \\ -2 & 4 \end{pmatrix}$	Cox & Jones, 2000
C_{60}/Cu(111)	$\begin{pmatrix} 4 & 0 \\ 0 & 4 \end{pmatrix}$	Hashizume *et al.*, 1993
DMe-DCNQI/Ag(110)	$\begin{pmatrix} 3 & 1 \\ -2 & 2 \end{pmatrix}$ phase I	Seidel *et al.*, 1999
	$\begin{pmatrix} 3 & 0 \\ 1 & 2 \end{pmatrix}$ phase II	
DPP-PTCDI/Ag(110)	$\begin{pmatrix} 6 & 0 \\ 0 & 3 \end{pmatrix}$ phase A	Nowakowski *et al.*, 2001
	$\begin{pmatrix} 7 & 0 \\ 0 & 3 \end{pmatrix}$ phase B	
	$\begin{pmatrix} 5 & 2 \\ -3 & 6 \end{pmatrix}$ phase C	
DP5T/HOPG	$\begin{pmatrix} 12 & 6 \\ 0 & 12 \end{pmatrix}$	Müller *et al.*, 1996
EC4T/Ag(111)	$\begin{pmatrix} -3 & -5 \\ 5 & 1 \end{pmatrix}$	Soukopp *et al.*, 1998
heptahelicene/Cu(111)	$\begin{pmatrix} 5 & 1 \\ -1 & 4 \end{pmatrix}$	Fasel *et al.*, 2001
HBC/GeS(010)	$\begin{pmatrix} 0 & 4 \\ 3 & -2 \end{pmatrix}$	Karl & Günther, 1999
NTCDA/MoS_2(0001)	$\begin{pmatrix} 4 & 2 \\ 0 & 5 \end{pmatrix}$	Karl & Günther, 1999
PTCDA/Ag(111)	$\begin{pmatrix} 6 & 1 \\ -3 & 5 \end{pmatrix}$	Umbach *et al.*, 1998
PTCDA/Ag(110)	$\begin{pmatrix} 2 & 3 \\ -2 & 3 \end{pmatrix}$	

Note: $\begin{pmatrix} 3 & -1 \\ 1 & 3 \end{pmatrix}$ and $\begin{pmatrix} 2 & 2 \\ -2 & 4 \end{pmatrix}$ can be also expressed as $\sqrt{10} \times \sqrt{10}R \pm 18.4°$ and $\sqrt{12} \times \sqrt{12}R30°$, respectively.

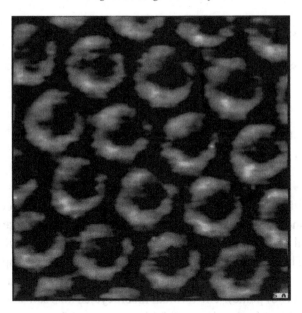

Figure 4.11. STM image of a SAM of cyclotrimeric terthiophenediacetylene on HOPG. $V_t = -0.88$ V, $I_t = 50$ pA: 10 nm × 10 nm. Courtesy of Dr E. Mena-Osteritz.

2001) with $\begin{pmatrix} 1 & 7/4 \\ 5 & 0 \end{pmatrix}$. Examples of *coincident-IB* systems are perylene/Cu(110) (Chen *et al.*, 2002), which exhibits eight-fold periodicity along the [110] azimuth and incommensurate order along [001] and HBC/Au(111) (Sellam *et al.*, 2001) with $\begin{pmatrix} -3 & 6 \\ -6.17 & 3 \end{pmatrix}$.

The formation of MLs of organic molecules at the liquid/HOPG interface can be *in situ* characterized with a STM. Depending on the molecule–solvent combination used, the MLs form spontaneously, giving rise to well-defined packings. HOPG surfaces are extremely popular because they can be easily prepared by cleavage and exhibit micrometre-sized terraces. However, because of the graphite basal plane, substrate–overlayer $\pi-\pi$ interactions are expected to strongly influence the packing of the overlayer molecules. The solvent used plays an important role in stabilizing the MLs.

Let us first explore the selected STM example concerning the molecular system cyclotrimeric terthiophenediacetylene shown in Fig. 4.11 (Mena-Osteritz & Bäuerle, 2001). The synthesis of this molecule was discussed in Section 2.5 (Fig. 2.17). When a solution of the cyclotrimeric terthiophenediacetylene macrocycle in $C_6H_3Cl_3$ is brought onto a freshly cleaved HOPG substrate, spontaneous formation of a SAM is found with honeycomb ordering at the solution/HOPG interface.

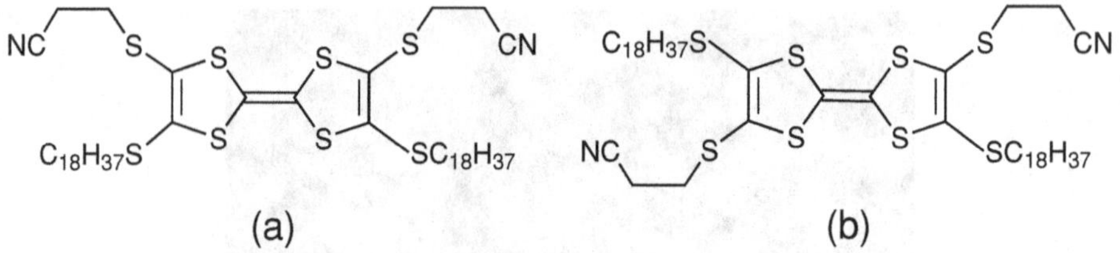

Figure 4.12. Isomers of bis-cyanoethylthio-bis-octadecylthio-TTF: (a) *cis*, C_{2v} and (b) *trans*, C_{2h}.

As measured with a low-current STM at RT the determined 2D lattice constants are $a_o = 2.46$ nm and $b_o = 2.52$ nm with $\alpha_o = 62°$. The arrangement and dimensions of the individual macrocycles indicate interdigitation of some butyl side chains in a regular way. This 2D lattice ordering corresponds to a more compressed molecular arrangement than that found in the bulk material ($a = 2.58$ nm, $b = 2.58$ nm, $\alpha = 60°$ for the (100) plane). The toroidal macrocycles thus self-assemble in a six-fold symmetry indicating that the three-fold symmetry of the molecules is increased in the adsorbate lattice. In fact in the ML coverage regime the crystal structure can be quite different from that of the bulk. The STM image reveals brighter and darker parts. The brighter regions correspond to higher I_t through the macrocyclic π-conjugated system. The alkyl side chains are supposed to lie in the dark regions.

Our second selected example concerns the isomers of bis-cyanoethylthio-bis-octadecylthio-TTF (Gomar-Nadal *et al.*, 2003), shown in Fig. 4.12. The synthesis of bis-cyanoethylthio-bis-octadecylthio-TTF is achieved starting from a bis-cyanothioethyl dithiole-thione molecule and using the mono-deprotection strategy discussed in Section 2.5 followed by alkylation. After conversion of this intermediate compound to the corresponding ketone and coupling in trimethyl phosphite (strategy S2) a mixture of both possible isomers is isolated.

The solutions were prepared in 1-octanol or 1-phenyloctane and a drop was placed on a freshly cleaved HOPG surface. A ML of *trans*-bis-cyanoethylthio-bis-octadecylthio-TTF in 1-octanol spontaneously physisorbs at the liquid/HOPG interface as revealed by scanning tunnelling microscopy (Fig. 4.13(a, c)).

Two polymorphs are observed, termed α and β, where for both structures the molecules are organized in a lamellar structure. The TTF cores in the middle of the lamellae (dark grey arrows in Fig. 4.13(a)) appear as bright bands. In the α polymorph TTF cores are lying flat on the surface, forming an angle of 31° with respect to the lamellar axis. The octadecyl chains appear with a darker contrast and form an angle of 60° with respect to the lamellar axis. The intermolecular distance within lamellae is 0.83 nm. The β polymorph (Fig. 4.13(b) and (c)) has the molecules organized in alternating double-core (dark grey arrows) and

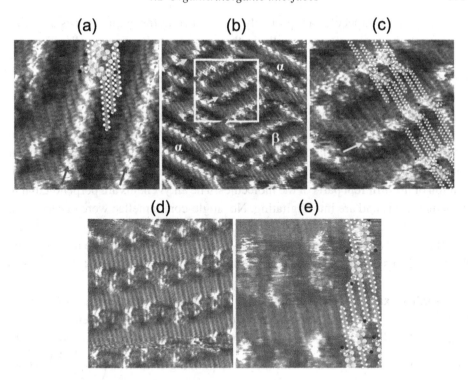

Figure 4.13. STM images at the liquid/HOPG interface. (a) A ML of a *trans*-bis-cyanoethylthio-bis-octadecylthio-TTF deposited from 1-octanol (10×10 nm², $I_t = 0.5$ nA, $V_t = -0.226$ V). (b) A ML of a *trans*-bis-cyanoethylthio-bis-octadecylthio-TTF deposited from 1-octanol (19.5×19.5 nm², $I_t = 0.5$ nA, $V_t = -0.294$ V) showing the two different polymorphs. (c) Zoom-in on the area indicated in (b). Dark grey and white arrows indicate double-core and single-core lamellae, respectively. Nine molecules are superimposed on the image for clarity (9×9 nm², $I_t = 0.5$ nA, $V_t = -0.294$ V). (d) A ML of *cis*-bis-cyanoethylthio-bis-octadecylthio-TTF (13.8×13.8 nm², $I_t = 0.9$ nA, $V_t = -0.31$ V). No single-core lamellae are observed. Alkyl chains are interdigitating. (e) ML of *cis*-bis-cyanoethylthio-bis-octadecylthio-TTF with superimposed molecules for clarity (6.9×6.9 nm², $I_t = 0.8$ nA, $V_t = -0.288$ V). Gomar-Nadal *et al.*, 2003. Reproduced by permission of the Royal Society of Chemistry.

single-core (white arrows) lamellae. In addition, the octadecyl chains are perpendicular with respect to the lamellar axes and are interdigitating. The TTF cores form an angle of 27° with respect to the lamellar axis. The intermolecular distance between equivalent molecules in a lamella is 1.52 nm. The alternation between double and single rows can be explained as a result of the constraints imposed by the alkyl chains in order to form a dense 2D packing with optimized van der Waals interactions between the alkyl chains. In a double-core lamella (as is the case with the first polymorph) the cyanoethyl groups cannot fully adsorb on the surface. Solvent molecules are probably coadsorbed in the empty spaces available in between the molecules especially in the case of single-core lamella. Thus the solvent is expected to have a strong impact on the formation of the ML. When 1-phenyloctane is used as a solvent, no MLs were observed by scanning tunnelling microscopy at the liquid/solid interface.

Upon applying a drop of *cis*-bis-cyanoethylthio-bis-octadecylthio-TTF in either 1-octanol or 1-phenyloctane a ML is spontaneously formed at the liquid/solid interface (Fig. 4.13(d)). A lamellar structure is again evident. Images with similar contrast to the *trans* isomer are again submolecularly resolved, allowing the identification of the different parts of the molecules. The alkyl chains are perpendicular to the lamella axis and are interdigitating. No single-core lamellae were observed for MLs of *cis*-bis-cyanoethylthio-bis-octadecylthio-TTF. Figure 4.13(e) shows an image of a ML of *cis*-bis-cyanoethylthio-bis-octadecylthio-TTF, where the cores and the alkyl chains can be clearly distinguished. The intermolecular distance between equivalent molecules in a lamella is 1.78 nm.

A further example of the influence of the solvent on the organization of the SAM is given by the self-organization of steroid-bridged thiophenes (Vollmer *et al.*, 1999). MLs of a particular complex molecule[1] in phenyloctane on HOPG substrates commensurably order with the corresponding R-matrix $\begin{pmatrix} 1 & 4 \\ -13 & 8 \end{pmatrix}$. However, in dodecane the molecular arrangement is similar but with slightly varied crystallographic parameters. In this case the transformation matrix is $\begin{pmatrix} 1 & 4 \\ -11 & 11 \end{pmatrix}$.

Phase diagrams

The ϑ vs. T phase diagrams of heterostructures based on molecular materials are usually very rich and interesting. However, only a few examples are known in detail. To start with, we rescue our forgotten guiding nitrogen molecule and consider the system N_2 on graphite. The phase diagram of N_2 physisorbed on graphite is presented in Fig. 4.14 and exhibits a high degree of complexity. 1 ML is defined here as the coverage of N_2 forming a complete $\sqrt{3} \times \sqrt{3}$ commensurate single layer.

We highlight here some relevant features important for later discussions. A detailed study can be found in Marx and Wiechert, 1996. Above 85 K only a homogeneous fluid phase (F) is stable for $\vartheta < 1$ ML. In the coverage range 0.2–0.8 and for temperatures below about 50 K the centre of mass of the molecules shows positional ordering in the commensurate solid phase $\begin{pmatrix} 1 & 1 \\ -1 & 2 \end{pmatrix} \equiv (\sqrt{3} \times \sqrt{3})R30°$ with disordered molecular orientations (CD) in coexistence with the fluid phase. The

---

[1] 17-[(2,2'-bithienyl-5-yl)methylidene]-3-(2-thienyl)-5α-androst-2-ene

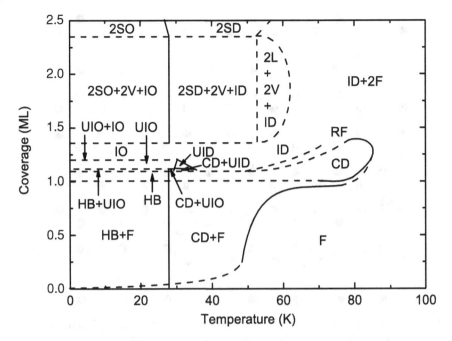

Figure 4.14. Phase diagram, coverage vs. temperature, of N_2 physisorbed on graphite. Symbols used: fluid without any positional or orientational order (F), reentrant fluid (RF), commensurate orientationally disordered solid (CD), commensurate herringbone ordered solid (HB), uniaxial incommensurate orientationally ordered (UIO) and disordered (UID) solid, triangular incommensurate orientationally ordered (IO) and disordered (ID) solid, second-layer liquid (2L), second-layer vapour (2V), second-layer fluid (2F), bilayer orientationally ordered (2SO) and disordered (2SD) solid. Solid lines are based on experimental results whereas the dashed lines are speculative. Adapted from Marx & Wiechert, 1996.

ideal $(\sqrt{3} \times \sqrt{3})R30°$ complete commensurate $\vartheta = 1$ ML structure is represented by the centres of the adsorbed molecules in Fig. 4.15. This superstructure consists of an equilateral triangular 2D array where the superlattice spacing of 0.426 nm is $\sqrt{3}$ larger than the lattice constant a of the underlying graphite (0001) basal plane. In addition, the lattice is rotated by 30° with respect to the graphite lattice. Such overlayer commensurate ordering is also found in many systems such as H_2/graphite ($\vartheta = 1/3$ ML), Xe/Ru(001) (Narloch & Menzel, 1997), H_2O/Pd(111) ($\vartheta < 1$ ML at 40 K) (Mitsui *et al.*, 2002) and methyl-terminated alkanethiol SAMs absorbed on Au(111) surfaces (Barrena *et al.*, 2004). This last case will be discussed later in some detail.

An orientational transition of the molecular axes is observed near 28 K independent of coverage up to the full ML and slightly beyond. This type of low-temperature orientational ordering is the 2×1 in-plane herringbone orientationally ordered

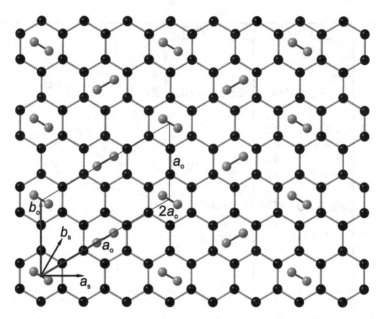

Figure 4.15. N_2/graphite. 2×1 in-plane herringbone orientational ordered super-structure within the positionally ordered commensurate $(\sqrt{3} \times \sqrt{3})R30°$ phase.

superstructure (HB) within the positionally ordered commensurate $(\sqrt{3} \times \sqrt{3})R30°$ solid phase as sketched in Fig. 4.15. An example of an orientationally disordered structure was discussed for H_2 in Chapter 1 (phase I). This commensurate ordering can also be viewed as two rectangular sublattices, with different molecular orientations with a transformation matrix $\begin{pmatrix} 3 & 0 \\ -1 & 2 \end{pmatrix}$, both lattices shifted by a vector $\boldsymbol{a}_s + \boldsymbol{b}_s$.

The commensurate $\sqrt{3}$ solids (HB and CD) still exist slightly above $\vartheta = 1$ ML coverage. Further increase of coverage (compression) leads first to a uniaxial incommensurate orientationally ordered phase (UIO) in coexistence with the commensurate phase (HB+UIO) slightly above $\vartheta = 1$ ML completion. In this regime both phases are orientationally ordered at low temperatures in terms of a herringbone structure. HB disorders at about 28 K evolving towards the CD phase, and UIO transforms to the uniaxial incommensurate orientationally disordered phase UID near 29 K, hence there is coexistence between a commensurate disordered and a uniaxial incommensurate ordered phase (CD+UIO). Further compression results in the pure uniaxial incommensurate phase UIO, which disorders orientationally around 30 K to UID. As shown in the phase diagram of Fig. 4.14 there are more phases than those discussed here for $\vartheta \geq 1.3$ ML, with a high degree of coexistence (see caption for details).

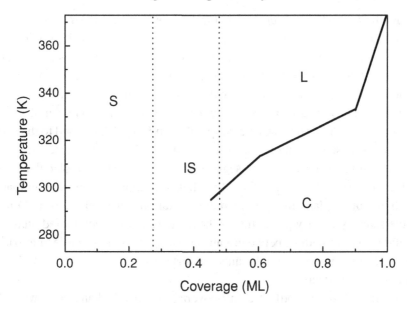

Figure 4.16. Schematic T-ϑ phase diagram of decanethiol on Au(111). The different regions and phases are denoted as S (stripes), IS (intermediate structures), C [$c(4 \times 2)$] and L (liquid). The broken lines indicate phase boundaries of the IS, which are not yet fully established. The solid curve between C and L (melting transition) exhibits a sharp rise near full coverage. Adapted from Schreiber *et al.*, 1998.

When adsorbed on the atomically highly corrugated, anisotropic Cu(110) surface N_2 orders commensurably in a complex seven-sublattice pinwheel structure with a transformation matrix $\begin{pmatrix} 4 & 1 \\ 1 & 3 \end{pmatrix}$ (Zeppenfeld *et al.*, 1997). At this point we should conclude that at the ML and sub-ML level the registry of even a simple model molecule such as N_2 on a simple substrate such as graphite exhibits commensurism of different types and incommensurism depending on the nitrogen coverage and temperature. This should warn us about the structural complexity that organic/inorganic interfaces can achieve.

Let us move now to the case of alkanethiol SAMs on gold surfaces. The stability of these SAMs originates from the high affinity of the thiol groups (-SH, also called mercapto groups) to gold. Strictly speaking, SAMs are not pure 2D systems due to the chains extending out-of-plane, and in fact the 3D character is essential for the molecular packing. Although the phase diagram of thiols on Au(111) is obviously a very fundamental issue, and several groups have studied it, important questions remain unresolved. Figure 4.16 shows the schematic phase diagram of decanethiol on Au(111) in temperature and coverage space.

Decanethiol adopts six discrete structural phases with increasing surface coverage. These are a 2D gas (α), three striped phases (β, χ and δ), a 2D liquid (ε) and the upright phase (ϕ) (Poirier *et al.*, 2001). The 2D gas comprises a low density of weakly interacting, highly mobile, surface-bound molecules. The striped phases correspond to lying-down structures, with the alkyl chains lying parallel to the substrate surface. The β-phase has a molecular packing area of 0.83 nm^2 $molec^{-1}$, and the alkyl chains in adjacent rows are antiparallel and strictly confined to the surface plane. The χ-phase has a molecular packing area of 0.65 nm^2 $molec^{-1}$ and comprises alternating β- and δ-like row-segments. The δ-phase has a molecular packing area of 0.54 nm^2 $molec^{-1}$, and the alkyl chains in adjacent rows are antiparallel and with an out-of-plane interdigitated configuration. The ε-phase is a 2D liquid. It has the same symmetry properties as the α-phase but is a condensed phase with molecular area intermediate between δ and ϕ. The ϕ-phase has a molecular packing area of 0.22 nm^2 $molec^{-1}$, and the alkyl chains are oriented upright and tilted 30° from the surface normal.

The basic structural motif of a full-coverage layer of alkanethiols with chain lengths between C10 and C22 on Au(111) is a $\sqrt{3} \times \sqrt{3}R30°$ lattice. This is expected based on simple packing arguments, which include the tilt angle of the hydrocarbon chains with respect to the surface normal, with values around 30°. However, $c(4 \times 2)$ superstructures have also been observed. Intermediate structures (IS in Fig. 4.16) are characterized by intermediate tilt angles or chains lying over each other but their structures are not yet completely understood. One has to keep in mind that some reported structures cannot be considered strictly as phases in a thermodynamic sense because they correspond to non-equilibrium phases and that one has to be aware of the potential structural changes on long time scales.

Since *ab initio* calculations are non-trivial for alkanethiols with long chains, methylthiols and other short-chain thiols on Au(111) have recently received considerable attention. While longer chains with their flexibility and non-negligible chain–chain interactions are essentially a defining feature of SAMs, methylthiols at least supposedly exhibit the same headgroup–substrate interaction. However, due to the different balance of the interactions (chain–chain versus headgroup–substrate), it is not possible to extrapolate the results from short-chain thiols to longer-chain thiols.

The determination of phase diagrams of MOMs on substrates is rare in the literature, and never to the level of detail of N_2 discussed above. Just recall here that for single crystals of MOMs, detailed phase diagrams only exist for BFS and for other TTF-derivatives (see Section 1.5). Among the few existing examples, let us discuss the *in situ* real-time evolution of the formation of OMBD-grown films of DMe-DCNQI/Ag(110) in the ML regime characterized by LEED (Seidel *et al.*, 1999). One of the limiting experimental factors of such experiments is the low sublimation temperature of DMe-DCNQI, about 320 K, and the fact that DMe-DCNQI polymerizes at 340 K, which forces the UHV chamber to be pumped for a prolonged period of time without being baked. During film growth three structural transitions are observed. At the beginning, the DMe-DCNQI film causes a diffuse diffraction pattern. At a higher dose, sharp diffraction spots occur corresponding to a single ML (structure I). 1 ML is defined here as 1.1×10^{14} cm^{-2}. In matrix notation this commensurate structure is described by $\begin{pmatrix} 3 & 1 \\ -2 & 2 \end{pmatrix}$, with $a_{oI} = 0.958$ nm, $b_{oI} = 1.000$ nm and $\alpha_{oI} = 100°$. Before the diffuse LEED images change to that of structure I, the LEED images show that the development of structure I starts from an incommensurate phase with unit vectors $a_o = 1.02$ nm, $b_o = 1.2$ nm and $\alpha_o = 87°$ and transformation matrix $\begin{pmatrix} 3 & 1.3 \\ -2 & 2.6 \end{pmatrix}$. The compression of this structure results in phase I. After the first ordered structure is complete, a second type of structure forms (structure II), which is totally different from the first one. The transition occurs at 1.18 ML. The transformation matrix is described by $\begin{pmatrix} 3 & 0 \\ 1 & 2 \end{pmatrix}$, with $a_{oII} = 0.867$ nm, $b_{oII} = 0.866$ nm and $\alpha_{oII} = 70.5°$. Structures I and II exist in two domain orientations. After structure II is completed, it undergoes another structural change upon increase of coverage. Structure II becomes compressed in the substrate $[1\bar{1}0]$ direction. The final structure ends with a smaller unit cell with unit cell vectors $a_o = 0.78$ nm, $b_o = 0.86$ nm and $\alpha_o = 72.5°$ and transformation matrix $\begin{pmatrix} 2.7 & 1 \\ 0.9 & 2 \end{pmatrix}$.

Interface-induced homochiral assemblies

We saw before for the NN/Au(111) system how the surface can induce chirality on non-chiral molecules because of the confinement to two dimensions. Chirality is an extremely interesting phenomenon with significant implications in biology and pharmacology and chiral recognition still remains mysterious (not for Flatland inhabitants!). The interactions involved in chiral recognition are indeed the same as those involved in the formation of crystalline phases of MOMs: hydrogen bonding, π–π and dipole-induced interactions. In addition, crystalline surfaces may induce segregation of enantiomers by adding molecule–substrate interactions, thus inducing homochiral assemblies, either in the form of dimers or as SAMs. We are going to discuss next two relevant examples of vapour-deposited organic layers on metallic substrates involving chirality: cysteine/Au(110) (Kühnle *et al.*, 2002) and heptahelicene/Cu(111) (Fasel *et al.*, 2003).

Cysteine, HS-CH$_2$-CH(NH$_2$)-COOH, binds covalently to gold due to the thiol group. When depositing enantiomeric pure (L-cysteine or D-cysteine) on Au(110) at

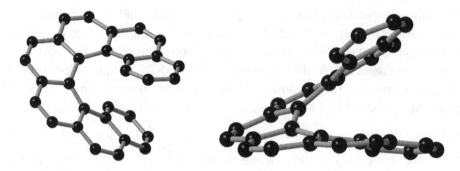

Figure 4.17. Two perspective views of heptahelicene corresponding to the mono-clinic $P2_1/c$ polymorph (Hark & Beurskens, 1976). Heptahelicene belongs to the low symmetry C_1 point group (only E applies).

low coverages ($\ll$ 1 ML) cysteine molecules form dimers exhibiting the same enan-tiomeric form: either LL or DD. When depositing a racemic mixture (L-cysteine and D-cysteine) dimers are only of the LL or DD forms, but never heterochiral (LD). Dimerization of cysteine molecules on Au(110) surfaces is thus highly stere-oselective.

The second example concerns heptahelicene. Figure 4.17 shows the structure of left-handed M-heptahelicene, where M stands for minus. On Cu(111) surfaces the M-heptahelicene molecules are found to adsorb in a geometry with their terminal phenanthrene group (the first three carbon rings) oriented parallel to the (111) faces and to successively spiral away from the surface from the fourth ring on, as determined by X-ray photoelectron diffraction experiments (Fasel *et al.*, 2001).

Figure 4.18 shows STM images of the two close-packed structures observed for M-heptahelicene. Each bright dot represents one molecule. At $\vartheta = 0.95$ ML a long-range ordered structure, apparently built-up from clusters containing six molecules and from clusters containing three molecules (6&3-structure), is observed. The six-membered clusters appear as an anticlockwise pinwheel, thus showing hand-edness. At $\vartheta = 1$ ML the unit cell of the adsorbate lattice contains a group of three molecules (3-structure) exhibiting a particular cloverleaf shape. Two rotational do-mains, rotated by 180° with respect to each other, are observed in both structures.

The observed adsorbate lattice structures show enantiomorphism, that is, adsorp-tion of the right-handed P-heptahelicene (P stands for positive) leads to structures which are mirror images of those observed for M-heptahelicene. This effect can be clearly observed in the high-resolution STM images of Fig. 4.19. Furthermore, the enantiomeric lattices form opposite angles with respect to the [1$\bar{1}$0] substrate surface direction. The combined molecule–substrate systems thus exhibit extended

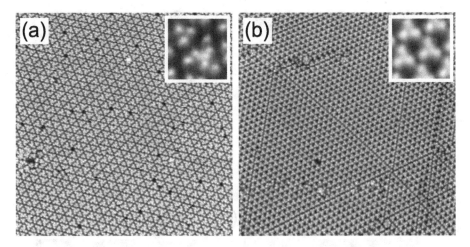

Figure 4.18. Constant-current STM images (70 nm × 70 nm) acquired at 50 K of long-range ordered MLs of M-heptahelicene on Cu(111). (a) 6&3-structure at $\vartheta = 0.95$ ML and (b) 3-structure of $\vartheta = 1$ ML. 1 ML is defined as the saturation coverage. The insets show the molecular cluster units of the structures. Reprinted with permission from Fasel *et al.*, 2003.

surface chirality. Remarkably, the unit cells are not only mirror images of each other, but also of the arrangements of molecules within the unit cells. This is most clearly seen for the pinwheel-clusters of the 6&3-structures shown in Figs. 4.19(a) and (b). The pinwheel's wings point either anticlockwise, as in the M-heptahelicene 6&3-structure (Fig. 4.19(a)), or clockwise, as in the P-heptahelicene 6&3-structure (Fig. 4.19(b)). In the case of the 3-structures illustrated in Figs. 4.19(c) and (d), the mirror symmetry is expressed by tilts of the three-molecule cloverleaf units into opposite directions with respect to the adsorbate lattice vectors.

If instead of enantiomeric heptahelicene a racemic mixture is sublimated onto the surface, the molecules self-organize in enantiomorphous mirror domains. Thus, as in the case of cysteine on Au(110), the surface is stereoselective.

Organic/inorganic superlattices

Superlattices result from the periodic infinite repetition of heterostructures. MBE-grown superlattices of III-V semiconductors exhibit sharp interfaces and high carrier mobilities of the resulting 2D carrier gas at low temperature (Ando *et al.*, 1982). To date no superconductivity has been found for such engineered solids, although some expectations were raised in 2000 after some reports on the obtention of superconductivity in semiconductor/insulator interfaces by field-induced

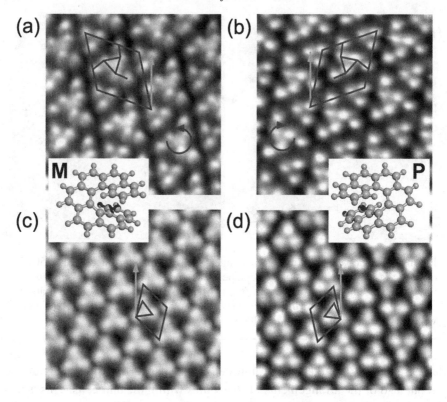

Figure 4.19. High-resolution STM images of M- and P-heptahelicene structures (10 nm × 10 nm). (a) M-heptahelicene at $\vartheta = 0.95$ ML, (b) P-heptahelicene at $\vartheta = 0.95$ ML, (c) M-heptahelicene at $\vartheta = 1$ ML, and (d) P-heptahelicene at $\vartheta = 1$ ML. The M- and P-heptahelicene structures are mirror images of each other. Unit cells and their basic building blocks are outlined by dark grey lines, the substrate [$1\bar{1}0$] direction is indicated by light grey arrows. Reprinted with permission from Fasel *et al.*, 2003.

charge injection. Such reports were found to be false and will be discussed in Section 6.3.

In Section 1.5 we saw that MgB_2, a superconductor with $T_c \simeq 39$ K, consists of alternating (intercalated) sheets of honeycomb boron layers and hexagonal magnesium layers and that graphite also becomes superconducting when doped with alkali metals forming GICs, with $T_c < 2$ K. Binary Li-GICs and K-GICs exhibit commensurate 2×2 and $(\sqrt{3} \times \sqrt{3})R30°$ as well as incommensurate superstructures (Lang *et al.*, 1992) while ternary KHg-GICs show commensurate 2×2 in addition to incommensurate superstructures (Kelty *et al.*, 1991). Figure 4.20 shows the 2×2 superstructure of idealized binary GICs.

We can also classify BFS, which can be regarded as chemically constructed multilayers (Day, 1993), in terms of the HFW framework. The projection of the

(a)

(b)

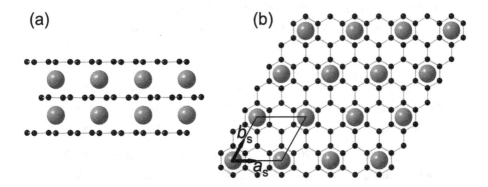

Figure 4.20. (a) Cross-sectional view of an idealized binary alkali-metal GIC and (b) top-view illustrating the 2×2 superstructure. C and alkali-metal atoms are represented by black and medium grey balls, respectively.

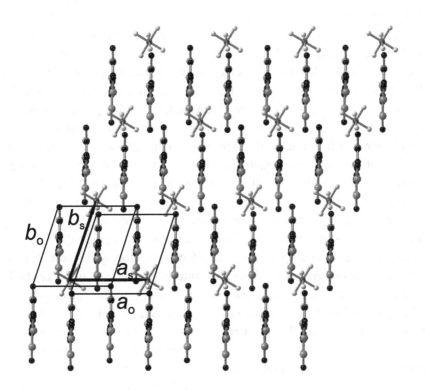

Figure 4.21. Projection of the *ab*-plane of $(TMTSF)_2PF_6$. C, Se, P and F are represented by black, medium grey, dark grey and light grey balls, respectively.

Table 4.2. *Examples of BFS salts containing non-centrosymmetric anions exhibiting anion ordering transitions*

Formula	T_{ao}	q_r	References
$(TMTSF)_2ClO_4$	24	$(0, \frac{1}{2}, 0)$	Pouget *et al.*, 1983
$(TMTSF)_2ReO_4$	177	$(\frac{1}{2}, \frac{1}{2}, \frac{1}{2})$	Moret *et al.*, 1982
$(TMTSF)_2NO_3$	41	$(\frac{1}{2}, 0, 0)$	Moret *et al.*, 1983
$(TMTTF)_2ClO_4$	70	$(\frac{1}{2}, \frac{1}{2}, \frac{1}{2})$	Pouget *et al.*, 1982
$(TMTTF)_2ReO_4$	160	$(\frac{1}{2}, \frac{1}{2}, \frac{1}{2})$	Parkin *et al.*, 1983b
$(TMTTF)_2NO_3$	50	$(\frac{1}{2}, 0, 0)$	Moret *et al.*, 1983

ab-plane of $(TMTSF)_2PF_6$ is depicted in Fig. 4.21. Along c alternate organic (TMTSF) and inorganic (PF_6) planes are stacked. The periodicity is OIOIOI..., where O and I stand for organic and inorganic, respectively. If the lattice vectors are defined by the PF_6 sublattice, there are two independent TMTSF sublattices, which order according to the identity 1×1 superstructure (or identity transformation matrix E). The examples given in Figs. 1.9 and 1.13, corresponding to herringbone structures, can also be described by two independent 1×1 superstructures.

Note that PF_6 is centrosymmetric. If the anions were non-centrosymmetric, an additional degree of freedom would be available in the case that such anions order. In fact BFS with non-centrosymmetric anions exhibit anion ordering below a given temperature, which can be classified as an order–disorder transition. Above the transition temperature T_{ao} anions exhibit random orientations, but for sufficiently low temperatures (low activation energy) they become ordered.

Table 4.2 shows some examples of BFS exhibiting such order–disorder transitions with the measured T_{ao} and the reduced wave vectors q_r (Moret & Pouget, 1986). The anion ordering transitions have important consequences for the low-temperature properties and ground states of these materials. This can be understood by realizing that when the a^* component of q_r is equal to 1/2, which means that the period along a is doubled, the anion potential can open an energy gap at E_F thus inducing a metal–insulator transition, as discussed in Section 1.7. This is the case for many of the transitions given in Table 4.2. As shown in Fig. 1.30 the quasi 1D character of the BFS is expressed by the preferential band dispersion along the Γ–X direction, which corresponds to the real-space a-direction. Doubling a induces a gap opening at $\pi/2a$. On the other hand, when the a^* component of q_r is zero there is no change of periodicity along a and the metallic character can be kept below the transition.

Examples of these two behaviours are provided by $(TMTSF)_2ReO_4$ and $(TMTSF)_2ClO_4$. In $(TMTSF)_2ReO_4$ the anions order at c. 180 K with a reduced wave vector $(\frac{1}{2}, \frac{1}{2}, \frac{1}{2})$ and undergo a sharp metal–insulator transition with $E_a = 0.16$ eV (Jacobsen *et al.*, 1982). The anion ordering transition of $(TMTSF)_2ClO_4$ occurs at 24 K with $q_r = (0, \frac{1}{2}, 0)$. This q_r value implies that the ClO_4 tetrahedra have the same orientation in a given ac-plane while the orientation alternates in the b-direction. Since the period along a is preserved, no gap opening is expected at T_{ao}, as experimentally verified (Gubser *et al.*, 1982). This is the reason why $(TMTSF)_2ClO_4$ becomes a superconductor in spite of the anion ordering transition.

$(TMTTF)_2BF_4$ shows $T_{ao} \simeq 40$ K at ambient pressure with $q_r = (\frac{1}{2}, \frac{1}{2}, \frac{1}{2})$ with the corresponding energy gap opening at E_F. However, we saw in Table 1.8 that it becomes superconducting below 1.4 K, which is ascribed to the application of an external pressure that modifies the intermolecular distances and thus the interaction strength and geometry. This represents an additional example of the importance of specifying the physical parameters when discussing the physical properties of materials.

A phase transition of a different origin is found for Fabre salts constituted of centrosymmetric anions (Chow *et al.*, 2000). No structural modifications have been observed for these salts along the transition, hence deserving the term structureless, and primarily the charge degrees of freedom are involved. The phase transition is ascribed to charge ordering, where the electronic equivalence of the TMTTF molecules is removed below a critical temperature making the charge disproportionate. This phase transition is ubiquitous for the TMTTF family and should be included in the generic phase diagram of Fig. 1.19.

Experiments based on ^{13}C NMR spectroscopy performed on ^{13}C spin-labelled $(TMTTF)_2PF_6$ and $(TMTTF)_2AsF_6$ samples reveal changes in the hyperfine coupling signal below a given temperature T_{co} (Chow *et al.*, 2000). At RT, each molecule is equivalent, but the two ^{13}C nuclei in each molecule have inequivalent hyperfine coupling, giving rise to two spectral lines separated by $\simeq 10$ kHz. Upon cooling, the NMR spectrum of $(TMTTF)_2AsF_6$ remains unchanged down to $T_{co} \simeq 105$ K, below which both the peaks split. From each molecule there is a signal from the nucleus with a stronger hyperfine coupling and a signal from the nucleus with a weaker hyperfine coupling. The doubling arises from two different molecular environments of roughly equal number, one with slightly greater electron density and one with a reduced electron density. For $(TMTTF)_2PF_6$, $T_{co} \simeq 65$ K.

Charge ordering below 220 K has also been found, in the organic conductor $(DI\text{-}DCNQI)_2Ag$ (Hiraki & Kanoda, 1998).

LB films also form well-defined engineered organic/inorganic superlattices. There are many examples in the literature and we just recall here some of them:

metal phosphonate LB films discussed in Chapter 3, built from layers of a variety of metal ions sandwiched within bilayers of the organophosphonate (Seip *et al.*, 1997), and LB films of polyoxometallates (Clemente-León *et al.*, 1997) and of single-molecule nanomagnets (Clemente-León *et al.*, 1998). Organic/inorganic superlattices of polyoxometallates using the LB technique have been prepared using the adsorption of Keggin heteropolyanions on positively charged MLs of dimethyl-dioctadecylammonium. Chemically stable Keggin polyoxometallates with general formula $[X^{+n}W_{12}O_{40}]^{-(8-n)}$, with $X^{+n}=2H^+$, P^{+5}, Si^{+4}, B^{+3} and Co^{+2}, have been used. These anions have the same structure but, depending on the nature of X, their charge varies between -3 and -6.

In the latter example of LB-based organic/inorganic superlattices acetate and benzoate derivatives of Mn_{12} were used. Upon spreading a pure sample at the gas/water interface, the clusters do not form stable Langmuir films. Therefore, a mixture of these complexes and a lipid is needed to obtain a ML. A homogeneous ML is formed upon use of behenic acid as matrix. The Langmuir film is stable over time if the lipid–cluster ratio is 5 or higher. Furthermore, the transfer of the ML onto a solid hydrophobic substrate is easily achieved. In these examples the periodicity is OIOOIOOIO . . . .

Semiconductor/metal interfaces

Electronic structure

Kroemer's Lemma of Proven Ignorance states that *if, in discussing a semiconductor problem, you cannot draw an energy band diagram, this shows that you don't know what you are talking about.* The associated corollary is also quite clear: *if you can draw one, but you don't, then your audience won't know what are you talking about* (Kroemer, 2001). With this advice in mind let us discuss the energy band diagram of semiconductor/metal interfaces starting from their energy diagrams, shown in Fig. 4.22.

Figure 4.22(a) represents the case where the interface dipole Δ, the energy difference between the vacuum levels of both organic and metal samples, is zero, thus indicating that they are aligned. Such a condition is known as the Schottky–Mott limit and is expected to be fulfilled for weakly interacting interfaces. However, this is not always the case as will be discussed later. The terms ionization energy I_E, electron affinity E_A and metal work function ϕ_M were defined in Section 1.5. The HOMO and LUMO energies referred to E_F are represented by E_{HOMO} and E_{LUMO}, respectively.

The case of non-vanishing Δ is shown in Fig. 4.22(b). From this figure we observe that $\Delta = \phi_M + E_{HOMO} - I_E$. Here we have assumed a common energy level, E_F, for both metal and semiconductor. This is not quite exact since the metal induces electronic states in the semiconductor gap that in turn may induce CT and

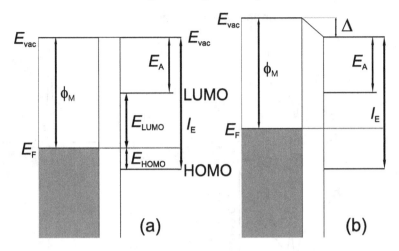

Figure 4.22. Schematic of an organic semiconductor/metal interface energy diagram: (a) $\Delta = 0$ and (b) $\Delta \neq 0$.

thus electronic dipoles. Taking into account these gap states, semiconductor electronic levels are usually referred to reference states derived from charge neutrality conditions such as the charge neutrality level (CNL) (Tejedor *et al.*, 1977) or effective midgap levels (Tersoff, 1984). In fact E_F corresponds to the charge neutrality in the metal. In a semiconductor/metal heterojunction the charge neutrality level is aligned to E_F, while for semiconductor/semiconductor systems both CNLs should have the same energy. The fundamentals of the formation of energy barriers are only partially understood and will be a matter of intense research in the coming years, mainly boosted by the industrial application of the associated devices.

ϕ_M, E_{HOMO} and I_E can be experimentally determined with UPS for sufficiently thin organic films ($\sim$0.5–1.0 nm), in order to prevent electrostatic effects inducing charging or band bending, associated with the semiconducting or insulating character of the films caused by the removal of photoelectrons. Figure 4.23 shows a scheme of the photoemission process.

Depending on the energy $\hbar\omega$ of the incident photons, valence band states and even core level electrons can be excited. UPS is a surface-sensitive technique since electrons have a very short inelastic mean free path, λ_i, which depends on the kinetic energy E_K, and has a minimum value of $\sim$0.5 nm for $E_K \sim 100$ eV. The leading edge of the valence band is taken as the VBM or HOMO maximum and has to be referred to E_F, which has to be determined from a clean inorganic metal surface. Those electrons with $E_K > 0$ are removed from the sample and transmitted to the detector. The fundamental equation of the photoemission process is (Einstein, 1905):

$$\hbar\omega = E_B + \phi_M + E_K. \tag{4.3}$$

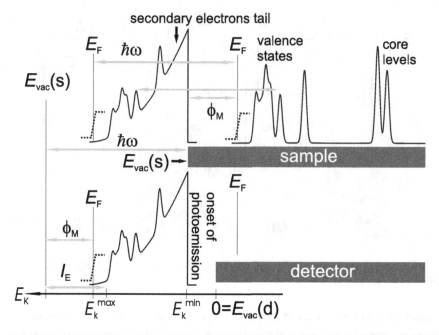

Figure 4.23. Photoemission process and UPS spectrum of a semiconducting material. $E_{vac}(s)$ and $E_{vac}(d)$ represent the sample and detector vacuum levels, respectively. The Fermi edge of the metal is represented by the discontinuous line.

From Fig. 4.23 we observe that $\phi_M = \hbar\omega - (E_K^{max} - E_K^{min})$, where E_K^{min} is obtained from the onset of photoemission also known as the secondary electron cutoff and E_K^{max} from E_F. The secondary electron tail originates from the inelastic scattering of photoexcited electrons on their way to the surface. Analogously, $I_E = \hbar\omega - (E_K^{max} - E_K^{min})$, where E_K^{max} is obtained from the leading edge of the VBM or HOMO maximum. $E_K^{max} - E_K^{min}$ represent the total width of the measured UPS spectra.

As an example let us consider the pentacene/samarium interface (Koch *et al.*, 2002). Samarium has a low work function ($\phi_M \simeq 2.7$ eV), which is comparable to E_A from pentacene ($\simeq 2.7$ eV). Thus, if $\Delta \simeq 0$, the condition $\phi_M \simeq E_A$ should provide efficient electron injection because in this case E_F and LUMO are nearly aligned. In order to avoid contamination that may alter the instrinsic ϕ_M, I_E and E_{HOMO} values, such heterostructures have to be prepared in ideally clean conditions, imposing the use of UHV. The UPS experiments performed with synchrotron radiation are shown in Fig. 4.24. After measuring ϕ_M of the clean samarium surface (2.7 eV) as described above, increasing amounts of pentacene are controllably deposited onto the samarium surface. The survey spectra of the valence states and a close-up view of the energy region near E_F are shown in Figs. 4.24(a) and (b),

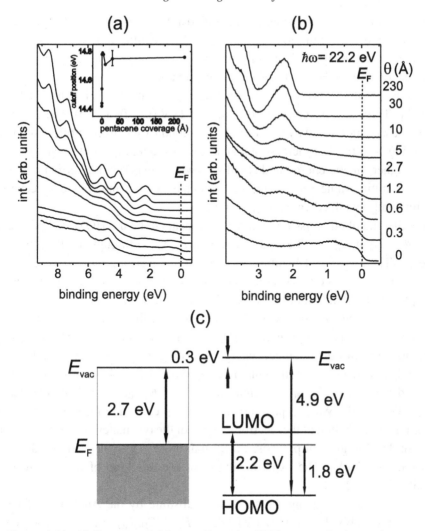

Figure 4.24. (a) Survey and (b) near E_F region UPS spectra for increasing pentacene coverage on samarium. θ stands for ϑ. Inset: kinetic energy position of the secondary electron cutoff. Reprinted with permission from Koch *et al.*, 2002. Copyright (2002) American Chemical Society. (c) Schematic energy level diagram of pentacene/samarium.

respectively. All spectra were obtained with $\hbar\omega \simeq 22.2$ eV. The nominal thickness of the pentacene layer is given on the right side of the figure. The survey spectra show that small amounts of pentacene effectively attenuate the emission arising from the substrate and that distinct emission features of the organic material become evident above 0.5 nm. The close-up spectra of Fig. 4.24(b) show that the first evaporation step (0.03 nm) produces a new photoemission feature at $E_B \simeq 2.3$ eV, where E_B is referred to E_F. As more pentacene is deposited, this feature grows in intensity at constant E_B, and finally is identified as the emission from the pentacene. Conversely, the photoemission features from the metallic samarium substrate are eliminated for coverages of 3.0 nm and above, leaving the energy gap region of pentacene (above the HOMO) free of valence features.

The onset of the HOMO, E_{HOMO} (leading edge of the peak), is found at 1.8 eV below E_F. This value is very close to $E_{opt} \simeq 2.2$ eV (see Table 1.6). The change in the sample ϕ_M, monitored by the shift of the secondary electron cutoff, is shown as inset in Fig. 4.24(a). For increasing pentacene coverage, the onset moves abruptly toward higher kinetic energy and stabilizes (within experimental error) $c.$ 0.3 eV above the initial value for a pentacene thickness of 0.5 nm. For pentacene the experimentally obtained value of I_E is 4.9 eV.

Note that the photoemission feature that appears to be the HOMO of pentacene in Fig. 4.24(b) at $E_B \simeq 2.3$ eV has a constant E_B value throughout the deposition sequence. The absence of bonding-related shifts is a strong indication of a weak interaction between metal substrate and organic overlayer and this should favour the application of the Schottky–Mott rule. The injection barrier Δ for electrons from samarium into pentacene is thus 0.3 eV, revealing that the interaction of pentacene with samarium is mainly physisorptive. Finally, following Kroemer's suggestion, we draw the schematic energy level diagram of the investigated organic semiconductor/metal structure, which is depicted in Fig. 4.24(c). Given are the measured metal ϕ_M, E_{HOMO} of pentacene and Δ. E_{LUMO} is evaluated as $E_{opt} - E_{HOMO}$. The van der Waals type interaction at the present interfaces does not allow for a strong perturbation of the electronic states of the two materials. Thus, basically a vacuum level alignment prevails, except that there is a small interface dipole Δ measured ($\sim$0.3 eV). The experimental error for this kind of measurements is typically about 0.2 eV.

However, when replacing the samarium substrate by the Au(111) surface, Δ becomes noticeably larger (1.0 eV) while $E_F - E_{HOMO} \simeq 0.5$ eV (Schroeder et al., 2002). In fact, for many organic/metal interfaces, $\Delta \sim$ 0.5–1.0 eV, indicating adsorbate-induced modification of ϕ_M, relatively strong bonding, charge exchange and perhaps defective interfaces. Further examples are PTCDA, Alq$_3$, α-NPD, CBP on magnesium, indium, tin and gold (Hill et al., 1998). Actually, weak molecule–substrate interactions do not automatically imply vanishing values of Δ.

Theoretical calculations based on the induced density of interface states (IDIS) model show that, although the molecule–metal interaction is weak, the metal-IDIS in the organic HOMO–LUMO gap can be sufficiently large so that a CNL of the organic molecule can be defined (Vázquez et al., 2004). CT between both solids can induce an electrostatic dipole at the interface and trigger the alignment of E_F and CNL. In the calculation the molecule–molecule interactions can be neglected, as they introduce only a small broadening of the molecular levels ($\sim$0.2 eV), and do not create any electron DOS in the molecular energy gap. The organic layer is thus treated as a non-interacting gas and only the molecule–substrate interactions are taken into account. The CNL is calculated by integrating the induced local density of states (LDOS) and imposing charge neutrality conditions: the total number of electrons up to the CNL equals that of the isolated molecule. For PTCDA/Au the CNL is located 2.45 ± 0.1 eV above the centre of the HOMO level of the molecule. The important outcome of the analysis is that there is a significant DOS at the PTCDA/Au interface, in spite of the weak interaction between the two materials. The implication is that the interface E_F is close to the CNL. Thus, the formation of the interface barrier is related to the transfer of charge across the interface. The charge transfer is associated with the IDIS, and creates an electrostatic interface dipole, which tends to align the metal E_F and the PTCDA CNL.

These calculations are in line with experiments performed on PTCDA/Ag(111), where it is shown that the central PTCDA carbon ring is the part of the molecule most effectively coupled to the substrate via CT from the metal surface to the LUMO, as evidenced by measurements using the vibrational spectroscopies *in situ* Raman (Wagner, 2001) and HREELS (Eremtchenko *et al.*, 2003). The vibrational frequencies depend on the bond strength and thus contain direct chemical information. EELS is most sensitive to infrared-active modes (dipole-induced) and with Raman spectroscopy dipole-inactive vibrations as well as in-plane vibrations can be analyzed. However, intrinsically Raman-active modes of the free molecule become infrared-active at the surface because of CT. Raman spectra of sub-ML coverages of PTCDA on Ag(111) show a doublet at 1297 and 1310 cm^{-1}. The 1297 cm^{-1} component is attributed to PTCDA molecules strongly bonded to the silver surface. This strong bonding induces a down-shift of the A_g-mode originally located at 1310 cm^{-1} for the free molecule, essentially involving the centre of the molecule. Therefore, this mode shift indicates that the central carbon ring constitutes the molecular reaction centre. That carbon atoms are mainly involved in the bonding to the silver substrate for a flat molecule is expected, because of the π-electronic structure. At the same time, the local nature of the reaction centre allows the molecular orientation to adjust to the intermolecular lateral interactions, inducing the stability of the PTCDA layers.

The comparison of PTCDA with its parent perylene molecule is extremely interesting. For perylene MLs on Ag(111), electron diffraction suggests an orientational liquid, in which the molecules are positionally ordered in an incommensurate close-packed superlattice but orientationally disordered and mobile. The same activated Raman peaks as for PTCDA are observed but they are, however, orders of magnitude weaker, indicating that, while a molecular reaction centre may still exist in the perylene backbone, its residual activity would be too small for the molecule to recognize a preferred site.

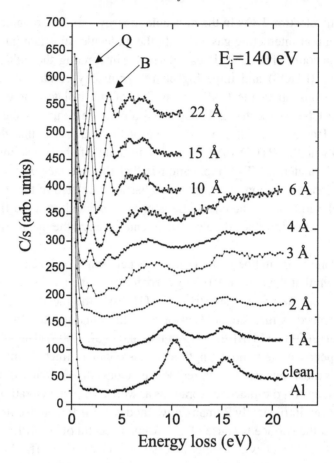

Figure 4.25. EELS spectra as a function of CuPc film thickness performed in specular geometry with a primary energy of 140 eV. Reprinted with permission from A. Ruocco, M. P. Donzello, F. Evangelista and G. Stefani, *Physical Review B*, **67**, 155408 (2003). Copyright (2003) by the American Physical Society.

The capabilities of the EELS technique can be further demonstrated for the CuPc/Al(100) system (Ruocco *et al.*, 2003). Figure 4.25 shows the EELS spectra measured in specular reflection as a function of CuPc coverage, which ranges from 0, the clean Al(100) surface, up to 2.2 nm. The clean aluminium spectrum is dominated by two structures at 10.5 eV and 15 eV that correspond to the surface $\hbar\omega_s$ and bulk $\hbar\omega_b$ plasmon, respectively. The intensity of $\hbar\omega_b$ drops with increasing coverage and almost disappears for a coverage of 0.6 nm. However, the $\hbar\omega_s$ feature shows a differentiated evolution. For the lowest coverages the centroid of this structure shifts towards lower loss energies. This shift increases as the thickness grows up to 0.3 nm. In this situation it is no longer possible to resolve the surface plasmon from the features appearing in the 5–8 eV region with similar intensity.

The $\hbar\omega_s$ peak is ascribed to an interface plasmon. Assuming that the aluminium conduction electrons are well described by a free-electron gas, the interface plasmon frequency is related to the relative dielectric constant ϵ of the molecular film through the relation:

$$\omega_s = \frac{\omega_b}{\sqrt{1 + \epsilon}}. \tag{4.4}$$

Equation (4.4) stresses that the surface plasma wave propagates within the aluminium substrate whose frequency is modified by the dielectric response of the molecular adlayer. From the measurements reported, with $\hbar\omega_s = 8.5$ eV and $\hbar\omega_b = 15$ eV, one obtains $\epsilon = 2.1$ for CuPc, a value in agreement with those measured for other planar organic molecules, with an extended delocalization of π-electrons (Alonso *et al.*, 2003).

Electronic transitions due to the CuPc molecule become evident in the EELS spectrum starting from a coverage of 0.3 nm. For this coverage two weak peaks centred at 1.9 and 3.7 eV (Q and B, respectively, in Fig. 4.25) appear together with the broad structure between 5 and 8 eV. Increasing the coverage, the Q and B transitions are always present: their intensities increase as a function of the coverage while their shape, energy position and relative intensity remain essentially unperturbed. On the contrary the broad structure at 5–8 eV shows a modest evolution reaching its final shape at 0.6 nm. Above this coverage two features located at 5.8 and 7.1 eV appear that are weaker than the B and Q transitions. It is interesting to note that starting from 1.0 nm the EELS spectrum does not show significant modifications, thus suggesting that the molecular film has reached a bulk-like configuration.

The transitions at lower energies (< 5 eV) that appear in Fig. 4.25 are assigned mostly to $\pi \rightarrow \pi^*$ electronic transitions of Pc molecules (Schaffer *et al.*, 1973). In particular the peak at 1.9 eV, corresponding to the Q band, is associated to $a_{1u}(\pi) \rightarrow e_g(\pi^*)$ and $b_{2g}(Np_\sigma) \rightarrow b_{1g}(d_{x^2-y^2})$ transitions and related to the formation of singlet excitons (S_0 and S_1 transitions) in the Pc ring. Although the two transitions are almost degenerate in energy, the former has a dipole moment perpendicular to the molecular plane while the latter, mostly from the copper d orbitals, has a dipole moment in the plane of the molecule. The Q band energy coincides with E_{opt}, as indicated in Table 1.6. The peak at 3.7 eV, associated to the B band, is related to the $a_{2u}(\pi) \rightarrow e_g(\pi^*)$ transition, thus having a well-defined symmetry with respect to the plane of the molecule. Peaks at 5.8 eV and 7.1 eV correspond to the C and X_1 bands, respectively, both assigned to $\pi \rightarrow \pi^*$ transitions.

From Fig. 4.25, the Q and B transitions appear for coverages of 0.3 nm and higher. This threshold value corresponds, in the hypothesis of a flat-lying adsorption geometry, to the saturation of the surface with one ML. In this framework molecules

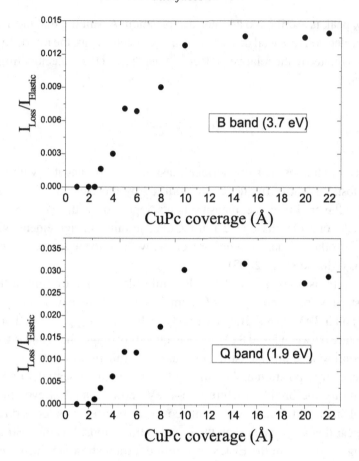

Figure 4.26. B and Q band intensities, normalized to the elastic peak intensity as a function of CuPc coverage. Reprinted with permission from A. Ruocco, M. P. Donzello, F. Evangelista and G. Stefani, *Physical Review B*, **67**, 155408 (2003). Copyright (2003) by the American Physical Society.

from the first layer do not contribute to Q and B bands. In Fig. 4.26 the B and Q band intensities are plotted as a function of CuPc coverage.

The appearance of the optical absorption bands (Q and B) has a clear threshold at a low non-zero coverage, implying that the electronic structure of the first adsorbed molecules is different from that of the bulk ones. Thus, a clear distinction between molecules directly bonded to the aluminium substrate ($\vartheta < 0.3$ nm) and molecules not directly bonded to the substrate ($\vartheta > 0.3$ nm) can be made. In the latter case the electronic structure, as revealed by EELS, is identical to that of bulk CuPc, while in the former case modification of the electronic structure prevents transitions toward the LUMO orbital. Above 1.0 nm the Q and B band intensities saturate. The optical transitions are inhibited for molecules directly bonded to the aluminium substrate due to CT. Electrons from the substrate fill up the molecular π^* LUMO and the $3d_{x^2-y^2}$ orbitals.

STM-based charge injection spectroscopy

In contrast to STM current–voltage ($I_t - V_t$) spectroscopy, which was discussed in Section 4.1, where the feedback loop has to be open during operation and where the energy dependence of the LDOS is obtained, STM $z - V_t$ spectroscopy probes the DOS via the voltage-dependent tip displacement z at constant I_t. Figure 4.27(a) shows an idealized $z - V_t$ spectrum of a thin film, which typically consists of three regions, indicated by A, B and C.

In the absence of tip–film mechanical contact (region A), the tunnelling resistance, $R_t = V_t/I_t$, is much larger than the electrical resistance of the organic layers. Charge-carrier injection takes place through the vacuum barrier into the organic material, as depicted in Fig. 4.27(b). This means that V_t drops completely at the tunnelling barrier, and the charge-injection energy can be tuned by adjusting the polarity and the magnitude of V_t. Reducing V_t under constant-current conditions forces the tip to approach proximity and finally make direct physical contact with the organic thin film. When this contact mode has been attained (region B), the tip E_F typically will be at an energy level within the forbidden energy gap of the semiconducting organic material. The high electrical field across the organic layer leads to band bending and the formation of a Schottky-like diode (Fig. 4.27(c)).

The gradient of the electrical field strongly depends on the curvature of the injecting electrodes. Thus, the highest electrical field drop appears at the STM tip apex, which, under ideal conditions, will be the predominant site for charge-carrier injection into the organic thin film. For example, with a typical STM tip apex radius $R \sim 50$ nm, one can obtain continuous injection current densities in the range of 10 to 10^4 A cm^{-2} at a V_t of a few volts, which is orders of magnitude higher than current densities at the planar interface between the organic material and the metal electrode. The injection of charge carriers into the organic material is possible until V_t is reduced to the value at which the tip E_F enters the forbidden energy gap at the interface with the metal electrode (region C). The point at which this transition occurs determines the position of the lowest electron polaron state E_P^- for an applied $V_t < 0$ or the highest polaron state E_P^+ for $V_t > 0$. In this phase, a characteristic logarithmic z–V_t curve is observed when either tunnelling directly into the metal electrode or non-resonantly through a monomolecular organic layer in contact with the electrode.

Let us discuss the example of CuPc thin films grown on Au(111) surfaces (Alvarado *et al.*, 2001a). For the β_2 polymorphic phase $E_P^- \simeq 0.1$ eV and $E_P^+ \simeq -0.2$ eV, which results in $E_{gsp} \simeq 0.3$ eV, according to Eq. (1.2). For the other

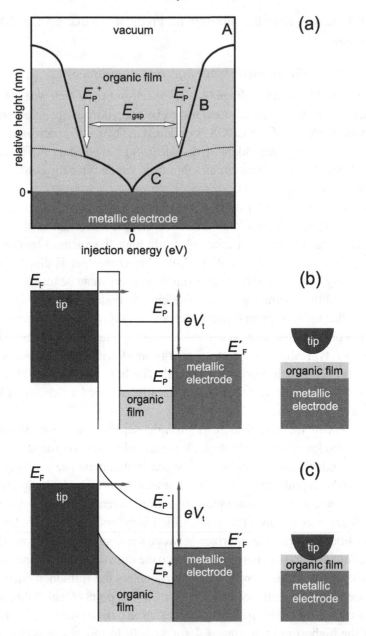

Figure 4.27. (a) Schematic of a STM $z - V_t$ injection spectrum (solid curve). The dashed curves represent typical STM tip displacements observed at a clean metal surface. (b) Energy band diagrams for STM tunnelling through a vacuum barrier into the organic thin film and (c) through a Schottky-like barrier with the tip in contact. In both cases, $V_t < 0$ relative to E_F is shown. Adapted from Müller *et al.*, 2001.

polymorphic phases found in the CuPc samples $E_{\text{gsp}} \simeq 0.2$–$0.8$ eV. For disordered samples $E_P^- \simeq -0.55$ eV and $E_P^+ \simeq 0.55$ eV, which yields $E_{\text{gsp}} \simeq 1.1$ eV. Considering that for CuPc $I_E \simeq 5.1$ eV (see Table 1.6) and that for the clean Au(111) surface $\phi_M \simeq 5.3$ eV, one would expect, applying the Schottky–Mott limit, the HOMO level of CuPc to lie just at or above the E_F. However, the HOMO level lies clearly below the substrate E_F, which indicates the presence of a dipole layer or a net charge at the interface. Thus, the Schottky–Mott approximation does not hold for the interface between the different CuPc polymorphs and Au(111). For all the polymorphic phases, E_{gsp} is remarkably smaller than $E_{\text{opt}} \simeq 1.7$ eV (see Table 1.6). Thus, the energy gap for charge-carrier injection into thin films of CuPc depends strongly on the intermolecular packing and crystallographic order of the molecules. The results suggest a tendency of decreasing gap energy and improving charge-transport properties with increasing degree of crystalline order in the thin film. Hence the importance of controlling the crystallinity of the films, an issue not always taken into account when determining the electronic structure of semiconductor/metal heterostructures and that may, at least partially, explain differences encountered in different experiments involving *identical* systems.

For thin Alq$_3$ films grown on Au(111) $E_P^- \simeq 1.2$ eV and $E_P^+ \simeq -1.8$ eV, with $E_{\text{gsp}} \simeq 3.0$ eV leading to $E_{\text{ex}} \simeq 0.2$ eV since $E_{\text{opt}} \simeq 2.8$ eV (Müller *et al.*, 2001).

Semiconductor/semiconductor interfaces

After discussing semiconductor/metal heterostructures let us now consider the case of semiconductor/semiconductor heterojunctions. The semiconductors will be characterized by the band gaps E_g^1 and E_g^2, where we arbitrarily assume that $E_g^2 > E_g^1$, and with valence–HOMO band offset ΔE_V, defined as the difference between VBM and HOMO, and the conduction–LUMO band offset ΔE_C, defined as the difference between CBM and LUMO (see Fig. 4.28). Both ΔE_V and ΔE_C are defined as positive.

There are basically two different band lineups possible: straddling lineup (Fig. 4.28(a)), which is the most common one, where $E_g^2 - E_g^1 = \Delta E_C + \Delta E_V$ and staggered lineup (Fig. 4.28(b)), where $E_g^2 - E_g^1 = \Delta E_C - \Delta E_V$ for $\Delta E_C < E_g^2$ and $\Delta E_V < E_g^1$. When $\Delta E_C > E_g^2$ and $\Delta E_V > E_g^1$ we obtain the exotic and rare case of broken band gaps. When extended to semiconductor/semiconductor heterostructures, the Schottky–Mott limit corresponds to the electron affinity rule, which states that $\Delta E_C = E_A^2 - E_A^1$, where E_A^1 and E_A^2 stand for the electron affinities of semiconductors 1 and 2, respectively. This rule breaks down for heterojunctions formed by inorganic IV, III-V and II-VI semiconductors, due to the interfacial dipoles caused by the strong covalent bonding (Niles & Margaritondo, 1986).

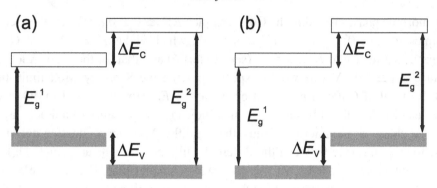

Figure 4.28. (a) Straddling and (b) staggered band lineups.

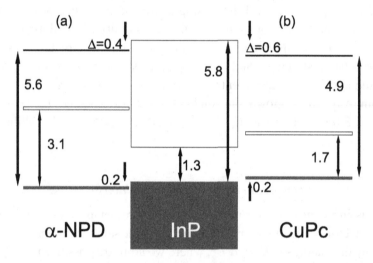

Figure 4.29. Band diagrams of (a) α-NPD/InP(110) and (b) CuPc/InP(110).

Let us illustrate all this with the α-NPD/InP(110) and CuPc/InP(110) heterostructures (Chassé *et al.*, 1999). Their band diagrams are shown in Fig. 4.29. Analogously to the methodology described for semiconductor/metal interfaces, I_E of the semiconductor surfaces, I_E^{InP} and I_E^{org}, where org stands for α-NPD and CuPc, can be determined via UPS by subtracting the total width of the UPS spectra from the energy of the incident photons. The top of the valence level E_V for the inorganic semiconductor and E_{HOMO} for the organic semiconductor are determined by linear extrapolation of the low-energy edges of their corresponding valence spectra.

Thus, $I_E = E_{vac} - E_V = \hbar\omega - (E_V - E_{onset})$. E_V and E_{onset} correspond to E_K^{max} and E_K^{min} in Fig. 4.23, respectively. $I_E^{InP} \simeq 5.8$ eV and $I_E^{\alpha\text{-NPD}} \simeq 5.6$ eV as experimentally determined. The band offset between the top of the InP valence band and the top of the α-NPD HOMO at the interface is determined by measuring E_V

of the clean substrate and $E^{\alpha\text{-NPD}}_{\text{HOMO}}$ of a thin film (4.5 nm). For α-NPD/InP(110) it turns out that $\Delta \simeq 0.4$ eV and for CuPc/InP(110) $\Delta \simeq 0.6$ eV (see Fig. 4.29). α-NPD/InP(110) and CuPc/InP(110) represent examples of straddling and staggered band lineups, respectively.

In general the clean surfaces of inorganic semiconductors are reactive due to the presence of dangling bonds in the UHV environment (large γ_s values). It is thus expected that $\Delta \neq 0$. Furthermore, when the molecule–substrate interaction is strong enough, drastic changes in the electronic structure of the molecule or of the substrate's surface can be achieved. For PbPc thin films grown on InAs(100) surfaces the oxidation state of Pb changes at the interface, $Pb^{+2} \rightarrow Pb^0$, as has been shown with UPS/XPS (Papageorgiou *et al.*, 2001).

In the case of C_{60} molecules on Ge(111) surfaces we saw in Fig. 4.7 that individual molecules can strongly perturb the surface reconstruction by inducing shifts of the adatom rows. In the ML coverage regime C_{60} exhibits different reconstructions as a function of T_{ann} intimately related to the perturbation induced at the underlying surface. For $520 < T_{\text{ann}} < 770$ K, C_{60} orders with a $(3\sqrt{3} \times 3\sqrt{3})R30°$ reconstruction and for $770 < T_{\text{ann}} < 970$ K the $(\sqrt{13} \times \sqrt{13})R14°$ superstructure is generated. A view of the $(\sqrt{13} \times \sqrt{13})R14°$ reconstruction as deduced from grazing incidence X-ray diffraction (GIXRD) measurements is shown in Fig. 4.30. Because X-rays have a large penetration depth, the surface to bulk signal ratio can be increased by illuminating the sample at very small (grazing) angles, thus reducing the probed depth.

The crystallographic structure of this interface reveals the formation of periodic defects in the unit cell created by the removal of six germanium atoms, large enough to accommodate individual C_{60} molecules. In fact, such an induced defect prioritizes the orientation of the C_{60} molecules with one of their hexagonal rings parallel to the substrate's surface plane. It appears that this reconstruction is achieved by the generation of surface holes rather than by strong rehybridization.

Reconstructed surfaces of metals are also perturbed by the C_{60} molecules, as has been shown for bare Au(110)-$p(1 \times 2)$ where a Au(110)-$p(6 \times 5)$ superstructure is induced (Pedio *et al.*, 2000). Again, the adsorption of C_{60} is accompanied by important displacements of underlying gold atoms.

This kind of study is gaining interest among the scientific community and will provide deeper insight into the understanding of the generation of interfaces involving organic materials.

Semiconductor/insulator interfaces

The most representative element belonging to such interfaces is the OFET. A simplified scheme of an OFET based on an organic thin film is given in Fig. 4.31.

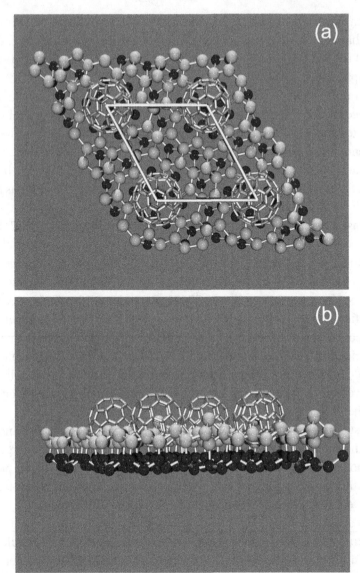

Figure 4.30. Projected views of the C_{60}/Ge(111)-$(\sqrt{13} \times \sqrt{13})R14°$ structure: (a) top and (b) lateral. Ge atoms are represented by both medium grey and black atoms. Courtesy of Dr X. Torrelles.

It mainly consists of a capacitor where one plate is a semiconductor film and the second plate consists of an insulating material with the gate electrode. Two additional electrodes, the source and the drain, are deposited on the semiconductor. When a voltage is applied between source and gate, charges of opposite sign are induced at both sides of the insulating film. The charge at the semiconductor/insulator

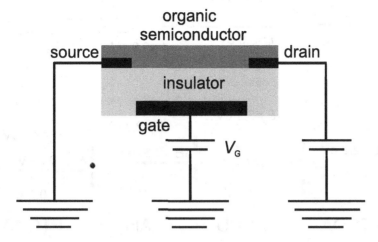

Figure 4.31. Schematic of the basic construction of an OFET.

interface can be driven by applying a second, independent, bias between source and drain. As a result, the device behaves as a variable resistor, the magnitude of which may be adjusted through the gate voltage. Because of the relevance of this structure, Section 6.3 will be devoted to OFETs.

4.3 Organic/organic interfaces

Organic/organic heterostructures

Contrary to organic/inorganic interfaces, organic/organic interfaces can be regarded as boundary regions of weakly interacting molecules, with no interface active bonds since γ_s is very low. Hence it is expected that $\Delta = 0$ and indeed this is experimentally observed in the majority of cases. However, there are a few examples where the Schottky–Mott rule is no longer valid (Rajagopal *et al.*, 1998). For instance, the staggered bands of Alq$_3$/PTCDA exhibit $\Delta E_V \simeq 0.3$ eV and $\Delta \simeq 0.5$ eV, while for α-NPD/Alq$_3$ $\Delta E_V \simeq 0.5$ eV and $\Delta \simeq 0.2$ eV, and for PTCDA/α-NPD $\Delta E_V \simeq 1.1$ eV and $\Delta \simeq 0.1$ eV, as shown in Fig. 4.32. The estimated experimental error in such measurements is about 0.2 eV.

Within the experimental error these organic/organic heterostructures exhibit transitivity, in the sense that

$$E_{\text{HOMO}}^{\text{Alq}_3} - E_{\text{HOMO}}^{\text{PTCDA}} + E_{\text{HOMO}}^{\alpha\text{-NPD}} - E_{\text{HOMO}}^{\text{Alq}_3} \simeq E_{\text{HOMO}}^{\alpha\text{-NPD}} - E_{\text{HOMO}}^{\text{PTCDA}}.$$

The question of commutativity and transitivity was previously raised for inorganic semiconductors of groups IV, III-V and II-VI (Katnani & Margaritondo, 1983). When prepared under controlled conditions, Ge/Si and Si/Ge heterostructures

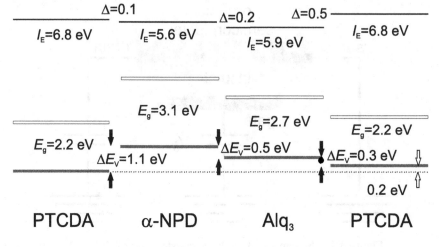

Figure 4.32. Band diagrams of PTCDA/α-NPD, α-NPD/Alq$_3$ and Alq$_3$/PTCDA. The experimental error, 0.2 eV, is indicated. Adapted from Rajagopal *et al.*, 1998.

exhibit the same band offset, $\Delta E_V \simeq 0.2$ eV and $|(E_{VBM}^{III-V} - E_{VBM}^{Ge}) - (E_{VBM}^{III-V} - E_{VBM}^{Si}) - (E_{VBM}^{Si} - E_{VBM}^{Ge})| \leq 0.2$ eV.

However, for organic/metal interfaces the commutativity rule is expected to break down, since the preparation sequence is quite different due to differentiated organic/metal and metal/organic diffusion profiles. F$_{16}$CuPc/Au shows $E_F - E_{HOMO} \simeq 1.2$ eV and $\Delta \simeq 0.2$ eV. However, when gold atoms are evaporated onto F$_{16}$CuPc they diffuse into the organic film increasing the electron injection barrier (Shen & Kahn, 2001). In general the interfacial properties of metal contacts on organic materials are strongly determined by the preparation conditions. Gold deposition at high rates and low substrate temperatures leads to well-defined interfaces with limited amounts of interdiffusion. However, low deposition rates combined with high substrate temperatures result in large interdiffusion (Dürr *et al.*, 2002b).

Many published results on electronic transport properties of organic materials, where metal contacts are usually made by evaporation of metals, do not describe the quality of the organic/metal interface, and some exotic observed features may perhaps be ascribed to extrinsic effects such as metal diffusion. The relatively simple contact lamination technique may become an alternative, since it provides a means for establishing electrical contacts without the potential disruption of the organic material associated with metal evaporation. The method consists in bringing the organic layer into mechanical contact with an elastomeric element coated with a thin metal film, which can also be patterned. The contacts are robust and reversible

(a)

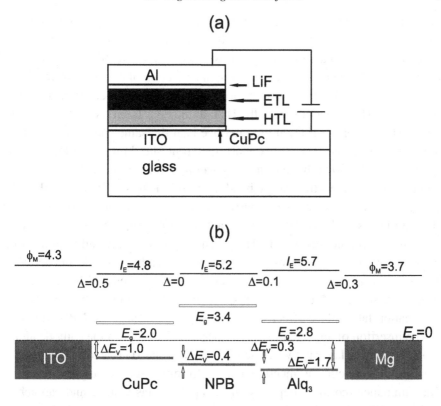

(b)

Figure 4.33. (a) Sketch of the basic construction of an OLED. (b) Schematic diagram showing the energy-level alignment for the OLED ITO/CuPc/NPB/Alq$_3$/Mg. Adapted from Lee *et al.*, 1999.

(Lee *et al.*, 2004). We shall see an example in Section 6.3, involving single crystals, where the highest reported μ_h values have been achieved thanks to the improved quality of the organic/metal and organic/organic interfaces.

The most extensively used devices based on organic/organic, as well as organic/inorganic, heterostructures are OLEDs. OLEDs are electroluminescent devices constructed using organic materials as the emitting elements (Tang & VanSlyke, 1987). A typical OLED consists of an ITO covered glass substrate coated with a hole transporting layer (HTL) like NPB or TPD and an electron transporting layer (ETL) such as Alq$_3$ (Fig. 4.33(a)). ITO, a highly doped wide band gap material ($E_{opt} \simeq 3.5$–4.0 eV), with a high work function, forms the transparent HTL anode, and a material with low work function, such as aluminium or magnesium, is used as a cathode to inject the electrons. ITO thin films in general exhibit high electrical conductivity, high optical transparency and smooth surface morphologies. Holes injected from the high work function anode into the HTL and electrons injected from the low work function cathode into the ETL are transported toward the organic/organic HTL/ETL interface. If the ETL is also the emissive layer, the organic/organic heterojunction must be designed to facilitate the injection of holes from the HTL into the ETL while blocking the injection of electrons from the ETL into the HTL, thus enhancing exciton formation and radiative recombination in the emissive layer. Hole injection from the HTL to the ETL is energetically favourable when the HOMO of the HTL at the interface is at the same energy as or at lower energy than that of the ETL. Conversely, electrons should remain blocked in the ETL when the LUMO at the interface is lower than that of the HTL.

Figure 4.33(b) shows the energy band diagram for the OLED ITO/CuPc/NPB/ Alq_3/Mg. Due to the large energy barrier for electrons of about 0.6 eV between the LUMO levels at the NPB/Alq_3 interface, electrons injected from the cathode are prevented from moving into the HTL. While electrons accumulate at this interface, the holes, once injected into CuPc, can travel readily towards and across the NPB/Alq_3 interface since the energy barriers for holes at CuPc/NPB and NPB/Alq_3 interfaces are small (0.4 and 0.3 eV, respectively). Accordingly, the electron–hole recombination takes place effectively in the Alq_3 layer near the NPB/Alq_3 interface. The insertion of an insulating layer (MgO, LiF) between the cathode and the ETL and of a CuPc layer between the ITO film and the HTL leads to a significant enhancement in the current injection and electroluminescence output.

The luminance (cd m^{-2}) depends on the applied bias voltage and can achieve values of $\sim 10^4$ with external quantum efficiencies (percentage of photons per electron) of 2–3%. The PTCDA/Alq_3 discussed above (Fig. 4.32) exhibits a low luminescence efficiency because the relative position of the LUMOs is inadequate to confine electrons in the emissive Alq_3 layer.

Organic/organic superlattices

Organic/organic superlattices can be defined as the regular alternate repetition in a given direction of different molecular organic planes. A MOM with generic formula A-B, where both A and B represent organic and organometallic molecules, is formed in a particular (but not necessarily unique) direction by stacking of alternate molecular planes containing either A or B molecules. The corresponding superlattice would be expressed by A/B. In the case of TTF-TCNQ, (100) planes (*bc*-planes) contain either TTF or TCNQ molecules and hence can be regarded as superlattices. In the two polymorphic modifications of TMTSF-TCNQ, (001) planes (*ab*-planes) are formed exclusively of TMTSF or TCNQ molecules. Figure 4.34 shows the case of the (101) planes of TTF-CA. The projection of the *ac*-plane with the *b*-direction pointing perpendicular to the plane is shown in Fig. 4.34(a) while Fig. 4.34(b) illustrates the CA (101) plane.

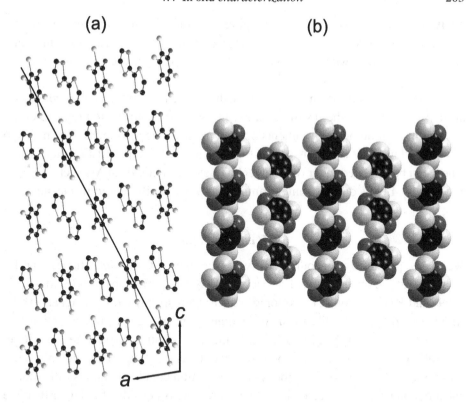

(a)   **(b)**

Figure 4.34. Crystal structure of the neutral phase of TTF-CA. Crystallographic
data from Le Cointe *et al.*, 1995. (a) Projection of the *ac*-plane with the *b*-direction
pointing perpendicular to the plane. The molecular (101) plane, which contains
the *b*-direction, is indicated by a straight line. (b) Space-filling model of the CA
(101) plane. C, O, S and Cl are represented by black, dark grey, medium grey and
light grey balls, respectively. H atoms are omitted.

Indeed, LB films can also be included in this classification. The most simple case
corresponds to the Y-type films, where the alternation is given by the head–head
and tail–tail contacts.

4.4 *In situ* characterization

The ability to *in situ* characterize the growth and properties of organic thin films
enables us to better understand the associated underlying mechanisms. Many techniques allow such studies. For films a few MLs thick the techniques should be sufficiently surface sensitive. *In situ* is too often confused with real-time techniques.
Real-time clearly means that the physical or chemical phenomena are studied
without interruption of the process. Indeed, this has to be done *in situ* during
growth. However, some *in situ* techniques, because of the experimental geometry,

imply perturbing the process, e.g., stopping the molecular flux, bringing the sample to RT, etc. In terms of the parameter hyperspace the properties are measured at a different point from where the sample was grown. In this case one can no longer talk about real-time measurements.

On the other hand, one has to distinguish between the experimental conditions of the sample and of the instrumentation in use. The samples are prepared either in vacuum or in liquid environments or even in air. Some instrumentation needs to be operated in UHV (those using electrons such as UPS, XPS, LEED, etc.), while others can be operated in air (optical techniques) or in liquid (SPMs such as STMs and AFMs). Let us discuss some examples employing different *in situ* techniques.

UHV environment

We have previously encountered two examples of *in situ* studies: DMe-DCNQI/Ag(110) measured with a specially designed LEED apparatus (Seidel *et al.*, 1999) and PTCDA/Ag(111) explored with Raman spectroscopy (Wagner, 2001). Another technique suitable for *in situ* studies is near-edge X-ray absorption fine structure (NEXAFS). NEXAFS refers to the absorption fine structure close to an absorption edge, about the first 30 eV above the actual edge. This region usually shows the largest variations in the X-ray absorption coefficient and is often dominated by intense, narrow resonances. This structure is considerably larger than the higher-energy extended X-ray absorption fine structure (EXAFS). EXAFS is due mostly to single scattering events of a high-energy photoelectron off the atomic cores of the neighbours, while NEXAFS is dominated by multiple scattering of a low-energy photoelectron in the valence potential set up by the nearest neighbours. NEXAFS is also called X-ray absorption near-edge spectroscopy (XANES). The term NEXAFS is typically used for soft X-ray absorption spectra and XANES for hard X-ray spectra although quite often they are indistinguishably used. In NEXAFS the X-ray energy is scanned and the absorbed X-ray intensity is measured, hence the mandatory use of synchrotron radiation. NEXAFS is element specific because the X-ray absorption edges of different elements have different energies, very sensitive to the bonding environment, and therefore has particular application to chemisorbed molecules on surfaces. This fine structure arises from excitations into unoccupied molecular orbitals.

Information concerning the orientation of the molecule can be inferred from the polarization dependence. NEXAFS is sensitive to bond angles, whereas EXAFS is sensitive to the interatomic distances. Linearly polarized X-rays are best suited for molecules possessing directional bonds. This is best exemplified for flat π-conjugated molecules lying flat on surfaces. When the electric field vector E is aligned along the surface normal, features due to the out-of-plane π orbitals are seen, and when E is parallel to the surface, resonances due to the in-plane σ orbitals are dominant. Transitions from C1s core levels to unoccupied π^* MOs can only be observed if a polarization component perpendicular to the molecular plane exists. Hence the light polarization vector E must be parallel to the p_z orbitals. By varying the angle between the incident beam and the substrate, the orientation of the molecules can be elucidated (Taborski *et al.*, 1995). An example of NEXAFS measurements performed on *ex situ* grown TTF-TCNQ thin films will be given in Section 6.1.

Photoelectron emission microscopy (PEEM) also allows the *in situ* characterization of materials. PEEM is a particluar mode of low-energy electron microscopy (LEEM) (Bauer, 1994; Tromp *et al.*, 1998). In a LEEM instrument high-energy electrons (several kV) are focused and decelerated to an energy of only a few eV, since the sample is held at nearly the same potential. The low-energy electrons are scattered at the sample surface, analogous to a LEED experiment, and reflected back into the objective lens. The electrons are again accelerated to the initial high voltage into the imaging column, where the projector lenses and the screen are located. Hence, when the low-energy electrons used for the illumination of the sample in the LEEM hit a crystalline surface, a diffraction pattern is obtained. For real-space imaging a diffraction spot has to be chosen with a contrast aperture located in the imaging column. The lateral resolution is a few nm but the vertical resolution can be much higher due to diffraction effects. The main differences with the SEM technique are that high-energy electrons impinge on the surface and that imaging is achieved after collection of secondary electrons in the detector. When operated in the PEEM mode, the electron beam is replaced by ultraviolet radiation. Contrast arises from differences in the work function of different materials.

PEEM has been used to characterize the real-time growth of pentacene films on cyclohexene-treated Si(100) surfaces (Meyer zu Heringdorf *et al.*, 2001). Pentacene was deposited by OMBD on the Si(100) passivated surfaces maintained at RT. By a combination of careful substrate preparation and surface energy control, thin films with single crystal grain sizes approaching 0.1 mm have been obtained. The large size of the crystals permits envisioning the microfabrication of a complete device in such crystals. However, the defect density has to be sufficiently low in order to avoid charge traps. The fractal character of the growth of this pentacene/Si(100) system will be discussed in Section 5.2.

Other examples of *in situ* studies use GIXRD (Fenter *et al.*, 1997), as previously discussed on page 197, and low-energy atom diffraction (LEAD) (Casalis *et al.*, 2003). The LEAD technique uses a monoenergetic atomic beam (e.g., helium with meV energy). The reciprocal space of the surface is mapped by detecting the angular distribution of the diffracted atoms at different azimuthal orientations of the substrate crystal. The techniques using scattering of helium atoms are extremely surface sensitive because the interaction of a helium atom with a surface is dominated by a long range van der Waals-like attraction and a shorter range repulsive interaction due to electronic overlap between the helium and the surface species.

An example of *in situ* real-time characterization of the growth of thin films of the radical *p*-NPNN in high vaccum will be discussed in detail in Chapter 5.

Liquid environment

The growth of ordered MLs in liquids can be *in situ* characterized with SPMs. One of the great advantages of SPMs is that they give real-space images with very high resolution in a variety of environments. In addition they can provide information on physical properties such as electrical conductivity, friction, mechanical resistance, etc. Operating under fluid allows imaging of samples in chemically active or electrochemical environments. For such studies, AFMs have become widely used, although many scanning tunnelling microscopy studies have been undertaken. The main reason lies in their working principle, force versus tunnelling, implying the absence or action of an electric field for the AFM or STM, respectively. In the case of the STM, the presence of such a field may strongly perturb the growth conditions. With an AFM, capillary forces, so important in air, are reduced when operating in a fluid.

In Section 4.2 we discussed the examples of the *in situ* characterization of MLs of cyclotrimeric terthiophenediacetylene (Mena-Osteritz & Bäuerle, 2001) and of the isomers of bis-cyanoethylthio-bis-octadecylthio-TTF (Gomar-Nadal *et al.*, 2003) at the liquid/HOPG interface with scanning tunnelling microscopy. We end this chapter with an example of *in situ* studies performed with an AFM also at a liquid/HOPG interface. The experiments were performed in CH_3CN with a single compartment fluid cell made of quartz with ports for fluid entry and exit. Freshly cleaved HOPG substrates were used as electrodes and platinum counter and silver reference electrodes were inserted through the outlet port of the fluid cell. Under these conditions layers of $(BEDT-TTF)_2I_3$ have been prepared and studied (Hillier *et al.*, 1994). In this case the choice of experimental conditions influences the selectivity toward the α and β polymorphs of $(BEDT-TTF)_2I_3$. High (low) overpotentials favour the formation of the α-(β-)phase and electrochemically etched HOPG favours the formation of the α-phase while the β-phase forms on pristine HOPG. AFM images allow a topographical view of the layer as well as details of the structure when operating in high resolution. In this case the unit cell of the layer can be obtained, within the experimental limitations of the technique, allowing in some favourable cases to differentiate between known crystallographic phases.

5

Thin-film growth: from 2D to 3D character

My growth is not your business, Sir!

Lewis Carroll, *Size and Tears*

As soon as the number of MLs increases, or equivalently, when we leave the realm of 2D heterostructures and enter our more familiar 3D space, the deposited materials tend to recover their intrinsic bulk properties, as already discussed in the previous chapter for CuPc, reducing the influence of the interface. For these increasingly thicker deposits the morphology becomes a relevant issue because films generally exhibit physical boundaries (steps, grains, etc.), so relevant that sometimes the physical properties of the films depend mostly on the particular morphology, rather than on their intrinsic properties. Some examples on this significant point will be given in Section 6.4. In fact, long-range ordered films as well as single crystals become thermodynamically stable only in the presence of perturbations such as defects and boundaries. The surface morphology depends not only on the nature of the intermolecular interactions but also on the selected experimental conditions. This situation is crucial. Depending on the chosen points or trajectories in the parameter hyperspace the morphology may substantially vary. Hence, when exploring the growth for a particular system the actual experimental conditions have to be described as accurately as possible.

The present chapter is devoted to several aspects of the growth of thin films of MOMs, such as growth modes, micrometre- and nanometre-scale morpohologies, control of orientation, polymorphism, etc., without pretending to be a classical canonical treatise of crystal growth. Among the host of works dedicated to the basics of crystal growth, Chernov (1984), Venables *et al.*, (1984) and Zhang and Lagally (1997) are suggested, and of course the seminal work of W. K. Burton, N. Cabrera and F. C. Frank (Burton *et al.*, 1951), where the Burton–Cabrera–Frank (BCF) model is introduced, is highly recommended.

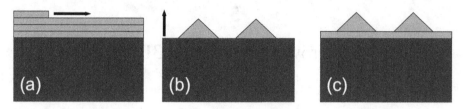

Figure 5.1. Scheme of the different mechanisms of growth: (a) Frank–van der Merwe, (b) Volmer–Weber and (c) Stranski–Krastanov. The substrate and over-layers are represented by dark grey and light grey shading, respectively.

5.1 Heterogeneous growth

Three modes of growth are known and they are termed after their discoverers: Frank–van der Merwe, Stranski–Krastanov and Volmer–Weber. A schematic of these basic mechanisms of growth is given in Fig. 5.1, which highlights the dimensionality of the growth: 2D (in-plane) or 3D (out-of-plane). The Frank–van der Merwe mode, also called the layer-by-layer mode, is of the 2D type, where a layer builds on top of the underlying one only after completion (Fig. 5.1(a)). This is the case of complete wetting. Contrary to this mode, the Volmer–Weber is purely 3D, where part of the substrate remains uncovered, representing the non-wetting case (Fig. 5.1(b)). In this 2D–3D competition, the Stranski–Krastanov mode occupies the intermediate position, since the first layers grow 2D but the 3D mode finally dominates (Fig. 5.1(c)). The growth is intrinsically a non-equilibrium process and the state obtained is not necessarily the thermodynamically most stable one but is instead kinematically determined. Herein lies the *essence* of the formation of different polymorphs.

Organic thin films have a tendency to adopt a preferential orientation with their molecular planes having the shortest molecule–molecule contacts (higher molecular density) parallel to the substrate surface in the case of weak molecule–substrate interactions. This well-known experimental fact can be understood within the framework of the heterogeneous growth model (Chernov, 1984), which is based on the minimization of the total Gibbs free energy G of the system. In order to evaluate the energy barrier G_c for the formation of stable crystalline aggregates or nuclei on a foreign surface, let us consider for simplicity nuclei having the shape of a parallelepiped of constant height h and sides L attached to a substrate single step also of height h, as depicted in Fig. 5.2. A step can be regarded as a line boundary at which the surface changes height by one or more molecular units.

In spite of its simplicity this approximation retains the necessary information and is justified because the vast majority of MOMs are layered and thus structurally anisotropic. Arrows in Fig. 5.2 aim to succinctly describe the basic atomistic mechanisms of step-flow growth, based on the BCF model (Burton *et al.*, 1951).

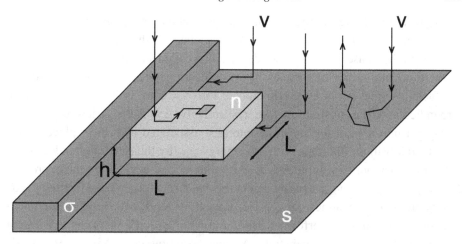

Figure 5.2. Scheme of a parallelepiped-shaped nucleus (n) on a substrate (s) and attached to a step (σ). Molecules coming from the vapour phase (v), represented by arrows exhibiting different trajectories, interact with the flat substrate, terrace or nucleus surfaces, diffuse and then desorb or attach to a step or start to nucleate. In the aggregation regime, the incoming material aggregates into the existing islands and finally they coalesce. The heterogeneous growth is symbolized by a different grey scale.

The major physical insight of the BCF model is that growth (the model can also be applied for sublimation) is strongly mediated by the presence of steps, acting as active and efficient sites for the attachment and detachment of molecules onto the terraces. During growth the concentration of the molecules exists in dynamic equilibrium with the step-edges by a balance between the rate of attachment and detachment from the step-edges, their rate of diffusion on the terraces governed by diffusion coefficient D_s, by desorption processes and deposition of molecules with a flux Φ_m. The diffusion is by far the most important kinetic process in film growth in the sense that sufficient surface mobility is mandatory in order to obtain smooth and uniform films. The BCF model assumes that the incorporation of molecules occurs at a time scale much faster than the motion of steps. In Section 5.5 we shall see that diffusion of molecules along the steps, the so-called periphery diffusion, terrace diffusion and attachment–detachment processes induce step fluctuations, that is the motion of steps.

The expression for G is given by (Molas *et al.*, 2000):

$$G = -G_{\mathrm{V}} + G_{\mathrm{S}} = -L^2 h \Delta\mu + L^2 \Delta\gamma_{\parallel} + L h \Delta\gamma_{\perp}, \qquad (5.1)$$

where G_{V} and G_{S} ($G_{\mathrm{V}} > 0$, $G_{\mathrm{S}} > 0$) represent the bulk and interface contributions to G and $\Delta\mu$ stands for the chemical potential per unit volume. $\Delta\mu = \mu_{\mathrm{v}} - \mu_{\mathrm{n}}$, where μ_v and μ_n represent the vapour and nuclei chemical potentials, respectively. The chemical potential expresses the work associated with changing the number of particles in the phase, vapour or solid. $\Delta\mu$, also called supersaturation, appears as the thermodynamic driving force of the crystallization process. $\gamma_\parallel$ and $\gamma_\perp$ denote the interfacial energies per unit area (in units of J m^{-2} = N m^{-1}) of planes parallel and perpendicular to the substrate surface, respectively. $\Delta\gamma_\parallel$ is the variation of the free energy per unit area of the substrate interface associated with the replacement of the substrate/vapour interface with free energy $\gamma_{\parallel sv}$ by the substrate/nucleus ($\gamma_{\parallel ns}$) and nucleus/vapour ($\gamma_{\parallel nv}$) interfaces. The value of $\Delta\gamma_\parallel$ is given by the expression $\Delta\gamma_\parallel = \gamma_{\parallel nv} + \gamma_{\parallel ns} - \gamma_{\parallel sv}$, making patent this balance, and can be expressed via the specific free energy of adhesion $\gamma_{\parallel ad}$. This magnitude represents the work per unit interface area that has to be performed in order to achieve reversible isothermal separation of the nucleus from the substrate. $\Delta\gamma_\parallel$ can be expressed as $\Delta\gamma_\parallel = 2\gamma_{\parallel nv} - \gamma_{\parallel ad}$ since $\gamma_{\parallel ad} = \gamma_{\parallel nv} + \gamma_{\parallel sv} - \gamma_{\parallel ns}$. When the adhesion is strong (weak), $\Delta\gamma_\parallel < 0$ ($\Delta\gamma_\parallel > 0$), leading to complete (incomplete) wetting. $\Delta\gamma_\perp = 3\gamma_{\perp nv} + \gamma_{\perp n\sigma} - \gamma_{\perp \sigma v}$, where $\gamma_{\perp nv}$, $\gamma_{\perp n\sigma}$ and $\gamma_{\perp \sigma v}$ represent the nucleus edge/vapour, nucleus edge/ step and step/vapour interface tensions, respectively. The factor 3 results from the choice of nucleus shape, in this case a parallelepiped, and is thus model-dependent.

Values of $\gamma_{\parallel nv}$ for several MOMs lie below 0.7 eV nm^{-2} ($\sim$ 0.1 J m^{-2}) (Northrup *et al.*, 2002; Verlaak *et al.*, 2003). According to first-principles pseudopotential DFT calculations (Northrup *et al.*, 2002) the (001) surface of pentacene is found to be the lowest in energy, having a formation energy of 0.15 eV per surface unit cell. This translates into a surface energy $\gamma_{\parallel nv}$ of 0.31 eV nm^{-2}. On the other hand the (010) surface of pentacene has a formation energy of 0.75 eV per surface unit cell, corresponding to 0.64 eV nm^{-2}. Inorganic solids exhibit surface energies of the order of 5–10 eV nm^{-2}.

Equation (5.1) describes the size effects associated with the presence of a finite nucleus and reflects the balance between bulk and surface contributions. The limiting case $L \rightarrow \infty$ represents the ideal formation of a film under complete wetting. The convention is that the system evolves towards negative values of G. When $G > 0$ the interface term dominates over the bulk term (low L values) while the opposite situation is found for $G < 0$ (higher L values). Under the condition $\partial G/\partial L = 0$, G_c and the critical length L_c of nucleation, are given by the expressions:

$$G_c = \frac{h^2}{4} \frac{\Delta\gamma_\perp^2}{h\Delta\mu - \Delta\gamma_\parallel}, \tag{5.2a}$$

$$L_c = \frac{h}{2} \frac{\Delta\gamma_\perp}{h\Delta\mu - \Delta\gamma_\parallel}. \tag{5.2b}$$

From Eq. (5.2a) it becomes clear that the stronger (weaker) the adhesion of the nucleus on the substrate, the smaller (larger) G_c. The role of inhomogeneities in the substrate, such as defects, impurities, etc., is not taken into account in the model but they may become relevant because they would reduce G_c. In Section 5.4 we shall deal with the case of a cylinder-shaped nucleus growing onto a substrate following the arguments described here. In this case L_c will be replaced by the critical 2D radius R_c. Note that $h\Delta\mu - \Delta\gamma_\parallel$ must be non-zero in order to avoid mathematical singularities and that $G_c = h\Delta\gamma_\perp L_c/2$.

All interface tensions are defined as positive and we consider $\gamma_{\perp nv} > \gamma_{\parallel nv}$ in order to intentionally account for anisotropy. This favours 2D growth but does not prioritize the Frank–van der Merwe mechanism over the others. In fact growth is *always* 2D, as indicated by the BCF model, but depends on the lateral size of the terraces. Small terraces necessarily imply 3D growth.

The chemical potential and interface tension terms can be defined in a consistent way if the dimensions of the nuclei are sufficiently large. The nuclei considered here should be large enough, well beyond the discrete nature of such nuclei, to be able to apply thermodynamic parameters. Estimates of the size of such nuclei range from a few molecules to relatively large R_c values. In the case of $(BEDT\text{-}TTF)_2I_3$, layers grown on HOPG surfaces, where single-layer deep pits have been created by thermal etching, reveal that nucleation is completely suppressed in pits with $R_c < 50$ nm (Hooks *et al.*, 1998) while for thin *p*-NPNN films $R_c < 25$ nm, as derived from the minimum distance between spiral centres of coupled spirals (Fraxedas *et al.*, 1998), a point that will be discussed later in Section 5.5. On the other hand, critical nucleus sizes i, obtained from the scaling of the nucleation density N as a function of the D_s/Φ_m ratio, give $i = 4$ for pentacene grown on chemically oxidized Si(100) (Ruiz *et al.*, 2003) and $i = 5$ for Alq$_3$ on H-terminated Si(100) (Brinkmann *et al.*, 2002b). N, D_s and Φ_m are related through the expression $N \propto (D_s/\Phi_m)^{-\chi}$, where $\chi = i/(i + 2)$. Here i represents the number of monomers, and for cluster sizes larger (smaller) than i the nuclei tend to grow (dissociate). The estimates are clearly compatible but the difference in range perhaps lies in the crystallinity of the nuclei.

In the absence of steps $\gamma_{\perp\sigma v} = 0$ and $\gamma_{\perp n\sigma} = \gamma_{\perp nv}$, giving rise to $\Delta\gamma_\perp = 4\gamma_{\perp nv}$ and Eqs. (5.2a) and (5.2b) transform into:

$$G_c = 4h^2 \frac{\gamma_{\perp nv}^2}{h\Delta\mu - \Delta\gamma_\parallel}, \tag{5.3a}$$

$$L_c = 2h \frac{\gamma_{\perp nv}}{h\Delta\mu - \Delta\gamma_\parallel}. \tag{5.3b}$$

In the homogeneous case $\Delta\gamma_\parallel = 0$ and $\Delta\gamma_\perp = 2\gamma_{\perp nv}$ since $\gamma_{\perp nv} + \gamma_{\perp no} - \gamma_{\perp ov} = 0$, thus:

$$G_c = \frac{h\gamma_{\perp nv}^2}{\Delta\mu}, \tag{5.4a}$$

$$L_c = \frac{\gamma_{\perp nv}}{\Delta\mu}. \tag{5.4b}$$

Finally, G_c for the homogeneous case in the absence of steps is obtained under the conditions $\Delta\gamma_\parallel = 0$ and $\Delta\gamma_\perp = 4\gamma_{\perp nv}$, which results in:

$$G_c = \frac{4h\gamma_{\perp nv}^2}{\Delta\mu}, \tag{5.5a}$$

$$L_c = \frac{2\gamma_{\perp nv}}{\Delta\mu}. \tag{5.5b}$$

As mentioned before, low adhesion is expressed by $\Delta\gamma_\parallel > 0$. In the case of growth $L_c > 0$, which imples that $h\Delta\mu - \Delta\gamma_\parallel > 0$, so that $\Delta\mu > 0$ if $\Delta\gamma_\parallel \geq 0$. Conditions $\Delta\gamma_\parallel > 0$ and $\Delta\mu > 0$ correspond to the Volmer–Weber mechanism of growth (Chernov, 1984). Strong adhesion or complete wetting (Frank–van der Merwe mode) implies that $\Delta\gamma_\parallel < 0$ and $\Delta\mu < 0$, while for the Stranski–Krastanov mode $\Delta\gamma_\parallel < 0$ and $\Delta\mu < 0$ for the first 2D layers and $\Delta\gamma_\parallel > 0$ and $\Delta\mu > 0$ for the 3D crystal growth.

The in-plane velocity v_{ip} can be approximated by:

$$v_{ip} \sim \frac{\partial L}{\partial t} \propto -\frac{\partial G}{\partial L}. \tag{5.6}$$

It can be easily shown that $v_{ip} \sim h\Delta\gamma_\perp$ at $L = L^*$ and that v_{ip} becomes independent of $\Delta\gamma_\perp$ for $L \gg L^*$, where L^* is obtained under the condition $G(L^*) = 0$ and is given by $L^* = 2L_c$. Hence, v_{ip} depends linearly on $\Delta\gamma_\perp$ at the earlier stages of growth. In the absence of steps $v_{ip} \sim 4h\gamma_{\perp nv}$ at $L = L^*$. In the case of 3D growth we observe from Eq. (5.2a) that for a given positive $\Delta\mu$ value, G_c decreases as $\Delta\gamma_\parallel \to 0$. Since $\Delta\gamma_\parallel = \gamma_{\parallel nv} + \gamma_{\parallel ns} - \gamma_{\parallel sv}$, in the low molecule–surface interaction limit of physisorption (low $\gamma_{\parallel sv}$), the $\Delta\gamma_\parallel \to 0$ condition will be fulfilled for low $\gamma_{\parallel nv}$ and $\gamma_{\parallel ns}$ values. Hence, nuclei with molecular planes of lower interface tension parallel to the substrate have higher probabilities of formation at the beginning of the nucleation process because of lower nucleation barriers and they grow faster since $\gamma_{\perp nv} > \gamma_{\parallel nv}$. Low $\gamma_{\parallel nv}$ and $\gamma_{\parallel ns}$ values imply that the molecular planes possess strong in-plane intermolecular interactions (the molecules are more strongly bound), which results in weak interaction between adjacent molecular planes.

PTCDA films OMBD-grown on Ag(111) single crystals show a transition from layer-by-layer growth with smooth and small-grained morphology towards the

(a) (b)

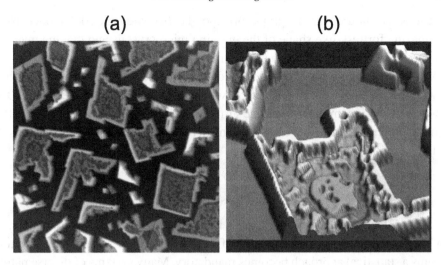

Figure 5.3. TMAFM images taken in ambient conditions of a thin film of EDT-TTF grown on KCl(100) by PVD. (a) 2D representation, 15 μm × 15 μm, (b) 3D representation, 3.5 μm × 3.5 μm. The walls in (b) are approximately 300 nm high and the plateaus inside exhibit heights of c. 100–150 nm.

formation of larger isolated microcrystals on complete coverage of two strained layers (Stranski–Krastanov mode) with increasing T_{sub} between 180 and 420 K (Chkoda *et al.*, 2003). This morphology transformation is observed for nominally $\sim$10 nm thick films with $D_t \sim 0.5$ ML min^{-1}. When grown under certain experimental conditions ($T_{sub} < 350$ K) the PTCDA/Ag(111) system shows a morphology transition as a function of T_{ann}. This transition, which in reality consists in a wetting–dewetting transformation, induces evolution from an initial (as-grown) pseudo-2D morphology (a smooth film) towards a final 3D morphology, where regions of the substrate become uncovered. This transition is thus of a different nature and simply reveals that the as-grown films are unstable against annealing (Krause *et al.*, 2003).

On the other hand, n-alkanes ($n = 4, 6, 7$) adsorbed from the vapour phase on Ag(111) surfaces also grow following the Stranski–Krastanov mechanism (Wu *et al.*, 2001).

Figure 5.3 gives an example of the Volmer–Weber mechanism of growth. The figure shows TMAFM images of an EDT-TTF film grown under a base pressure of $\sim$10^{-5} mbar on KCl(100) surfaces held at RT, taken after interruption of the growth process and exposure to air. It cannot be directly concluded that this corresponds to a snapshot of the growth process, although it is tempting to propose it, since the sample was exposed to air after stopping the molecular beam flux. From the images it becomes clear that the film is not continuous but instead formed by isolated microcrystals oriented along well-defined surface crystallographic directions. The non-covered regions correspond to the bare KCl(100) surface. Note the amazing fortress-like shape of the growing microcrystals and that the material retained inside exhibits round-shaped forms, indicative of incomplete growth.

And what about our long forgotten guiding N_2 molecule? It turns out that, when growing on HOPG at coverages above two layers in the temperature range $10 <$ $T_{sub} < 20\,K$, N_2 incompletely wets the surface (Seguin *et al.*, 1983). The system thus seems to prefer the Volmer–Weber mechanism of growth over the other mechanisms for the particular growth conditions.

5.2 Dynamic scaling theory

The 2D to 3D transition implies changes in morphology and thus in roughness, as can be inferred from Fig. 5.1. Because of the complexity of such time-fluctuating systems a statistical approach becomes mandatory. Many systems of diverse nature exhibiting rough interfaces in their evolution process, e.g., fluids invading porous media, fire fronts, crystal growth, growing tumours, etc., have been successfully described by means of scaling analysis, a powerful mathematical tool used in the study of fractal geometry (Family, 1990; Barabási & Stanley, 1995; Krug, 1997). The dynamics of such systems can be characterized by a set of critical exponents obtained from scale-invariant properties of certain physical quantities. Systems exhibiting the same critical exponents belong to the same universality class, sharing a common mathematical description.

Let us describe the essential trends of the dynamic scaling theory (DST) adapted to the field of film growth. We first assume that the surface morphology can be described by a continuous function $h(x, y)$, where h stands for the height and x and y represent the in-plane 2D Cartesian coordinates (a perfect flat surface would be represented by a constant h value). In many cases the surface morphology of thin films is, within certain spatial ranges, invariant under scale transformations. This invariance implies that there exist rescaling factors s_1 and s_2 such that $h(x, y)$ is statistically indistinguishable from $s_1 h(s_2(x, y))$. This property is known as self-affinity.

The magnitude most commonly employed to statistically characterize the surface morphology for a given growing system of size L observed at time t is the roughness, $\varsigma(L, t)$. This magnitude represents the height distribution of the surface and is a measure of the width or dispersion of the real surface. ς is defined as the mean square deviation of the local height with respect to the mean height, $\varsigma(L, t) = [\langle (h(x, y, t) - \langle h(x, y, t) \rangle)^2 \rangle]^{1/2}$, where $\langle \cdots \rangle$ represents the average value of the quantity within brackets for the length scale L. The surface roughness can be described by the dynamic scaling expression $\varsigma(L, t) = L^\alpha f(t/L^z)$, where the function $f(v)$ scales as v^β for $v \ll 1$ and is constant for $v \gg 1$. The exponents α, β and z are known as the roughness, growth and dynamic exponents, respectively. They are characteristic of the mechanism governing the system of growth dynamics and are related by the expression $z = \alpha/\beta$.

This scaling law implies the existence of a characteristic time interval, $\tau \propto L^z$, which defines a transition for the roughness behaviour. For short enough periods, ς is independent of the system size and increases with time as indicated in Eq. (5.7a). From this expression it is possible to experimentally obtain the β exponent, provided that ς is measured over large enough length scales. For long enough periods ς reaches a stationary regime, that is, it becomes time independent and scales with the system size as indicated in Eq. (5.7b):

$$\varsigma(L, t) \propto t^\beta \qquad t \ll \tau, \tag{5.7a}$$

$$\varsigma(L, t) \propto L^\alpha \qquad t \gg \tau. \tag{5.7b}$$

In spite of the fact that real systems are finite, it is not always clear which is the appropriate L value. In practice L is taken as the observation size L_S, provided that L_S is smaller than the actual system size. Thus, the value of α can be determined experimentally from Eq. (5.7b). Note that within this choice ς depends on the size L_S of the observation window. In the case of measurements performed with an AFM L_S is taken as the image size.

Another magnitude that characterizes the surface morphology is the lateral correlation length, $\xi(L, t)$, which represents the typical width of these structures. DST applies to systems in which, depending on the temporal range, $\xi(L, t)$ follows the expressions:

$$\xi(L, t) \propto t^{1/z} \qquad t \ll \tau, \tag{5.8a}$$

$$\xi(L, t) \propto L \qquad t \gg \tau. \tag{5.8b}$$

The $1/z$ exponent describes the coarsening process of the structures generated during the system growth and is also known as the coarsening exponent. The exponents α, β and z can be experimentally measured, e.g., from AFM images. One way to analyse the scaling behaviour from the AFM images is by using the power spectral density (PSD) function. The PSD function provides information relative to the lateral correlations of the surface, and hence to $\xi(L, t)$, as well as to ς. The PSD function is defined as $\text{PSD}(k) = (2\pi/L_S)^2 \langle |\tilde{h}(k)|^2 \rangle_k$, where $k^2 = k_x^2 + k_y^2$ and where $\tilde{h}(k)$ represents the 2D Fourier transform of the surface topography function $h(x, y)$. Thus, the PSD is a 1D representation of the surface in the reciprocal space and gives the intensity with which each spatial frequency contributes to ς. This applies only for the case of isotropic film surfaces in the substrate plane. In practice the lowest and highest frequencies in reciprocal space are taken as $1/L_S$ and N_{pix}/L_S, where N_{pix} stands for the number of pixels used in the AFM images. The surface roughness can be obtained from $\varsigma^2(L_S) = 2\pi \int \text{PSD}(k)k \, dk$. For a self-affine system $\text{PSD}(k) \propto k^{-(2\alpha+d)}$, where d is the (fractal) dimension. For instance, the fractal dimension of pentacene layers grown on Si(100) surfaces for coverages between 0.05 and 0.2 ML is 1.6–1.7 (Meyer zu Heringdorf *et al.*, 2001). In this case 1 ML is defined as a single molecular layer of crystalline pentacene with a thickness of ~1.5 nm. At higher coverages the coalescence of islands makes the fractal dimension tend towards 2, as expected for a compact layer.

To illustrate the utilization of DST for thin films of MOMs we start with the change of morphology observed in films of α-6T (Biscarini *et al.*, 1997). In fact only a few examples are found in the literature. The α-6T films have been grown on ruby mica by high-vacuum sublimation (base pressure 10^{-6}–10^{-7} mbar). During deposition T_{sub} was varied, in order to elucidate the effect of the deposition temperature on the film roughness scaling behaviour, while the film thickness (~100 nm) and $D_t \simeq 0.02$–0.1 nm s^{-1} were kept constant. As characterized with an AFM in contact mode, the films consist of grains at $T_{sub} < 473$ K while above this temperature the aggregate is made of large anisotropic lamellae. The films evolve towards a self-affine topology as the morphology changes from grains to lamellae. The roughness exponent α decays with temperature from about 1 at RT down to 0.7 at 473 K as derived from PSD curves. Thus, α maps the range within the predictions of diffusion-limited aggregation ($\alpha = 1$) and OMBD-growth controlled by adsorption at kink sites ($\alpha = 0.66$). In this system, the apparent scaling behaviour exhibits a crossover from linear (diffusion) to non-linear (adsorption) regimes.

In the case of DIP films grown in UHV on atomically smooth SiO$_2$ substrates the values of the scaling coefficients are $\alpha = 0.68$, $\beta = 0.74$ and $1/z = 0.92$ as derived from atomic force microscopy, X-ray reflectivity and diffuse X-ray scattering measurements (Dürr *et al.*, 2003b). DIP films with thickness between 7 and 900 nm were prepared on oxidized (~ 400 nm) Si(100) substrates at $T_{sub} \simeq 418$ K and for $D_t \simeq 1.2$ nm min^{-1} under UHV conditions ($\leq 5 \times 10^{-10}$ mbar). The derived exponents point to an unusually rapid growth of vertical roughness and lateral correlations.

Finally, LB films consisting of eight bilayers of cadmium-substituted tricosenoic acid (C$_{22}$H$_{45}$COOH), where the two hydrogen atoms belonging to adjacent carboxylic heads are replaced by a doubly charged Cd^{+2} ion, exhibit a roughness exponent $\alpha \simeq 0.5$ as derived from diffuse X-ray scattering measurements (Gibaud *et al.*, 1995).

5.3 Control of orientation

The relative orientation of the molecules with respect to the substrate surface or, in other words, the molecular planes parallel to the substrate can be controlled, to a certain extent, by varying external variables such as D_t, T_{sub}, ϑ, the chemical and physical nature of the substrate, etc. In the following lines we find examples of the selectivity of orientation by leaving only one variable free while maintaining all the rest fixed for $\vartheta > 1$ ML and we start with variable T_{sub}.

Films formed by either lying or standing p-6P molecular orientations can be achieved on the (100) cleavage plane of KCl at $T_{sub} \simeq 293$ or $T_{sub} \simeq 423$ K, respectively, at a deposition rate of 1 nm min^{-1} at 5×10^{-6} mbar (Yanagi & Okamoto, 1997). Alternatively, NTCDA is oriented on Ag(111) with its molecular plane mainly parallel to the substrate for $T_{sub} \simeq 150$ K but essentially perpendicular to the substrate when $T_{sub} > 285$ K, at 1 ML min^{-1} (Gador et al., 1998). In the case of octaethylporphyrinplatinum (+2) films grown at $D_t \leq 0.04$ nm s^{-1} on KCl and KBr substrates, when the films are grown at $T_{sub} = $ RT the molecules adopt the standing up position while the lying position is obtained for $T_{sub} \simeq 323$ K (Noh et al., 2003).

When T_{sub} is kept constant and D_t is varied, similar selectivity in the film orientation is observed. Thin CuPc films prepared on amorphous substrates using PVD at $T_{sub} = $ RT and low D_t (0.1 nm s^{-1}) are oriented with their (100) crystallographic planes preferentially parallel to the substrate surface (standing). At higher D_t (10 nm s^{-1}), an additional second type of preferred orientation is observed with (110) planes oriented preferentially parallel to the substrate surface (lying) (Resel et al., 2000).

After these few examples let us analyse the case of varying the chemical nature of the substrate. Figures 5.4(a) and 5.4(b) show the $\theta/2\theta$ XRD patterns of thin films (thickness ~ 1 μm) of the neutral π-donor TMTSF molecule grown by sublimation in high vacuum ($\sim 10^{-6}$ mbar) on KCl(100) and on a ML of a BaAA LB-film transferred to Si, respectively (Fraxedas, 2002). The substrates were held at 320 K for the KCl(100) and at RT for the BaAA substrates. We observe that the TMTSF films are preferentially oriented with their ($0\bar{2}1$) and (100) molecular planes parallel to the surface for KCl(100) and BaAA surfaces, respectively. The films are highly oriented as demonstrated by the low value of the FWHM of the rocking curves. Figures 5.4(c) and (d) depict views of the ($0\bar{2}1$) and (100) planes, respectively. This is an example of low $\gamma_{\parallel ns}$ and $\gamma_{\parallel nv}$ values. Both ($0\bar{2}1$) and (100) planes are the most densely packed ones, where both a and c parameters define stacking directions induced by short Se$\cdots$CH$_3$ (0.387–0.393 nm), Se$\cdots$Se (0.391–0.404 nm) and CH$_3 \cdots$CH$_3$ (0.397 nm) contacts.

The crystallographic parameters for the thin films derived from the reflections ($0\bar{2}1$), ($0\bar{4}2$), (010), ($01\bar{1}$), ($2\bar{2}1$), (002), (011), ($2\bar{2}\bar{1}$), ($1\bar{2}\bar{1}$) and ($1\bar{1}0$) give $a = 0.696 \pm 0.003$ nm, $b = 0.809 \pm 0.001$ nm, $c = 0.632 \pm 0.001$ nm, $\alpha = 105.62 \pm 0.07°$, $\beta = 95.435 \pm 0.109°$ and $\gamma = 108.934 \pm 0.193°$, in good agreement with the values obtained from single crystals ($P\bar{1}$, $a = 0.693$ nm, $b = 0.809$ nm, $c = 0.631$ nm, $\alpha = 105.51°$, $\beta = 95.39°$, $\gamma = 108.90°$) (Kistenmacher et al., 1979).

Substrate-induced orientation selectivity has also been observed for thin films of F$_{16}$ZnPc grown in UHV from ML coverages to average thickness of about 20 nm.

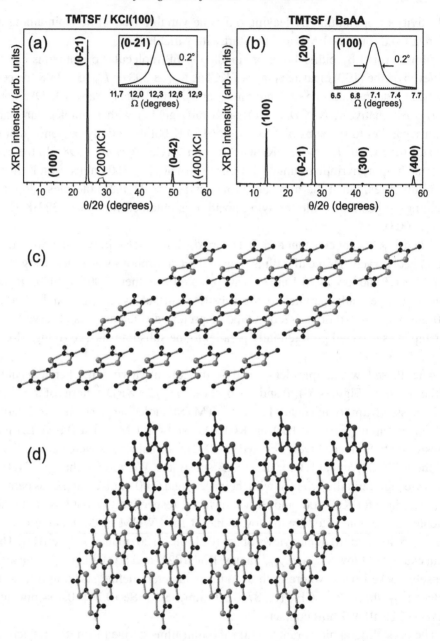

Figure 5.4. XRD patterns of ~1 μm thick TMTSF films grown on (a) KCl(100) and (b) a BaAA LB film. The insets show the rocking curves (Ω) of the (a) $(0\bar{2}1)$ and (b) (100) reflections. Views of the TMTSF molecular planes (c) $(0\bar{2}1)$ and (d) (100).

KCl and KBr substrates induce a distribution where the $F_{16}ZnPc$ molecules lie parallel to the surface as opposed to NaCl and amorphous SiO_2 where growth occurs in cofacial stacks of molecules predominantly with the molecular plane vertical to the surface (Schlettwein *et al.*, 2000).

In contrast, the modification of gold surfaces with phenyl-terminated thiols not only leads to a greater efficiency in transfer of the CuPc-based molecule shown in Fig. 3.11 but also influences the orientation of such molecules (Zangmeister *et al.*, 2001). Thin CuPc films grown on technical substrates such as ITO, oxidized silicon and polycrystalline gold are almost as highly ordered as on single crystal substrates, but the molecular orientation in the two cases is radically different: the CuPc molecules stand on the technical substrates and lie on the single crystalline substrates (Knupfer & Peisert, 2004). The relative orientation has been determined by means of NEXAFS (see Section 4.4). For thin CuPc films (20–50 nm thick) grown on Au(110) and ITO substrates a very clear angular dependence of the $N1s \rightarrow \pi^*$ resonances (402–408 eV) is observed: the maximum intensity of the π^* resonances for CuPc on Au(110) is observed at grazing incidence, whereas for the ITO substrate the highest intensity is found at normal incidence. Furthermore, the angle dependence of the $N1s \rightarrow \sigma^*$ intensities (408–430 eV) behaves in an analogous, but opposite manner for both substrates. These observations demonstrate that the organic molecules are well ordered on both substrates, but that the adsorbate geometry in each case is completely different: for Au(110) the CuPc molecular plane is parallel to the substrate surface and for ITO it is perpendicular.

The chemical functionalization of TTF-derivatives, such as the case of the asymmetric EDT-TTF molecule (see Chapter 2), permits control over the strength of molecule–molecule and molecule–substrate interactions, and thus on the growth. Among the large number of available monofunctionalized EDT-TTF molecules we select here the carboxylic acid derivative EDT-TTF-COOH in order to explore the role of hydrogen bonding. EDT-TTF-COOH is synthesized by hydrolysis with LiOH of the corresponding ester EDT-TTF-CO_2Me (Heuzé *et al.*, 1999).

This acid crystallizes in the monoclinic system, space group $P2_1/n$, with lattice parameters $a = 0.637$ nm, $b = 2.714$ nm, $c = 1.424$ nm and $\beta = 92.29°$ (Heuzé *et al*, 1999). Two crystallographically independent molecules (A and B) are found in general positions in the unit cell and are associated into AB pairs by two OH $\cdots$ O hydrogen bonds to form the eight-membered cyclic motif $R_2^2(8)$, following Etter's nomenclature (Etter, 1990). Details of the crystallographic structure of the dimers are given in Fig. 5.5. The associated synthon of this structure was shown in Fig. 1.8(a). Within this nomenclature, $R_d^a(n)$ describes a cyclic motif made of one type of hydrogen bonds involving d hydrogen atoms, a hydrogen bond acceptors within a ring of n atoms (including H atoms). Two AB pairs are further associated through the inversion centre into face-to-face dimers, a motif often encountered

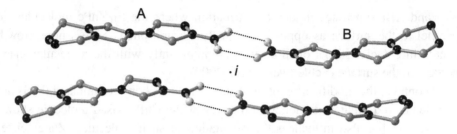

Figure 5.5. Inversion-related dimers of hydrogen-bonded pairs of EDT-TTF-COOH. C, S and O atoms are represented by black, medium grey and light grey balls, respectively. H atoms are omitted.

with such sulfur-rich molecules and stabilized by S$\cdots$S van der Waals interactions, as observed in the structures of the neutral forms of EDT-TTF or BEDT-TTF. In the solid state these pairs of dimers are further organized in a herringbone pattern, with the ethylene end groups pointing toward the sulfur atoms of a neighbouring EDT-TTF moiety, as will be shown later.

When sublimated onto *ex situ* cleaved alkali halide (100) surfaces in high vacuum ($\sim 10^{-6}$ mbar) with the substrates held at RT, the obtained films are polycrystalline, with typical microcrystal long dimensions of *c*. 7 μm, and exhibit no texture, as evidenced by the bright field images given in Fig. 5.6(d–f). The relatively large sizes of the microcrystals (something quite common for vapour-deposited neutral TTF-derivatives, where lengths of 15–20 μm are routinely found) allow selected area electron diffraction (SAED) to be performed, whose patterns are shown in Fig. 5.6(a–c). Alkali halide substrates are ideal for this kind of experiment since they can be removed by dissolving them in water.

The analysis of such patterns reveals that the microcrystals are preferentially oriented with their (021) planes, the contact planes, parallel to the substrate's surface. The interesting point is that, in order to satisfy such orientation, the hydrogen bonds of the dimers at the interface have to be broken and in addition some reorganization of the molecules is needed (see Fig. 5.6(g)). In conclusion, the molecule–substrate interactions are sufficiently strong (larger $\gamma_{\parallel ns}$ and $\gamma_{\parallel nv}$ values) to induce COO$\cdots$Alk bonds, where Alk represents sodium and potassium, but the growing crystals adapt their structure in order to crystallize in the known monoclinic bulk phase.

5.4 *In situ* studies of growth

Let us start this section with real-time measurements of the crystallization process of highly ordered thin films of the molecular organic radical *p*-NPNN grown from the vapour phase on glass substrates (Caro *et al.*, 2000). These films exhibit 2D

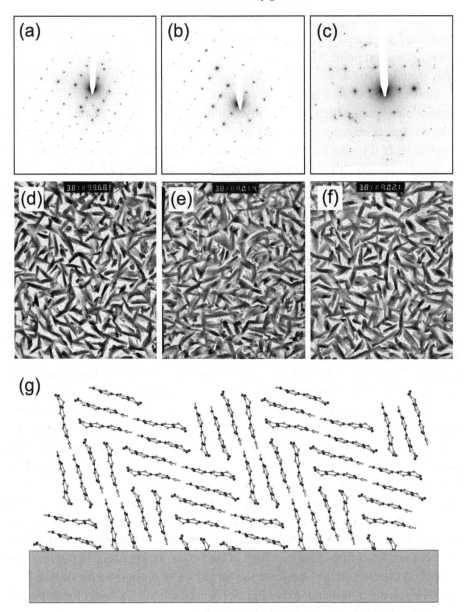

Figure 5.6. SAED patterns of EDT-TTF-COOH films grown on (a) NaCl(100), (b) KCl(100) and (c) KBr(100) substrates. The bright field images of the films corresponding to (d) NaCl(100), (e) KCl(100) and (f) KBr(100) substrates are shown after dissolution of the substrates in distilled water. Courtesy of Dr M. Brinkmann. (g) Structure of EDT-TTF-COOH projected along the *a*-axis and with the (021) plane parallel to the substrate. Note the herringbone distribution mentioned above.

spherulitic morphology with a low density of nucleation centres. The spherulitic morphology is characterized by the 'radial growth of lamellar crystals around a central nucleation point. Most of the knowledge on spherulitic growth arises primarily from investigations performed on polymers. The usual method of obtaining thin films of pure or of a mixture of polymers is the isothermal crystallization of a supercooled melt, which is generally confined between two glasses. The same behaviour has been reported for thin films of polymethine dye salts obtained by spin coating. In this case the films remain amorphous below 30 nm in thickness but crystallize with spherulitic morphology above 30 nm (Dähne, 1995).

Video microscopy with crossed polarizers permits the direct and non-invasive observation of the nucleation and growth process for many substances, and thus the study of the time evolution of the spherulite radius $R(t)$. When the growth is controlled by diffusion the radius of the spherulites increases as $R(t) \propto t^{1/2}$, while when the growth is determined by a nucleation-controlled process (incorporation of atoms or molecules to the surface of the crystalline part) the radius increases linearly with time, $R(t) \propto t$.

The thin films of p-NPNN reported here were grown on vacuum glass viewports by evaporation of the p-NPNN precursor in high vacuum (low 10^{-6} mbar range) with $D_t \sim 0.5$ μm h^{-1} and with the viewport held at RT. Under these conditions the films crystallize in the monoclinic α-phase and are (002)-oriented (ab-planes parallel to the substrate surface) exhibiting a high degree of orientation (see Fig. 3.19). Figure 1.23(a) shows the bulk molecular distribution of α-p-NPNN oriented with the ab-planes parallel to the substrate. The real molecular arrangement of the film/substrate interface is unknown and tentatively assumed to be the bulk distribution. The ab-plane is the most densely packed plane and the stacking along the a-direction might favour the radial in-plane growth.

Figure 5.7 shows an optical microscope image (crossed polarizers) of a thin film of p-NPNN grown on a glass substrate exhibiting the characteristic Maltese cross associated with 2D spherulitic morphology. Note the clearly defined boundaries between spherulites. The curvature of the boundaries between two spherulites depends on the elapsed time between the formation of those two spherulites. Both spherulites on the right-hand side of Fig. 5.7 are separated by a nearly linear boundary because they formed roughly at the same time.

The time evolution of the formation of the film during evaporation is shown in Fig. 5.8 as recorded with a CCD camera assembled to an optical microscope with crossed polarizers. The 2D spherulites are labelled by order of formation. The in-plane spherulitic crystallization is completed ~ 300 s after the observation of the first spherulite, **1** in Fig. 5.8(a). From Fig. 5.8 we observe that the nucleation centres do not appear simultaneously and that the density of spherulites is rather low (~ 1.6 mm^{-2}). This low density permits a detailed analysis of the time evolution of

Figure 5.7. Optical microscope image of a thin film (thickness $\sim$2 μm) of α-p-NPNN grown on a glass substrate (1.6 × 1.0 mm^2, crossed polarizers). Reprinted from *Journal of Crystal Growth*, Vol. 209, J. Caro, J. Fraxedas and A. Figueras, *Thickness-dependent spherulitic growth observed in thin films of the molecular organic radical p-nitrophenyl nitronyl nitroxide*, 146–158, Copyright (2000), with permission from Elsevier.

the radii of the spherulites in a wide time period. We also observe that when two spherulitic fronts meet, the boundaries have a well-defined curvature and remain static. Another important point is that the vertical resolution improves with deposition time: **8** and **9** in Fig. 5.8(f) are better resolved than **1** and **2** in Fig. 5.8(a) despite having similar values for the radii. This implies that the thickness of the 2D spherulites is independent of the radius and that it increases with deposition time.

Figure 5.9 shows the time evolution of the radii of selected 2D spherulites from Fig. 5.8. We observe that the process is non-linear and accelerated, $d^2R/dt^2 > 0$. It is also interesting to notice that, at a given time, the radial growth velocity $v_R = dR/dt$ (slope) is nearly the same for all spherulites, which implies that it depends on the deposition time and certainly not on the radius of the spherulites. In the case discussed here the thickness of the film is increasing with time because of continuous exposure to the molecular beam. The non-linearity is more pronounced at the beginning of the experiment (roughly between 250 and 350 s) and the velocity nearly tends towards an asymptotic value, so that 2D spherulites that are formed last show almost linear growth.

The mechanism of growth of the films may be modelled by considering that the vapour phase transported to the substrate initially condenses in an amorphous state

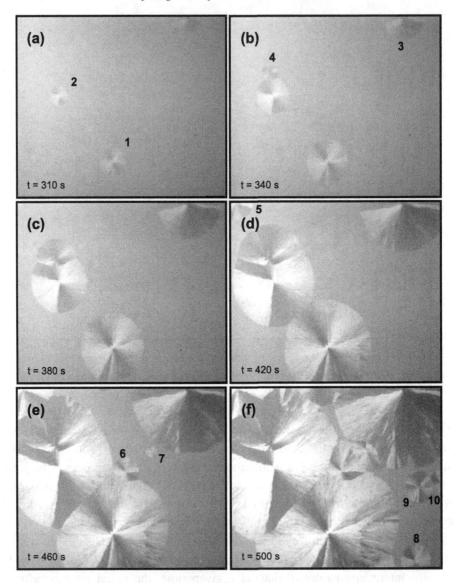

Figure 5.8. Time evolution of the 2D spherulitic crystallization of a thin film of *p*-NPNN during evaporation. In each picture (2.65 × 2.20 mm², polarizers not completely crossed in order to avoid excessive loss of light) the deposition time is indicated. The 2D spherulites discussed in the text are labelled by order of formation. Reprinted from *Journal of Crystal Growth*, Vol. 209, J. Caro, J. Fraxedas and A. Figueras, *Thickness-dependent spherulitic growth observed in thin films of the molecular organic radical p-nitrophenyl nitronyl nitroxide*, 146–158, Copyright (2000), with permission from Elsevier. Details on the determination of the origin of the time scale ($t = 0$) can be found in the original article.

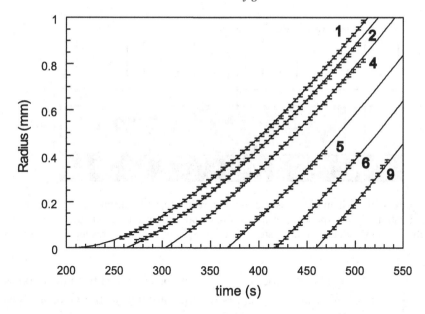

Figure 5.9. Time evolution of the radii of selected 2D spherulites from Fig. 5.8. The curves correspond to $R(t)$ obtained by integration of Eq. (5.9). Reprinted from *Journal of Crystal Growth*, Vol. 209, J. Caro, J. Fraxedas and A. Figueras, *Thickness-dependent spherulitic growth observed in thin films of the molecular organic radical p-nitrophenyl nitronyl nitroxide*, 146–158, Copyright (2000), with permission from Elsevier.

up to a certain critical thickness, h_c, and that the film crystallizes with spherulitic morphology when the increasing film thickness overcomes h_c. The condensed phase corresponds to a positive balance between the incoming (evaporation) and the outgoing (desorption) material.

The observed non-linear radial growth of the spherulites can be described in terms of a thickness-dependent growth law given by the expression:

$$v_R(t) = v_\infty \left\{ 1 - \frac{h_c}{h(t)} \right\}, \tag{5.9}$$

where v_∞ is the asymptotic value of the radial growth velocity when the film thickness $h(t) \to \infty$. The dependence of v_R on $h(t)$ given by Eq. (5.9) can be derived by taking into account the contribution of the interfacial energies to G. Equation (5.9) can be derived assuming that v_R has a linear dependence on the chemical potential change, $\Delta\mu$, involved in the phase transformation:

$$v_R = K_c \frac{\Delta\mu}{k_B T}, \tag{5.10}$$

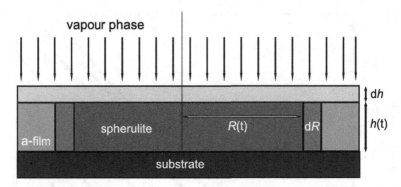

Figure 5.10. Schematic cross-sectional view of a thin film during spherulitic crystallization. The perpendicular axis corresponds to a rotation axis conferring cy-
clindrical symmetry on the system. The amorphous film is represented by a-film.

where K_c stands for a kinetic coeffcient. $\Delta\mu$ can be evaluated from the expression
$\Delta\mu = -\Omega^{-1}\mathrm{d}G/\mathrm{d}N$, where $\mathrm{d}G$ represents the variation of the G of the system
when a number $\mathrm{d}N$ of molecules transfer from the nutrient (mother) phase to the
crystal.

In the following derivation we will assume an almost complete wetting of the
substrate by the material, in such a way that a continuous amorphous condensed
film is formed at a thickness h smaller than the critical size of nucleation. In order to
evaluate $\mathrm{d}G/\mathrm{d}N$ of the process of incorporation of molecules from the amorphous
condensed film to the spherulite, that is the ordered phase, we will hypothesize that
the thickness of the amorphous film increases linearly with time, $h(t) = v_h t$, where
the velocity v_h is a constant, and that the spherulite has a cylindrical shape of radius
R and height h, as illustrated in Fig. 5.10.

The height and growth rate from the vapour phase are assumed to be identical
for both amorphous and crystalline phases. Taking into account only the in-plane
radial growth resulting from the lateral incorporation of amorphous material, the
number $\mathrm{d}N$ of molecules incorporated into the crystalline spherulite contained in a
volume $\mathrm{d}V$ is $\Omega^{-1}2\pi Rh(t)\,\mathrm{d}R$, where $\mathrm{d}R$ represents the increase of the radius and
Ω stands for the molecular volume. The increase in the spherulite volume involves
an increase in both the basal, A_1, and perimeter, A_2, interface areas: $\mathrm{d}A_1 = 2\pi R\,\mathrm{d}R$
and $\mathrm{d}A_2 = 2\pi h\,\mathrm{d}R + 2\pi R\,\mathrm{d}h$. The incremental increase of height is represented
by $\mathrm{d}h$. The G variation of the system, $\mathrm{d}G$, is given by:

$$\mathrm{d}G = -\Delta\mu_0\,\mathrm{d}V + \Delta\gamma_\parallel\,\mathrm{d}A_1 + \gamma_{\perp\mathrm{ca}}\,\mathrm{d}A_2, \qquad (5.11)$$

where $\Delta\mu_0 = \mu^a - \mu^c$ and $\Delta\gamma_\parallel = \gamma_{\parallel\mathrm{cv}} + \gamma_{\parallel\mathrm{cs}} - \gamma_{\parallel\mathrm{av}} - \gamma_{\parallel\mathrm{as}}$. μ^a and μ^c stand for
the chemical potential of the amorphous and crystalline phases, respectively, and
$\gamma_{\parallel\mathrm{cv}}, \gamma_{\parallel\mathrm{cs}}, \gamma_{\parallel\mathrm{av}}, \gamma_{\parallel\mathrm{as}}$ and $\gamma_{\perp\mathrm{ca}}$ represent the isotropic specific interfacial free energies

of the crystal/vapour, crystal/substrate, amorphous/vapour, amorphous/substrate and crystal/amorphous interfaces, respectively.

The first term in Eq. (5.11) reflects the gain in the bulk energy while the second term accounts for the variation in the total free energy associated with the replacement of the substrate/amorphous and vapour/amorphous interfaces (dA_1) by the substrate/crystal and vapour/crystal interfaces. The last term represents the increase in the total free energy due to the increase in the crystal/amorphous interface (dA_2). Taking into account Eq. (5.11) and the expressions for dV, dA_1 and dA_2 given above, dG/dN can be expressed as:

$$\frac{1}{\Omega}\frac{dG}{dN} = -\Delta\mu_0 + \frac{1}{h(t)}\left\{\Delta\gamma_\| + \gamma_{\perp ca}\frac{dh}{dR}\right\} + \frac{\gamma_{\perp ca}}{R(t)}$$

$$= -\Delta\mu_0 + \frac{1}{h(t)}\left\{\Delta\gamma_\| + \gamma_{\perp ca}\frac{v_h}{dR/dt}\right\} + \frac{\gamma_{\perp ca}}{R(t)}. \quad (5.12)$$

Taking into account that $R(t) \gg h(t)$ and that $dR/dt \gg v_h$, Eq. (5.12) can be approximated by:

$$\frac{1}{\Omega}\frac{dG}{dN} = -\Delta\mu_0 + \frac{\Delta\gamma_\|}{h(t)} = -\Delta\mu. \quad (5.13)$$

Combining Eqs. (5.10) and (5.13) we obtain:

$$v_R = \frac{K_c}{k_B T}\left\{\Delta\mu_0 - \frac{\Delta\gamma_\|}{h(t)}\right\} \quad (5.14)$$

and defining $v_\infty \equiv K_c\Delta\mu_0/k_B T$ and $h_c \equiv \Delta\gamma_\|/\Delta\mu_0$, Eq. (5.9) is obtained. Note that Eq. (5.9) is only valid for $h(t) \geq h_c$. Finally, the mathematical expression for $R(t)$ can be derived from Eq. (5.9) by integration. In this case:

$$R(t) = \int v_R(t)dt = v_\infty(t - t_c) - v_\infty t_c \ln\left\{\frac{t}{t_c}\right\}, \quad (5.15)$$

where t_c is defined as $t_c \equiv h_c/v_h$. Note that $R(t_c) = 0$. Figure 5.9 shows the fit to the experimental points obtained with Eq. (5.15). The predicted curves fit well the experimental points. The effects of non-linearity are more evident for the spherulites forming first because the factor $1 - h_c/h$ is rather small, while for spherulites appearing later $h_c/h \ll 1$, so that the velocity nearly reaches an asymptotic limit and the growth is thus approximately constant. h_c has been estimated from TMAFM measurements as 35 nm. This value is of the same order as those found for thin films of polymethine dye salts, $h_c \sim 30$ nm (Dähne, 1995), as previously discussed.

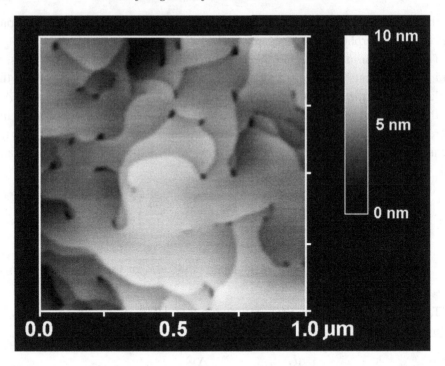

Figure 5.11. TMAFM image of a thin film (thickness ~1 μm) of α-p-NPNN on NaCl(100): 1 μm × 1 μm. Reprinted from *Surface Science*, Vol. 415, J. Fraxedas, J. Caro, A. Figueras, P. Gorostiza and F. Sanz, *Dislocation hollow cores observed on surfaces of molecular organic thin films: p-nitrophenyl nitronyl nitroxide radical*, 241–250, Copyright (1998), with permission from Elsevier.

5.5 Nanometre-scale surface morphology

The nanometre-scale morphology of the surfaces of the films can provide valuable information related to the growth mechanisms. Figure 5.11 shows a TMAFM image, measured at ambient conditions, of a 1 μm thin film of p-NPNN grown by PVD on an *ex situ* cleaved NaCl(100) substrate. Such p-NPNN/NaCl(100) films are also highly oriented, as when grown on glass substrates, with the molecular ab-planes parallel to the substrate's (100) planes (Caro *et al.*, 1998). The image reveals a random distribution of dislocations. These dislocations are spirals associated with screw dislocations coupled in pairs of opposite sign, each spiral emerging from a hollow core.

Hollow cores (black filled circles in the image) are empty tubes located at the dislocation core and they originate from the stress field generated during growth. F. C. Frank predicted the formation of such hollow cores along dislocation lines with Burger's vectors Λ with moduli larger than 1 nm (Frank, 1951). As you will soon see, he was right, for the height of the steps is 1.2 nm, the distance between two adjacent *ab*-planes. In the image, individual hollow cores are not found because the dislocation density is rather high ($\sim 3 \times 10^9$ cm^{-2}), which induces the interaction or coupling of spirals due to their close proximity. The origin of the high density of dislocations is due to the way in which the films grow (rapid in-plane crystallization as discussed in the previous section), under the given experimental conditions, rather than to an intrinsic property of the molecular solid. In fact it is well established that small distances between spirals of opposite sign, or equivalently a high density of dislocations, can be obtained if stress is present at some stage of the formation of the solid (Sutton & Balluffi, 1995).

According to F. C. Frank (Frank, 1949), growth from the vapour (or solution) of crystals of highly non-equiaxed organic molecules proceeds through the formation of an adsorption film differing in molecular orientation from the bulk crystal (liquid or liquid-crystalline) with a subsequent rearrangement. If the transition (or rearrangement) to the solid phase occurs at an elevated rate, the density of defects increases and the so-formed defects may act as spiral centres. Hollow cores have also been observed for PVD-grown thin pentacene films (Nickel *et al.*, 2004). The measured dislocation densities of 10–20 nm thick films grown on OTS-terminated silicon wafers, a silicon oxide surface and on H-terminated silicon by PVD are $0.5 \times 10^{11}, 0.9 \times 10^{11}$ and 2.1×10^{11} cm^{-2}, respectively. The films were prepared with $D_t = 1.25 \times 10^{-3}$ nm s^{-1} keeping T_{sub} at RT and with a background pressure of $\sim 1 \times 10^{-7}$ mbar.

Two spirals can be coupled only if they have opposite signs, because in this case they both define a common terrace. This is topologically not possible when two spirals have the same sign, because in this case no common step can be formed. In Fig. 5.12 a detailed example of two interacting spirals is illustrated. Here the coupling of pairs of screw dislocations of opposite sign is clearly observed, thus showing the out-of-plane mechanism of growth, following the BCF model referred to at the beginning of this chapter (Burton *et al.*, 1951). Once *p*-NPNN molecules reach the surface, they diffuse on it and those that do not re-evaporate interact with a step. In this way the step advances by incorporation of material into the ledge and the step tends to spiral around both static dislocations. This mechanism of growth is known as the Frank–Read source (Frank & Read, 1950) and constitutes a continuum way of growth, implying that no further surface nucleation is needed.

The measured step heights of the interacting spirals from the images shown in Figs. 5.11 and 5.12 are 1.2 nm, as pointed out before. From both figures we observe that either one step (single spiral) or two steps (double spiral) emerge from the hollow cores but never more than two spirals. Single spirals are associated with circular hollow cores with mean radii $R_{hc} \sim 7$–10 nm, whereas double spirals emerge normally from asymmetrically shaped hollow cores (elliptic) and when they appear circular (possibly due to insufficient lateral resolution), the measured radii

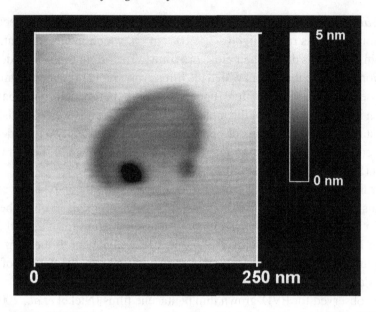

Figure 5.12. TMAFM image of a thin film (thickness ~1 μm) of α-*p*-NPNN on NaCl(100): 250 nm × 250 nm. Reprinted from *Surface Science*, Vol. 415, J. Fraxedas, J. Caro, A. Figueras, P. Gorostiza and F. Sanz, *Dislocation hollow cores observed on surfaces of molecular organic thin films: p-nitrophenyl nitronyl nitroxide radical*, 241–250, Copyright (1998), with permission from Elsevier.

are about 15–20 nm.[1] This is true for most of the spirals, but double spirals emerging from hollow cores with radii of about 7–10 nm and single spirals emerging from hollow cores with radii of about 15–20 nm are not uncommon.

As mentioned before and assuming the validity of the continuum elasticity theory at the dislocation core, F. C. Frank derived the expression for the characteristic radius of a hollow core (Frank, 1951):

$$R_F = \frac{S_M \Lambda^2}{8\pi^2 K \gamma},\tag{5.16}$$

known as the Frank radius, where S_M is the shear modulus of the crystal, γ the interfacial tension and K a constant that determines the character of the dislocation ($K = 1$ for screw dislocations). A modified version of this model (van der Hoek *et al.*, 1982) introduces a strain energy function $\tilde{u}(R)$ in order to account for the deviation at the core from the continuum theory. According to this more realistic model, a hollow core will be thermodynamically stable at a radius $R = R_{hc}$ if the

---

[1] The radius of the tip used for the TMAFM images was in the 5–10 nm range, thus the obtained R_{hc} values may in fact be smaller than 7–10 nm.

free enthalpy of the system has a minimum. In this case:

$$\left.\frac{dG}{dR}\right|_{R_{hc}} = 2\pi\gamma\left\{\frac{R}{R_c} + 1 - \frac{R\tilde{u}(R)}{\gamma}\right\} = 0, \tag{5.17a}$$

$$\left.\frac{d^2G}{dR^2}\right|_{R_{hc}} = 2\pi\gamma\left\{\frac{1}{R_c} - \frac{\tilde{u}(R)}{\gamma} - \frac{R}{\gamma}\frac{d\tilde{u}}{dR}\right\} > 0, \tag{5.17b}$$

where

$$\frac{\tilde{u}(R)}{\gamma} = \frac{R_F}{R_H^2}\frac{1}{1 + R^2/R_H^2}. \tag{5.18}$$

Using Eqs. (5.17a) and (5.17b) and assuming that R_{hc}, R_c and R_F are all positive, the following expression can be derived:

$$2R_{hc}\left\{1 + \frac{R_{hc}}{R_c}\right\}^2 > R_F > R_{hc}\left\{1 + \frac{R_{hc}}{R_c}\right\}. \tag{5.19}$$

From this expression we observe that $R_F > R_{hc}$. The terrace connecting two spirals, one left-handed and the other right-handed, will grow indefinitely if the diameter of the critical 2D nucleus, $2R_c$, is less than the distance between the emerging points of the two spirals (Frank, 1949). The minimum measured separation between coupled spirals is about 50 nm, so that $R_c < 25$ nm (see Section 5.1). Asymmetric hollow cores can be clearly seen in Fig. 5.13 and are in fact composed of two single hollow cores *of the same sign* separated by about $2R_{hc}$, as can be inferred from the figure.

The step height of each spiral is the modulus of Λ, and the superposition of the spirals, even if they originate from two different hollow cores, is probably facilitated by the layered structure of the α-phase. Positive/negative sign of curvature means that, after the emergence of the spiral, it maintains/changes the sign of rotation. Figure 5.12 is an example of spirals with positive sign of curvature: one right-handed and one left-handed. In fact, the Frank–Read mechanism of growth, which governs the evolution of the spirals, forces the change of sign of the curvatures of the spirals upon the advance of the step. The positive sign of curvature illustrated in Fig. 5.12 will turn to negative after the two hollow cores are connected by a straight terrace. With the advance of the step, the spirals will have negative sign (see e.g., centre of Fig. 5.11) which will turn back to positive when the situation shown in Fig. 5.12 is restored. In Fig. 5.13 we find examples of two spirals emerging from hollow cores, each with a different curvature.

Changes in curvature are related to the high elasticity of the material as well as to its layered structure. The high elasticity is suggested by several examples of screw dislocations reaching step heights of 1.2 nm very rapidly, that is very close to the hollow cores (about 30–40 nm away from the centre of them). At larger distances the steps adopt different shapes. This is due to the strong decay of the

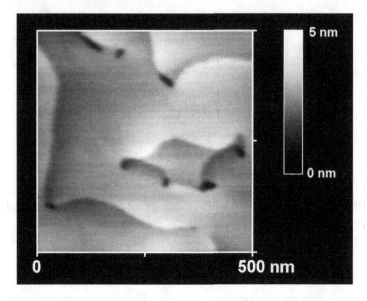

Figure 5.13. TMAFM image of a thin film (thickness ~ 1 μm) of α-p-NPNN on NaCl(100): 500 nm $\times$ 500 nm. Reprinted from *Surface Science*, Vol. 415, J. Fraxedas, J. Caro, A. Figueras, P. Gorostiza and F. Sanz, *Dislocation hollow cores observed on surfaces of molecular organic thin films: p-nitrophenyl nitronyl nitroxide radical*, 241–250, Copyright (1998), with permission from Elsevier.

stress field accumulated in the hollow cores at distances above R_{hc}, which is in turn a consequence of the weak intermolecular interactions and of the low rotational barrier of the radical around the carbon–carbon bond that bridges the nitrophenyl with the nitronyl nitroxide groups.

The decay of the stress field can be approximated by the following expression:

$$\frac{\tilde{u}(n R_{\text{hc}})}{\tilde{u}(0)} < \frac{1}{1 + n^2/(1 + 2R_{\text{hc}}/R_{\text{c}})}, \tag{5.20}$$

where $\tilde{u}(0)$ and $\tilde{u}(n R_{\text{hc}})$ represent the strain energy density in the centre and at a distance $n R_{\text{hc}}$ from the hollow core, respectively, and $n \in \mathbb{N}$.[2] In the case of α-p-NPNN $R_{\text{hc}} \sim 7$–10 nm and $R_{\text{c}} < 25$ nm, so that if we assume $R_{\text{hc}}/R_{\text{c}} \sim 1$, at a distance $n R_{\text{hc}} \sim 40$–50 nm ($n \sim 5$) from the centre of the hollow core, we obtain $\tilde{u}(n R_{\text{hc}}) < \tilde{u}(0)/9$, thus a drastic reduction at a short distance. Compared to an inorganic material such as $YBa_2Cu_3O_{7-\delta}$, $R_{\text{hc}} \sim 330$ nm and $R_{\text{c}} \sim 23$ nm for thin films (Klemenz, 1998) so that at distances as large as e.g., 2300 nm the stress field is only reduced by a factor 1/3, which is manifested by regular shapes of the steps at such large distances.

---

[2] This expression can be derived from Eqs. (5.17a), (5.17b) and (5.18) assuming that $R_{\text{F}} > 0$, $R_{\text{c}} > 0$ and $R_{\text{hc}} > 0$ ($R_{\text{H}}^2 > 0$) and replacing R by $n R_{\text{hc}}$.

In general, the shape of the step connecting two hollow cores at $R \gg R_{hc}$ shows a certain degree of modulation as the distance between them increases. The observed step shapes are due to thermal fluctuations, where molecules flow along the steps (periphery diffusion), move from steps to terraces and then reattach to the step at different sites (terrace diffusion), and move between the steps and terraces with no correlation between motion at different sites (attachment–detachment). Steps can in fact be viewed as 1D interfaces or massless strings that can vibrate with any wavelength larger than the molecular scale. The restoring force that keeps step fluctuations from growing indefinitely is the free energy cost of increasing the step length, which is governed by the step stiffness. Several examples of steps with node-like shape are observed, such as those given in Fig. 5.13. A detailed discussion of the properties of steps in thermal equilibrium (for inorganic materials) can be found in Jeong and Williams (1999).

Figures 5.11, 5.12 and 5.13 were acquired by reducing the AFM cantilever oscillation amplitude to a minimum in order to avoid tip-induced step movement while maintaining good image contrast. If the cantilever oscillation amplitude is intentionally increased the surface becomes slightly perturbed. Under these conditions tip-induced step movement is observed as manifested by the unwinding of the spirals, as illustrated in Fig. 5.14. Here is shown the time evolution of TMAFM images of the surface of a thin film of α-p-NPNN taken at ambient conditions. Tip-induced unwinding of screw dislocations has been reported for spirals in the layered inorganic material α-HgI$_2$ (Lang *et al.*, 1996). In this case etching depends on the mechanical contact between the tip and the α-HgI$_2$ step edges leading to vaporization of the material.

The unwinding of screw dislocations induces the exchange of spiral centres. Spiral centres necessarily exchange among each other at every turn because of their close proximity. The tip-induced etching observed in Fig. 5.14 can be viewed as a time-reversed growth. The tip removes material from the steps, so that the area of the terraces defined by two pairs of coupled spirals is reduced and as a consequence 2D menisci are formed, as illustrated in Fig. 5.14(a). All menisci shown in Fig. 5.14 have a height of 1.2 nm. The thinning of a meniscus down to its collapse can be clearly followed in the sequence Fig. 5.14(a–c). The neck dimension decreases down to *c.* 10–15 nm, where the meniscus breaks. This value is possibly limited by the tip diameter.

Figure 5.14(c) shows the case of the rupture of the meniscus during scanning. After rupture the remaining steps are highly unstable and relax by reducing their curvature (Fig. 5.14(d)). The relaxation is an intrinsic effect and involves sublimation of material, simply triggered by the tip, and is a consequence of the reduction of the line tension γ_l. The example of a 2D island of a given area can help in understanding this effect. The associated line tension has an energy $\gamma_l L_b$,

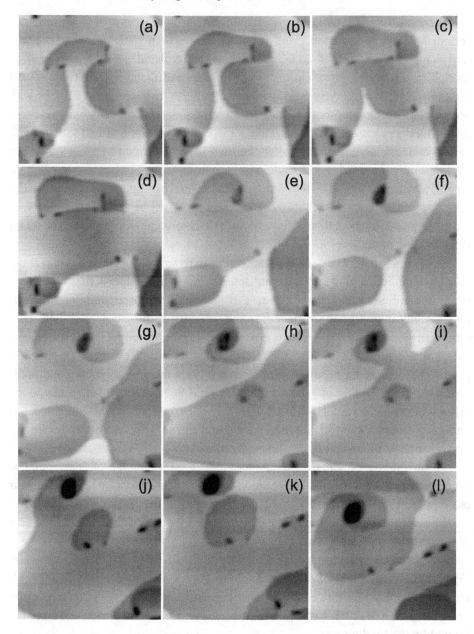

Figure 5.14. Time evolution of TMAFM images of a thin film (thickness ~1 μm) of α-p-NPNN on NaCl(100) taken at ambient conditions: 500 nm × 500 nm. The acquisition time is 87 s per image.

where L_b represents the length of the boundary and becomes minimized by shapes having the smallest perimeter consistent with the prescribed area (e.g., a circle). The menisci discussed here cannot be considered as a model 2D liquid (think of a 3D water meniscus) because of the interaction between adjacent molecular planes.

The exchange of spiral centres can be seen in the sequence Fig. 5.14(f–h) in the centre of the images. From two neighbouring hollow cores two spirals of opposite sign emerge and they evolve until they form a common terrace. Again the evolution leads to the formation of a new meniscus (Fig. 5.14(h)), which is driven to collapse (Fig. 5.14(i)) and then relaxation sets in (Fig. 5.14(j)). And so on. You can spend hours looking at this kind of images!

5.6 Polymorphism

We have already discussed in Section 1.6 that single crystals may exhibit different crystallographic phases by modifying the synthesis route or simply by varying the external variables such as temperature and pressure, and the example of TTF-CA will be studied in Section 6.4. We also referred to the apparently surprising situation where some materials show many polymorphs, such as the BEDT-TTF salts, but for some closely related molecules, as in the case of BFS, only recently have new polymorphs been discovered. The preparation of polycrystalline thin films can also be regarded as a synthesis route well-differentiated from other preparation techniques optimized for single crystals. We should thus not be at all surprised when new polymorphs are found when the materials are grown as thin films. However, and because of their relative stability, most of the crystallographic phases coincide with those obtained for single crystals. In this section we concentrate on polymorphic phases that have only been obtained with thin films and that have not (yet?) been reported for single crystals. It is of course not excluded that such phases can in the future be prepared as single crystals.

At this point it is important to realize that from single crystals the full crystallographic information is usually obtained by XRD methods (space group, lattice parameters *and* atomic coordinates) but that for thin films this is hardly the case, because of the physical limitation imposed by the substrate, in particular when the microcrystals are oriented, in which case only a reduced number of reflections are available. Hence only partial information is obtained with conventional diffraction methods. Removing the microcrystals, e.g., by scratching, in order to perform XRD measurements from the powder is not always possible and handling such small and fragile microcrystals with the usual tools is almost impossible.

We saw previously the example of TMTSF thin films where the lattice parameters could be obtained, and proof that the triclinic phase coincides with that found for

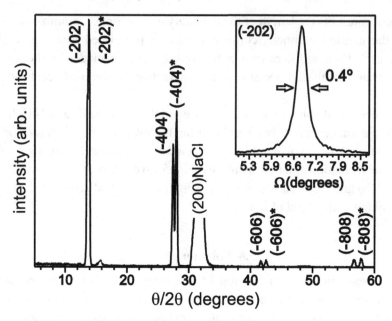

Figure 5.15. XRD pattern of a ~1 μm thick TMTTF/NaCl(100) film grown by PVD. The (200) reflection from the substrate is intentionally only partially shown.

single crystals. No extra reflections could be detected within the accuracy of the experimental system. In the case of thin TTF-TCNQ films grown by PVD the lattice parameters derived from XRD measurements are $a = 1.213$ nm, $b = 0.387$ nm, $c = 1.866$ nm and $\beta = 103.36°$, assuming a monoclinic lattice, in agreement with the known TTF-TCNQ parameters (Caro *et al.*, 1999). However, when the diffraction patterns cannot be compared to those derived from single crystals or powder samples, we can in principle hypothesize the existence of new phases, but keep in mind that they cannot be fully characterized. Let us illustrate all this with some examples, starting with TMTTF, the cousin of TMTSF.

Figure 5.15 shows the $\theta/2\theta$ XRD pattern of a TMTTF/NaCl(100) thin film grown at $T_{sub} = $ RT in HV (~10^{-6} mbar). In addition to the structure given by the $\{\bar{2}02\}$ family of planes, $(\bar{2}02)$ plane and homologues that coincide with the known monoclinic structure found in single crystals grown from solution ($C2/c$, $a = 1.614$ nm, $b = 0.606$ nm, $c = 1.428$ nm, $\beta = 119.974°$), we observe new diffraction reflections that are tentatively assigned as corresponding to the $\{\bar{2}02\}^*$ family, because they are compatible with a reduction of the distance between $(\bar{2}02)$ planes of about 2%. Note the different behaviour between TMTTF and TMTSF for the same kind of substrates. As mentioned before, and due to the experimental geometry used and to the high degree of orientation of the micro-crystals, the crystallographic information is limited. We can only conjecture on the presence of two phases: the well-known monoclinic one and an additional phase closely related to it. This new phase is welcome considering the fact that TTF exhibits at least two polymorphs as we saw in Section 1.6.

XRD patterns of TTF-TCNQ films grown by CVD on Si(100) substrates also show this kind of extra reflections (de Caro *et al.*, 2000a). This is perhaps the first evidence, albeit incomplete and thus questionable, of a new crystallographic phase of TTF-TCNQ. The conclusive observation of a new phase of TTF-TCNQ, e.g., with mixed-stacked structure as for the red phase of TMTSF-TCNQ, would be extremely interesting. The known and newly observed structures for both TMTTF and TTF-TCNQ might be tentatively ascribed to the thermodynamical and kinematical phases, respectively.

If we use different substrates under identical experimental conditions (samples grown in the same experimental run) we find interesting results that are summarized in Fig. 5.16. The figure shows the $\theta/2\theta$ XRD patterns for TMTTF films grown on three different substrates: (a) OTS films prepared on silicon wafers, (b) LB BaAA films and (c) thermally grown SiO_2. For OTS substrates the $\{\bar{2}02\}$ and $\{\bar{2}02\}^*$ families are clearly resolved and in addition some microcrystals are oriented with the (200) and ($\bar{2}04$) planes parallel to the substrate. What is relevant is that using either BaAA or SiO_2 substrates either the $\{\bar{2}02\}$ or the $\{\bar{2}02\}^*$ families are predominantly selected, respectively. The conclusion is that substrates can not only determine the orientation of the molecules but discriminate polymorphs.

Moving to different materials, for instance pentacene, two different crystallographic phases characterized by different interlayer distances $d_{(100)}$ have been identified for thin films. As mentioned above, the pentacene (100) plane exhibits the lowest-energy surface and therefore the films tend to be oriented with these planes parallel to the substrate surface. The deposition of pentacene on SiO_2 initially leads to the formation of dendritic domains with a substrate-induced structure (so-called thin film phase or α-phase), which is characterized by a layer separation of 1.55 nm (Dimitrakopoulos *et al.*, 1996). When exceeding a critical coverage, the XRD pattern points towards the formation of a second phase (β-phase) with $d_{(001)} = 1.45$ nm, which corresponds to the S-phase discussed in Section 1.6 (the V-phase is also known as the γ-phase). A comparison of the interlayer separation in these two phases, $d_{(100)}$, with the length of the pentacene molecule, ~ 1.6 nm, suggests that the molecules are inclined by different angles with respect to the layer planes. This critical thickness is T_{sub}-dependent, ranging from about 100 nm at RT down to 30 nm for $T_{sub} \simeq 360$ K (Bouchoms *et al.*, 1999). Films deposited at RT consist completely of the α-phase, while the β-phase appears for increasing substrate temperatures.

Pentacene and tetracene films can be oriented by using friction-transferred PTFE polymer substrates (see Section 3.3). The films consist of both crystallographic

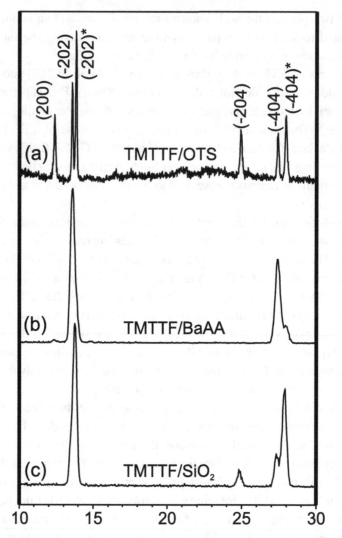

Figure 5.16. XRD patterns of ∼1 μm thin TMTTF films grown by PVD on (a) OTS, (b) BaAA and (c) SiO₂ substrates.

modifications, α- and β-phase, with the bulk (001)-axis oriented normal to the substrate surface. The highest structural quality of the pentacene and tetracene thin films is achieved when the deposition is carried out at $T_{sub} > 343$ K and $T_{sub} > 298$ K, respectively (Brinkmann *et al.*, 2003). Pentacene films deposited at $T_{sub} < 323$ K mainly consist of the α-phase, whereas films grown at $T_{sub} > 323$ K are dominated by the β-form.

In the case of vacuum-deposited α-6T films, X-ray studies based on meridional (00l) reflections show evidence for various crystalline phases depending on D_t

and T_{sub} (Servet *et al.*, 1994). At T_{sub} = RT the long interlayer spacing of films obtained under low D_t (0.1–0.5 nm s^{-1}) is about 2.37 nm, while for fast D_t this distance increases to 2.44 nm, corresponding to the most stable form (α-phase) and to the metastable kinetically favoured (β-phase), respectively. When $T_{sub} \simeq 463$ K the interlayer spacing is 2.24 nm, indicating a different structure, which has been designated as the γ-phase. At temperatures near the bulk melting point (about 560 K) additional reflections suggest the appearance of a new phase, the δ-phase, with a long spacing of 3.63 nm.

The deposition of DIP on atomically smooth SiO$_2$ surfaces at $T_{sub} \simeq 418$ K leads to the formation of organic thin films with an exceptionally high degree of molecular ordering (Dürr *et al.*, 2003a). Films 40 nm thick exhibit a layered structure with a characteristic layer spacing of 1.65 nm. The value of this distance suggests an upright orientation of the DIP molecules within the layers. A comparison with the length of the molecules of 1.84 nm yields an estimation of the tilt angle of 15–20° relative to the substrate normal. This thin-film phase is referred to as the σ-phase. When grown on polycrystalline gold surfaces the DIP films crystallize in a second phase (λ-phase). This polymorph is characterized by an interlayer distance of about 1.6 nm. In this case the tilt angle between the DIP molecular axis and the crystal planes is found to be 17°. In contrast to the situation for the SiO$_2$ substrate discussed above, the DIP molecules are oriented with their long axes almost parallel to the gold surface.

p-6P films deposited on isotropic substrates such as glass or ITO-coated glass reveal the existence of different growth regimes. When the deposition is carried out at $T_{sub} < 433$ K, the known bulk-like β-phase is found, whereas at $T_{sub} > 433$ K a new phase, denoted as the γ-phase, is observed. This monoclinic γ-phase is characterized by the lattice parameters $a = 0.798$ nm, $b = 0.554$ nm, $c = 2.724$ nm and $\beta = 99.8°$ (Resel, 2003).

Concerning Pc-based films, new crystalline phases have been suggested for H$_2$Pc (Heutz & Jones, 2002) and TiOPc (Brinkmann *et al.*, 2002a). In this last case the phase is triclinic, with space group $P\bar{1}$, and with lattice parameters $a = 1.40$ nm, $b = 1.40$ nm, $c = 1.06$ nm, $\alpha = 109.2°$, $\beta = 133.9°$ and $\gamma = 90°$. Recall here the case of CuPc thin films grown in microgravity briefly described in Section 3.2. When grown in the Space Shuttle Orbiter in low Earth orbit, CuPc crystallizes in the M-phase, a phase that has never been obtained under unity gravity (Debe & Kam, 1990).

Although the detailed information on the polymorphic phases has to be obtained with e.g., diffraction and scanning probe methods, our eyes are attracted by the beauty of optical microscope images often encountered with polymorphs under transformation. Figure 5.17 shows an optical microscopy image of a *p*-NPNN thin film (thickness ~2 μm) on a glass substrate, exhibiting a transformation at

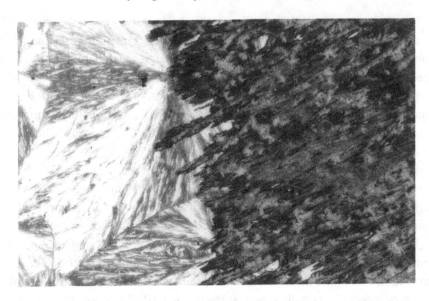

Figure 5.17. Optical microscopy image (1.7 mm × 1.1 mm, crossed polarizers) of a partially transformed thin film of *p*-NPNN grown on a glass substrate: α-phase (left) and β-phase (right). Reprinted with permission from J. Fraxedas, J. Caro, J. Santiso, A. Figueras, P. Gorostiza and F. Sanz, *Europhysics Letters* 1999, Vol. 48, 461–467.

ambient conditions from the original as-deposited isotropic spherulitic morphology (α-phase, left) to an anisotropic fibrilar and thus rougher morphology (β-phase, right).

To conclude this chapter, which has been devoted to the growth of *thick* films, the sequence of optical microscopy images as a function of time corresponding to a *p*-NPNN/glass film is shown. The time evolution is illustrated in Fig. 5.18 revealing the transformation in ambient conditions. The mean transformation velocity at ambient conditions along the directions defined by the fibres is $6.0 \pm 0.4 \ \mu m \ h^{-1}$.

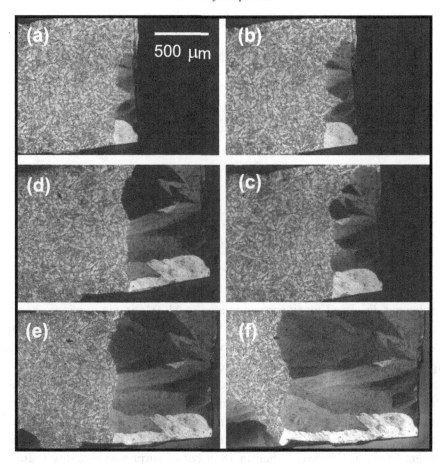

Figure 5.18. Optical microscope images (crossed polarizers) of the evolution of a transforming *p*-NPNN thin film grown on a glass substrate. The images were taken (a) 144, (b) 168, (c) 216, (d) 312, (e) 382, (f) 408 h after first exposure of the as-grown film to the atmosphere. Molas *et al.*, 2003. Reproduced by permission of the Royal Society of Chemistry.

6

A miscellany of physical properties

Forget it! Forget it! Everything I write is just so much bleating!

Gary Larson, *Last Chapter and Worse*

After having discussed many aspects related to MOMs in the previous chapters, from the synthesis of the building blocks, the molecules, to the assembling of such molecules in the solid state in the form of single crystals and thin films, this last chapter is devoted to the physical properties of MOMs based on selected examples. The reader should be aware that this selection, being personal and thus inevitably partial, does not cover the plethora of examples that can be found in the literature. I assume the risk of overlooking some important works. In order to apologize in advance for not including several important investigations I reproduce the wise comment from M. Faraday written in 1859 but of surprising actuality (Faraday, 1859):

I very fully join in the regret . . . that scientific men do not know more perfectly what has been done, or what their companions are doing; but I am afraid the misfortune is inevitable. It is certainly impossible for any person who wishes to devote a portion of his time to chemical experiment, to read all the books and papers that are published in connection with his pursuit; their number is immense, and the labour of winnowing out the few experimental and theoretical truths which in many of them are embarrassed by a very large proportion of uninteresting matter, of imagination, and of error, is such, that most persons who try the experiment are quickly induced to make a selection in their reading, and thus inadvertently, at times, pass by what is really good.

I wonder how many scientific journals were published in the middle of the nineteenth century. Certainly, not as many as nowadays. That's our present time paradox: the more information we can access the less time we have to process it. Or to express it in other words, scientists are led to spend more time writing than reading!

242

In MOMs dimensionality is a major issue. As discussed back in Chapter 1, although all materials are structurally 3D, some of them exhibit physical properties with lower dimensionality, 1D or 2D, mainly due to the pseudo-planar conformation of the molecules. In fact for bulk materials one cannot strictly use the terms 1D or 2D because intermolecular interactions build anisotropic but indeed 3D networks. Hence, one is led to using the prefixes pseudo or quasi when referring to 1D or 2D systems. However, ideal 1D and 2D systems can be artificially prepared exhibiting real 1D and 2D properties, respectively, and we will find some examples of this in the next sections.

6.1 1D molecular metals

Electronic structure of 1D metals

In order to study the electronic structure of 1D metals and compare it to the simple model discussed in Section 1.7, we start with the reference system Au/Si(111)-(5×1), an ideal artificial inorganic 1D metal patterned by a vicinal surface. We shall see here how the simple Fermi liquid scenario breaks down (Segovia *et al.*, 1999). In this system gold atoms build linear chains along the substrate [1$\bar{1}$0] direction, with a gold–gold spacing of 0.383 nm within the chains, and with an interchain spacing of about 2.0 nm, determined by the width of the substrate terraces. This long spacing guarantees negligible interchain interactions and that the system is truly 1D. The preparation of this and similar systems is extremely tedious because the experimental parameters are critical. These samples necessarily have to be prepared and measured *in situ* in UHV, in order to avoid contaminant-induced artifacts. Figure 6.1 shows the ARUPS spectra of Au/Si(111)-(5×1) taken at different emission angles θ_e at $T_{sub} \simeq 12$ K and with linearly polarized monochromatic 21.2 eV photons.

Since the gold chains are located at the surface of the sample, the wave vectors of the electronic states linked to the chains lie in the surface plane. This wave vector, $k_\parallel = \sqrt{2m_e E_K / \hbar^2} \sin \theta_e$, is strictly conserved in the photoemission measurements. The spectra from Fig. 6.1(a) show a band approaching E_F and finally crossing it, demonstrating the metallic character. It is important to notice that no emission from the substrate can be observed at E_F because the substrate is semiconducting.

Figure 6.1(b) displays the 2D band structure $E(k_\parallel)$, fitted to the free-electron parabola $E = \hbar^2 k_\parallel^2 / 2m_e^*$, following the physicists' most simplified approach. Although the bands clearly cross E_F, the intensity of the spectra at E_F, $N(E_F)$, is rather low, a property predicted within the Tomonaga–Luttinger scenario and incompatible with the Fermi liquid model (see discussion in Section 1.7). A closer examination of the ARUPS spectra near E_F reveals the dispersion of two separated

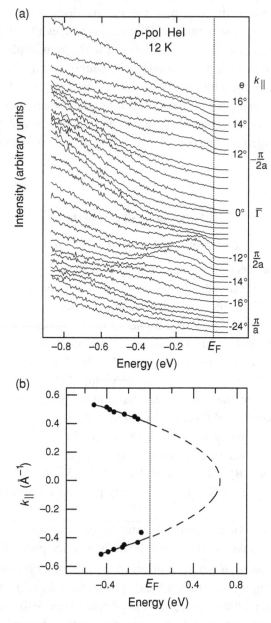

Figure 6.1. (a) ARUPS spectra of Au/Si(111) with $k_\parallel$ along the $[1\bar{1}0]$ chain direction. θ_e is referred to the surface normal. $k_\parallel$ values are given for emission at E_F, with a being the Au–Au spacing within a chain (0.383 nm). E is contained in the emission plane. (b) $E(k_\parallel)$ of the main structure plotted and fitted to a free-electron parabola. Energy is given relative to E_F. Reprinted with permission from Segovia *et al.*, 1999. Copyright (1999) by the Nature Publishing Group.

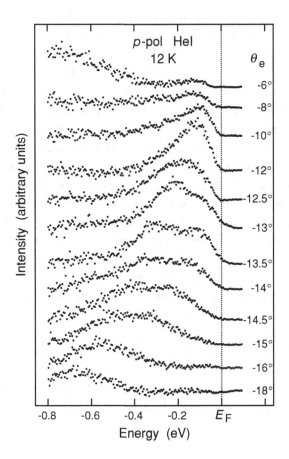

Figure 6.2. Detail of the energy dispersion shown in Fig. 6.1. Energy is given relative to E_F. Reprinted with permission from Segovia *et al.*, 1999. Copyright (1999) by the Nature Publishing Group.

bands as depicted in Fig. 6.2. Both bands have different effective masses and have been interpreted as originating from the separation of spin and charge, another consequence of the electronic correlation in 1D systems. The spinon band would correspond to the higher effective mass value (lower velocity and closer to E_F) while the holon band would correspond to the lower effective mass value (higher velocity). Using Eq. (1.34), even if it is only valid for weakly correlated systems, with $k_F \sim 4 \, \mathrm{nm}^{-1}$, as derived from Fig. 6.1(b) and $a = 0.383$ nm, we obtain that the one-electron band is half occupied ($N_{at} = 1$), clearly indicating a metallic system.

The suppression of the photoemission signal at E_F in inorganic quasi-1D metals has also been observed for several materials such as $TaTe_4$, $NbTe_4$ and $(NbSe_4)_3I$ (Coluzza *et al.*, 1993), $(TaSe_4)_2I$ and $K_{0.3}MoO_3$ (Dardel *et al.*, 1991), $Li_{0.9}Mo_6O_{17}$

(Smith, 1993) and $(DCNQI)_2Cu$ (Inoue *et al.*, 1993). However, one has to be extremely cautious when the photoemission data show that $N(E_F) \to 0$, because, since the photoemission process is extremely surface-sensitive, a 3D metal may exhibit low emission at E_F when the surface and bulk exhibit different stoichiometries. Perhaps the best example to illustrate this problem is $K_{0.3}MoO_3$. It is well-known that transition metal oxides desorb oxygen in UHV when irradiated with X-ray photons with energies greater than the absorption edges of the transition metal core levels. This photon stimulated desorption is due to an interatomic Auger process (Knotek & Feibelman, 1978). In the case of $K_{0.3}MoO_3$ for $\hbar\omega$ larger than the Mo3d threshold energy, $c.$ 230 eV, the surface becomes oxygen-deficient and thus semiconducting, and hence $N(E_F) = 0$. Emission at E_F thus strongly depends on the surface stoichiometry. The case of high-temperature superconductors is also revealing and edifying since the first photoemission experiments performed by 1988 showed an absence of emission at E_F and many articles were produced justifying this experimental fact as a result of strong electronic correlations. Those experiments were performed on polycrystalline samples. However, when single crystals were available a finite intensity at E_F was observed with ARUPS and the superconducting gap could be unambiguously measured (Imer *et al.*, 1989). In conclusion, special care has to be taken with the stoichiometry and morphology, and thus with the quality of the surface.

With these advisory ideas in mind let us spend some time on the electronic structure of one of the most extensively studied MOMs: the quasi-1D metal TTF-TCNQ.

Electronic structure of TTF-TCNQ

Theoretical band structure close to E_F

Because of the molecular nature and the relatively large unit cell, the correct description of the electronic structure of TTF-TCNQ, as for most of the molecular conductors, is still a challenge from the computational viewpoint even with current computational capabilities. Figure 6.3 shows the calculated band structure near E_F ($E_F = 0$) for the RT and ambient pressure structure of TTF-TCNQ (Kistenmacher *et al.*, 1974) along the a^*-(Γ–X), b^*-(Γ–Y) and c^*-(Γ–Z) directions of the reciprocal lattice. The calculations shown here were performed using a numerical atomic orbitals DFT approach (see Appendix B), using the generalized gradient approximation (GGA) with atomic orbitals and pseudopotentials, developed and designed for efficient calculations in large systems and implemented in the SIESTA code (Soler *et al.*, 2002). The predicted band structure is almost identical to those found in Ishibashi & Kohyama (2000) and Sing *et al.* (2003b) despite technical differences.

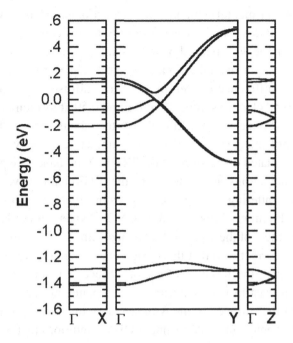

Figure 6.3. Band structure near E_F for the RT and ambient pressure structure of TTF-TCNQ (Kistenmacher *et al.*, 1974). $\Gamma = (0\ 0\ 0)$, $X = (1/2\ 0\ 0)$, $Y = (0\ 1/2\ 0)$ and $Z = (0\ 0\ 1/2)$ in units of the monoclinic reciprocal lattice vectors. Energy is given relative to E_F. Reprinted with permission from J. Fraxedas, Y. J. Lee, I. Jiménez, R. Gago, R. M. Nieminen, P. Ordejón and E. Canadell, *Physical Review B*, **68**, 195115 (2003). Copyright (2003) by the American Physical Society.

The local density approximation (LDA) and GGA within a plane-wave pseudopotential method was used in Ishibashi and Kohyama (2000) while DFT within the linearized augmented plane wave (LAPW) approach was employed in Sing *et al.* (2003b).

In the vicinity of E_F we observe two pairs of bands. The bands pair up at the Y and Z points because there are two symmetry-related chains of each type in the unit cell. The two upper bands at Γ, located at $+0.12$ and $+0.15$ eV, are built from the TTF HOMO. In contrast, the two lower bands at Γ are built from the TCNQ LUMO, and located at -0.08 and -0.20 eV. This 0.12 eV splitting points to the fact that the TCNQ chains are not completely isolated within the solid and interact along the *c*-direction (see Fig. 1.15). The two TTF HOMO bands are practically degenerate ($+0.03$ eV splitting) all over the Brillouin zone except for the region of the weakly avoided crossing around E_F. The bandwidth W of the TTF HOMO and TCNQ LUMO bands along the chain direction are not very different: 0.62 and 0.73 eV, respectively. The value for the TCNQ LUMO bands is in reasonably good agreement with experimental estimates which are around 0.7–0.8 eV. However, the value for the TTF HOMO bands does not seem to follow commonly accepted ideas according to which the dispersion should be smaller than that of the TCNQ LUMO bands. In this case experimental estimates suggest values around 0.4–0.45 eV.[1]

There is also a difference concerning ϱ. X-ray diffuse scattering studies have shown that $\varrho = 0.55$ electrons molec^{-1} at RT and 0.59 at low temperature near the Peierls transition (Pouget, 1988). In contrast, the calculated ϱ at RT is 0.74 and practically does not change with pressure or temperature. This overestimation of ϱ and the larger calculated dispersion for the TTF HOMO bands are perhaps related. Note that $4k_F$ fluctuations associated with the TTF chains have been observed in X-ray diffuse scattering experiments suggesting that electron correlations are important for TTF but less important for TCNQ (Pouget, 1988). However, a shift of the order of 50 meV of the TTF HOMO bands with respect to the TCNQ LUMO would lead to the correct ϱ so that this disagreement is not really significant.

When considering the possible effect of surface structural relaxation on the band structure, at least in what concerns the *ab*-plane, no noticeable changes in ϱ are expected at the surface, in line with scanning tunnelling microscopy results performed at cryogenic temperatures under UHV conditions that will be discussed later (Wang *et al.*, 2003).

A very important advantage of TTF-TCNQ is that its band structure can be compared to those of neutral TTF and TCNQ because both molecules exist as neutral *and* charged species (see Section 2.4). Figure 6.4 shows the band structure and the partial density of states (PDOS) of TTF-TCNQ and Fig. 6.5 the band structures and PDOS of neutral TTF and neutral TCNQ. Note that there are two molecules per repeat unit of the crystal structure for neutral TTF ($a = 0.736$ nm, $b = 0.402$ nm, $c = 1.392$ nm, $\beta = 101.42°$ (Cooper *et al.*, 1971)) but four for neutral TCNQ ($a = 0.891$ nm, $b = 0.706$ nm, $c = 1.639$ nm, $\beta = 98.54°$ (Long *et al.*, 1965)). TTF contains 1D stacks along the *b*-direction where every successive molecule slides along the long molecular axis. These stacks are similar to those found in TTF-TCNQ.

In contrast, neutral TCNQ does not really contain 1D stacks as in TTF-TCNQ but does contain 2D stacks in the *ab*-planes. While the TTF HOMO bands are practically degenerate in TTF-TCNQ, they are not in neutral TTF, e.g., the interchain interactions are larger in the neutral compound. The dispersion of these bands along the chain direction is not very different in TTF and TTF-TCNQ. In contrast, the structural origin of the spreading of the TCNQ LUMO levels is very different in the neutral and the salt compounds, and as a consequence the appearance of the TCNQ LUMO bands is very different in both cases.

---

[1] Computations using the crystal structure at 60 K, just above the Peierls transition, as well as at RT and 4.6 kbar show no relevant changes in the band structure except for a small expected increase in the dispersion along the chain direction (Fraxedas *et al.*, 2003).

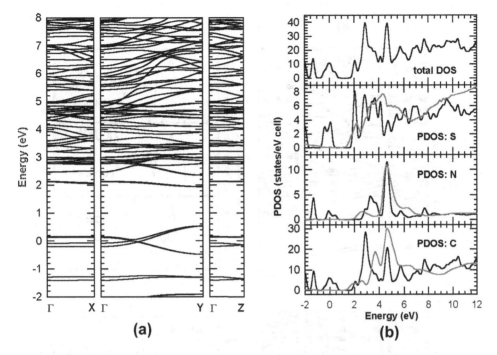

Figure 6.4. (a) Band structure calculated for the RT structure of TTF-TCNQ where $\Gamma = (0\,0\,0)$, $X = (1/2\,0\,0)$, $Y = (0\,1/2\,0)$ and $Z = (0\,0\,1/2)$ in units of the monoclinic reciprocal lattice vectors. (b) Total DOS calculated for TTF-TCNQ and PDOS for the sulfur, carbon and nitrogen atoms (black lines). The S2p, N1s and C1s NEXAFS spectra of TTF-TCNQ (grey lines) are superposed to the PDOS of S, N and C, respectively. Energy is given relative to E_F. Reprinted with permission from J. Fraxedas, Y. J. Lee, I. Jiménez, R. Gago, R. M. Nieminen, P. Ordejón and E. Canadell, *Physical Review B*, **68**, 195115 (2003). Copyright (2003) by the American Physical Society.

Electronic structure of occupied states

The electronic structure of the occupied states of TTF-TCNQ has been experimentally determined in UHV by means of ARUPS for *in situ* cleaved single crystals (Zwick *et al.*, 1998; Claessen *et al.*, 2002) as well as for thin films prepared *ex situ* with PVD (Rojas *et al.*, 2001). Let us start with the ARUPS spectra measured on high-quality single crystals of TTF-TCNQ along the *b*-axis (Γ–Y direction) taken at 150 K with 20 eV monochromatic radiation shown in Fig. 6.6 (Zwick *et al.*, 1998). A prominent feature dispersing symmetrically about the Γ-point ($k_\parallel = 0$) is observed in Fig. 6.6(a), and it approaches E_F. At larger wave vectors, it disperses away from E_F, to a maximum binding energy of 0.75 eV at the Y-point. The spectra measured perpendicular to the chains, along the *a*-axis (Fig. 6.6(b)), do not exhibit any dispersion, evidencing the 1D character. Figure 6.6(c) reproduces spectra from Fig. 6.6(a), at the Γ- and Y-points. According to the spectra the band crossing at E_F

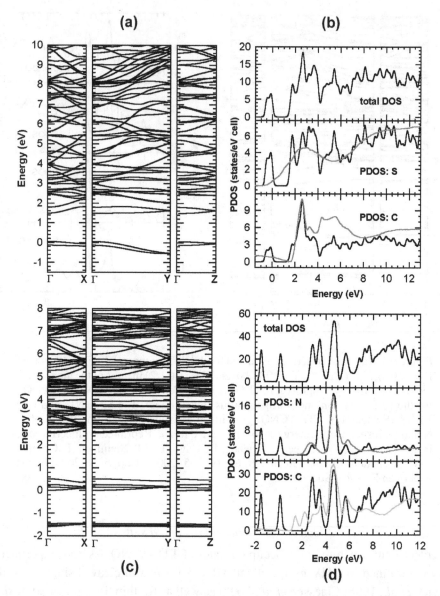

Figure 6.5. (a) Band structure and (b) total DOS calculated for neutral TTF and PDOS for the S and C atoms (black lines). The S2p and C1s NEXAFS spectra of TTF (grey lines) are superposed to the PDOS of S and C, respectively. Energies are referred to the HOMO maximum. (c) Band structure and (d) total DOS calculated for neutral TCNQ and PDOS for the N and C atoms (black lines). The N1s and C1s NEXAFS spectra of TCNQ (grey lines) are superposed to the PDOS of N and C, respectively. Energies are referred to the LUMO minimum. The Γ-, X-, Y- and Z-points are defined as in Fig. 6.4. Reprinted with permission from Fraxedas *et al.*, 2003. Copyright (2003) by the American Physical Society.

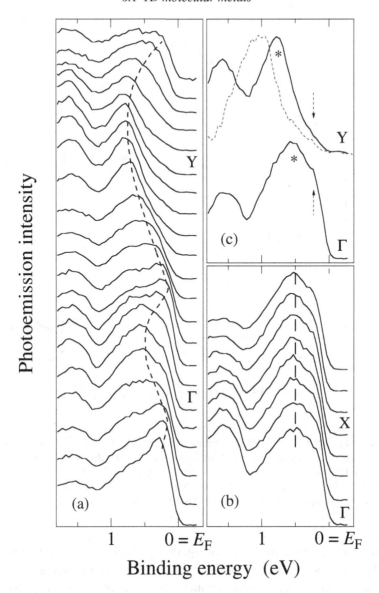

Figure 6.6. (a) ARUPS spectra ($T \simeq 150\,\mathrm{K}, \hbar\omega \simeq 20\,\mathrm{eV}$) of TTF-TCNQ along the chain direction. The dashed line outlines the dispersion of the main spectral feature. (b) ARUPS spectra along the perpendicular a-direction. (c) Selected spectra from (a), after background subtraction. The asterisks mark the main dispersive peak, and the arrows mark emission not accounted for by band theory. Energies are referred to E_F, determined to 1 meV accuracy on an evaporated, polycrystalline silver film. Reprinted with permission from F. Zwick, D. Jérome, G. Margaritondo, M. Onellion, J. Voit and M. Grioni, *Physical Review Letters*, **81**, 2974 (1998). Copyright (1998) by the American Physical Society.

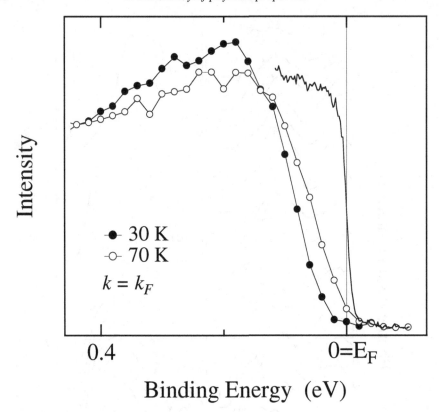

Figure 6.7. High-resolution ARUPS spectra measured at k_F ($\hbar\omega = 21.2$ eV) above (70 K) and below (30 K) the Peierls transition. The solid line spectrum corresponds to E_F, as determined for a silver film. Note that the spectrum follows the shape given by Eq. (1.27). Reprinted with permission from F. Zwick, D. Jérome, G. Margaritondo, M. Onellion, J. Voit and M. Grioni, *Physical Review Letters*, **81**, 2974 (1998). Copyright (1998) by the American Physical Society.

occurs at $k_F = 2.5 \pm 0.2$ nm^{-1}. Using again Eq. (1.34), as for the previously discussed Au/Si(111)-(5 × 1) case, for $N_{at} = \varrho = 0.59$ and $b = 0.382$ nm we obtain $k_F = 2.4$ nm^{-1}, in perfect agreement with the experiment.[2] Thus ϱ can be evaluated from ARUPS experiments.

High-resolution spectra taken at k_F and at 70 and 30 K, just above and below the CDW transition, respectively, are shown in Fig. 6.7. At 70 K the intensity at E_F is vanishingly small, characteristic of 1D systems. Taking the midpoint of the leading edge as an indicator, a pseudogap energy of ∼120 meV is derived. Between 70 and 30 K the leading edge of the spectrum shifts by 20 meV to higher binding energies, indicating the opening of a gap, the Peierls gap, of ∼40 meV.

---

[2] The a parameter from Eq. (1.34) corresponds to the stacking parameter b from TTF-TCNQ.

ARUPS spectra measured in an independent set of experiments on TTF-TCNQ single crystals along the b-axis are shown in Fig. 6.8 (Claessen *et al.*, 2002). Again, clean surfaces parallel to the ab-plane were obtained by *in situ* cleavage of the crystals at a base pressure of 10^{-10} mbar. As can be seen, the data are in excellent agreement with those from Zwick *et al.* (1998). At the Γ-point two peaks at 0.19 (a in Fig. 6.8) and 0.54 eV (b in Fig. 6.8) below E_F are clearly identified. For increasing wave vectors the a feature becomes more prominent up to $k = 2.4$ nm^{-1}, while the b feature seems to split contributing to both the a and d features. For even higher momenta a weak structure (c in Fig. 6.8) moves back again from E_F and displays a dispersion symmetric about $k = 8.7$ nm^{-1}, corresponding to the Z-point. Simultaneously, structure d in Fig. 6.8 disperses to higher binding energy and eventually becomes obscured by peak c. In addition, a dispersionless feature at about 1.5 eV binding energy is also observed. From Fig. 6.8 we obtain $k_F = 2.4$ nm^{-1}, in perfect agreement with the results from Zwick *et al.* (1998) and with the calculations performed using Eq. (1.34).

Astonishingly, the same ARUPS spectra have been observed for *ex situ* grown TTF-TCNQ thin films (Rojas *et al.*, 2001). The films were obtained by thermal sublimation in HV ($\sim 10^{-6}$ mbar) on cleaved KCl(100) substrates and consisted of highly oriented and strongly textured rectangular-shaped microcrystals as shown in the TMAFM image of Fig. 6.9. The molecular ab-planes are parallel to the substrate surface and the microcrystals are oriented with their a- and b-axis parallel to the [110] and [$\bar{1}$10] substrate directions, respectively, due to the cubic symmetry of the substrates. ARUPS spectra taken on the as-received films, measured along the substrate equivalent [100] directions at $T \simeq 100$ K, are shown in Fig. 6.10.

Due to the morphology of the films the bands are mapped along the directions bisecting both a- and b-directions, which does not correspond to any high symmetry direction in reciprocal space. In spite of this, band dispersion is clearly observed. The resemblance of the spectra to those obtained from single crystals is striking and a consequence of the high surface stability of the TTF-TCNQ films in air. Band dispersion is also observed, even with the overlap of both a- and b-directions in the films, because of the shape of the band surface in reciprocal space, which may be approximated by a sheet in E vs. Γ–X, Γ–Y space with $\partial E/\partial k_{a^*} = 0$ and $\partial E/\partial k_{b^*} \neq 0$, where E represents the band energy and k_{a^*} and k_{b^*} stand for the wave vectors along the a^* and b^* directions, respectively. In addition to the pseudogap already observed in the metallic phase in the single crystal samples an extra shift of about 50 meV is observed due to contamination and non-stoichiometry at grain boundaries. Taking into account this 50 meV shift, $k_F \sim 2.5$ nm^{-1} is obtained.

In spite of the fact that (001) surfaces are stable in air, one has to worry about surface contamination, stoichiometry and potential damage induced by VUV

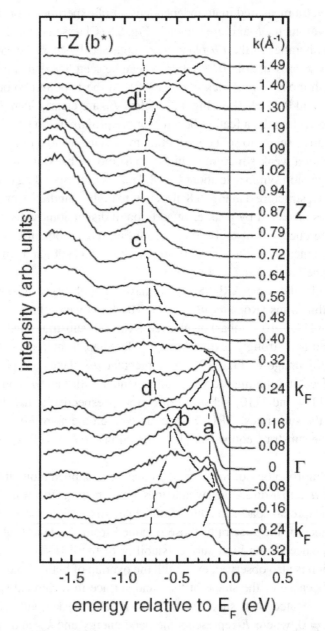

Figure 6.8. ARUPS spectra measured for k along the Γ–Z direction ($T \simeq$ 61 K, $\hbar\omega \simeq 25$ eV). The energy and angular resolution amounted to 60 meV and $\pm 1°$, respectively. The thin lines indicate the dispersion of the spectral features. Reprinted with permission from R. Claessen, M. Sing, U. Schwingenschlögl, P. Blaha, M. Dressel and C. S. Jacobsen, *Physical Review Letters*, **88**, 096402 (2002). Copyright (2002) by the American Physical Society.

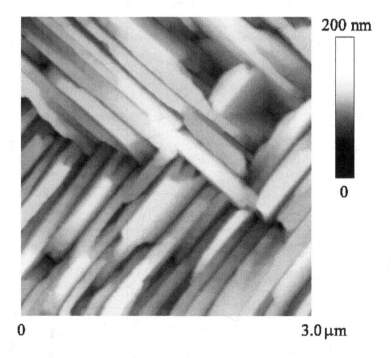

200 nm

0

0   3.0 μm

Figure 6.9. TMAFM image taken at ambient conditions of a TTF-TCNQ thin film (thickness ~1 μm) grown on KCl(100), revealing its polycrystalline morphology. Reprinted from *Surface Science*, Vol. 482–485, C. Rojas, J. Caro, M. Grioni and J. Fraxedas, *Surface characterization of metallic molecular organic thin films: tetrathiafulvalene tetracyanoquinodimetane*, 546–551, Copyright (2001), with permission from Elsevier.

irradiation. In fact great care has to be taken to avoid photon-induced surface damage by minimizing the exposure to the incident radiation, in particular when using synchrotron light. The effect is demonstrated in Fig. 6.11, which shows spectra taken at k_F. For a surface prepared immediately after cleavage of the crystal, an intense peak is observed close to E_F (continuous line). After two hours of exposure to VUV radiation its intensity strongly decreases and its peak position shifts by more than 0.1 eV away from E_F (discontinuous line). However, the original spectrum is recovered by taking data on another previously unexposed sample spot.

Let us now return to the Fermi liquid vs. Luttinger liquid competition. Figure 6.12 shows the negative second energy derivative of the photoemission intensity displayed in Fig. 6.8, as a grey-scale plot in the $E - k$ plane. This mathematical procedure enhances the visibility of the spectral structures and helps to visualize their dispersion. Also shown are DFT calculations of the bands using the GGA (Sing *et al.*, 2003b). The comparison of experiment and theory reveals significant

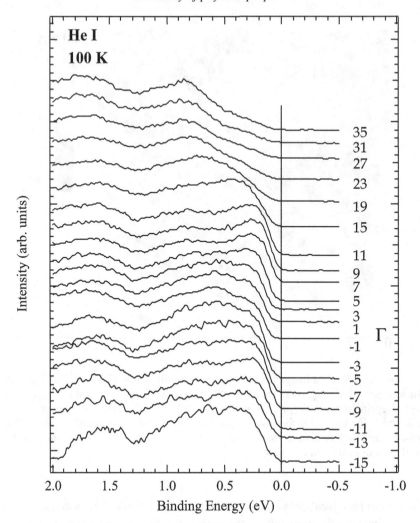

Figure 6.10. ARUPS spectra taken on a TTF-TCNQ thin film at $T \simeq 100$ K and with 21.2 eV photons along the directions bisecting both *a*- and *b*-directions (polar angles θ_e are shown). A parabolic background has been subtracted to enhance the features. Reprinted from *Surface Science*, Vol. 482–485, C. Rojas, J. Caro, M. Grioni and J. Fraxedas, *Surface characterization of metallic molecular organic thin films: tetrathiafulvalene tetracyanoquinodimetane*, 546–551, Copyright (2001), with permission from Elsevier.

discrepancies. According to this comparison we must conclude that band theory is not appropriate to predict the experimental band structure of TTF-TCNQ and, as a consequence, new theoretical approaches should be explored.

One approach consists of considering electronic correlations since it is well known that TTF-TCNQ undergoes instabilities of the $4k_F$ type (see

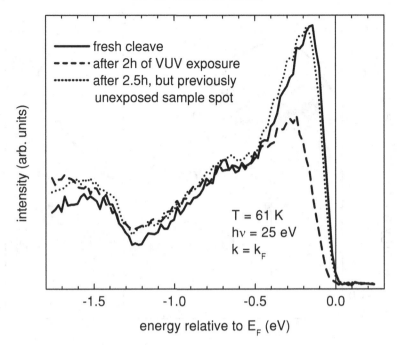

Figure 6.11. ARUPS spectra taken at k_F showing the effect of photon-induced surface degradation ($\hbar\omega \simeq 25$ eV, $T \simeq 61$ K). Reprinted with permission from M. Sing, U. Schwingenschlölgl, R. Claessen, P. Blaha, J. M. P. Carmelo, L. M. Martelo, P. D. Sacramento, M. Dressel and C. S. Jacobsen, *Physical Review B*, **68**, 125111 (2003). Copyright (2003) by the American Physical Society.

discussion in Chapter 1). By using a 1D Hubbard model at finite doping N_{at} (see Fig. 6.13), a U/W ratio of 1.2 with $U = 1.96$ eV and $W = 1.6$ eV for the TCNQ-related features has been derived (Sing *et al.*, 2003b). However, this seems to contradict the accepted idea that electronic correlations are weak for TCNQ (Pouget, 1988).

On the other hand, if we assume that photoemission is able to reveal both charged and neutral states of TTF and TCNQ, in the case where the timescale of the photoemission process is much shorter than that of CT, thus considering the CTSs as dynamical systems with regard to charge, then one could speculate whether the c, d and d′ features from Fig. 6.8 may in fact arise from the HOMOs from *neutral* TTF and TCNQ, as XPS spectra seem to suggest (see Fig. 1.31). From Fig. 6.5 we observe that both HOMOs are located *c.* 0.7–0.8 eV below E_F, assuming that E_F lies halfway between the HOMO–LUMO gaps. This second approach would render band theories valid, except for the suppression of spectral intensity at E_F. Thus, even for TTF-TCNQ, a satisfactory scenario has not yet been achieved. This is clearly an open field that will emerge in the coming years and *old* materials

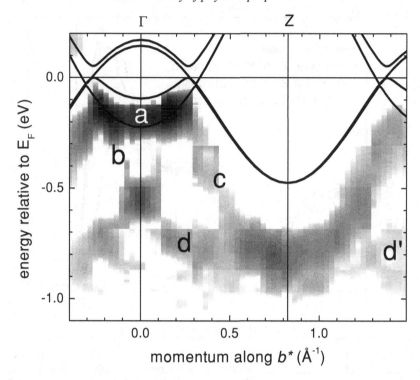

Figure 6.12. Grey-scale plot of the ARUPS in comparison to the band dispersion obtained by DFT calculations. Reprinted with permission from M. Sing,
U. Schwingenschlögl, R. Claessen, P. Blaha, J. M. P. Carmelo, L. M. Martelo,
P. D. Sacramento, M. Dressel and C. S. Jacobsen, *Physical Review B*, **68**, 125111
(2003). Copyright (2003) by the American Physical Society.

should perhaps be saved from oblivion and synthesized again in order to perform
ARUPS measurements.

High-resolution UPS experiments performed on $(TMTSF)_2PF_6$ and
$(TMTSF)_2ClO_4$ single crystals confirm the absence of intensity at E_F (Vescoli
et al., 2000). The surfaces of BFS are extremely sensitive to VUV irradiation,
so that photoemission data published for such materials have to be cautiously
examined. A fit of the leading edge of the spectra close to E_F to the Luttinger
liquid power-law expression $|E - E_F|^\upsilon$ leads to $K_\rho \sim 0.2$, corresponding to the
very strong repulsion limit. This is confirmed by measurements of the optical
conductivity $\sigma(\omega)$ of the three Bechgaard salts $(TMTSF)_2PF_6$, $(TMTSF)_2AsF_6$ and
$(TMTSF)_2ClO_4$, where $K_\rho \simeq 0.23$ is derived (Schwartz *et al.*, 1998). In this case
consistent deviations from the simple Drude response are found, with $\sigma(\omega)$ instead
consisting of two distinct features: a narrow mode at low energy containing a very
small part of the spectral weight ($\sim$1%), and a high-energy mode centred around

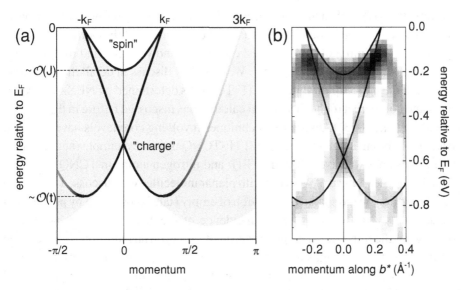

Figure 6.13. (a) Schematic electron removal spectrum of the doped 1D Hubbard model with band filling $1/2 < N_{at} < 2/3$. The shaded region denotes the continuum resulting from spin–charge separation. The solid curves indicate the dispersions of the spin and charge singularity branches. (b) Theoretical spin and charge branch dispersions of the 1D Hubbard model calculated for $U = 1.96$ eV, $W = 1.6$ eV, and $N_{at} = 0.59$ in comparison to the experimental ARUPS spectra. Reprinted with permission from M. Sing, U. Schwingenschlögl, R. Claessen, P. Blaha, J. M. P. Carmelo, L. M. Martelo, P. D. Sacramento, M. Dressel and C. S. Jacobsen, *Physical Review B*, **68**, 125111 (2003). Copyright (2003) by the American Physical Society.

200 cm^{-1}. These are the characteristics of a highly anisotropic interacting electron system, with either a half- or a quarter-filled band. This leads to a Mott gap and, at frequencies above the effective interchain transfer integral, to a Luttinger liquid state. On the other hand, evidence for the separation of spin and charge in BFS has been suggested based on thermal conductivity experiments, where it is shown that spin excitations have a much larger thermal conductivity than charge excitations (Lorenz *et al.*, 2002).

Electronic structure of unoccupied states

The unoccupied electronic states of a solid can be experimentally explored by different techniques. The most commonly used are inverse photoemission, where low-energy electrons impinge on the surface of the solid, and the photon-based techniques ellipsometry, NEXAFS and constant-initial-state spectroscopy. Results derived from inverse photoemission spectroscopy might be questionable unless low-energy electrons (*c.* 10–20 eV) and low beam currents are used as in LEED experiments. NEXAFS and constant-initial-state spectroscopy need synchrotron radiation, because the photon energy has to be scanned and they differ in the energy range of the initial occupied states: core levels and valence band for NEXAFS and constant-initial-state, respectively. We will now discuss in detail the electronic structure of the unoccupied states of TTF-TCNQ as determined by NEXAFS experiments and compare it to the theoretical calculations discussed before in this section (Fraxedas *et al.*, 2003). Absorption techniques involving core levels have the great advantage of being element-selective. TTF-TCNQ is a nice example (again!) since sulfur atoms are exclusively found in TTF and nitrogen atoms in TCNQ.

The further advantage of working with planar molecules was discussed in Section 4.4. Information on the spatial distribution of empty states is obtained by performing measurements at different angles of incidence of the X-ray radiation, θ_E, taking advantage of the intrinsic linear polarization of synchrotron radiation. The example of graphite is illuminating, since from initial s states, σ, symmetric with respect to the molecular plane, or π, antisymmetric with respect to the molecular plane, final states are selected when the polarization vector E lies parallel or perpendicular to the basal plane, respectively. Apart from this purely geometrical constraint, dipolar selection rules also apply. According to these selection rules atomic levels with an orbital quantum number l can be excited only to those levels with $l \pm 1$. Hence s initial states can only be excited to p final states, while from p the achievable final states are s and d. This selectivity is mainly accomplished when working with MOs, and usually simply replacing s by σ and p by π does the job. However, this approximation is insufficient as will be evident below.

The calculated total DOS and PDOS for the relevant atoms of TTF-TCNQ, TTF and TCNQ were introduced in Figs. 6.4(b), 6.5(b) and 6.5(d), respectively, and compared to experimental NEXAFS data (grey lines in the figures). The S$2p$, C$1s$ and N$1s$ NEXAFS spectra obtained from thin films of TTF-TCNQ, TTF and TCNQ are also shown in Fig. 6.14 where the relevant MOs of TTF and TCNQ associated with the different features of the spectra are indicated. The differences found in the energies of the peaks in the NEXAFS spectra compared with the peaks in the calculated PDOS are in part due to the fact that the calculated PDOS corresponds to the electronic structure of the ground state while the NEXAFS spectra correspond to the excited state with a core hole.

TTF-derived spectral features

Let us start with neutral and charged TTF. The experimental spectrum for S$2p$ (Fig. 6.14(a)) corresponding to neutral TTF is disappointingly unstructured, whereas that for charged TTF is clearly rich in structure. As shown in Fig. 1.31, the S$2p$ core levels are composed of two spin-orbit split components, $2p_{3/2}$ and $2p_{1/2}$, separated by 1.3 eV, so that the S$2p$ NEXAFS spectra consists of the superposition of both

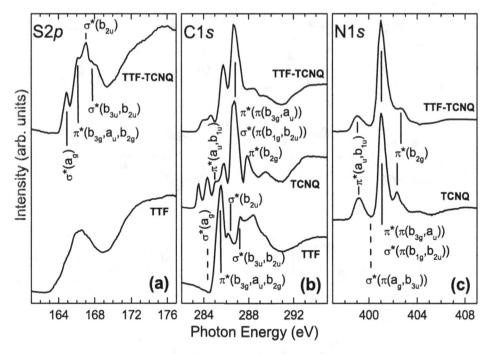

Figure 6.14. (a) S2p, (b) C1s and (c) N1s NEXAFS spectra for thin films of TTF-TCNQ, TTF and TCNQ. Reprinted with permission from J. Fraxedas, Y. J. Lee, I. Jiménez, R. Gago, R. M. Nieminen, P. Ordejón and E. Canadell, *Physical Review B*, **68**, 195115 (2003). Copyright (2003) by the American Physical Society.

contributions. N1s in TCNQ shows narrower spectral lines in the neutral state as compared to the TTF-TCNQ salt (Fig. 6.14(c)). On the other hand, C1s NEXAFS spectra show well-defined features for both neutral and charged TTF and TCNQ (Fig. 6.14(b)). The reason for this contrasting behaviour in the linewidths is the charge transferred upon formation of the TTF-TCNQ compound from the donor TTF to the acceptor TCNQ. The extra charge that is allocated within the TCNQ electronic structure broadens its electronic levels, whereas the loss of charge in TTF narrows the lineshape of the corresponding states. The narrowing of the TTF spectral width corresponds to an increase of the core-hole lifetime in the excited state that occurs when charge is transferred to form the salt.

Since the charge is transferred from the TTF HOMO, which mainly derives from the sulfur atoms as mentioned before, the line narrowing is especially important in the S2p NEXAFS. For the energy region of interest here, where clearly defined features are observed (c. 5–6 eV), the C PDOS is dominated by the p-type contributions with the s- and d-type contributions being much smaller. However, for the S PDOS the weight of the d-type contributions becomes considerably larger.

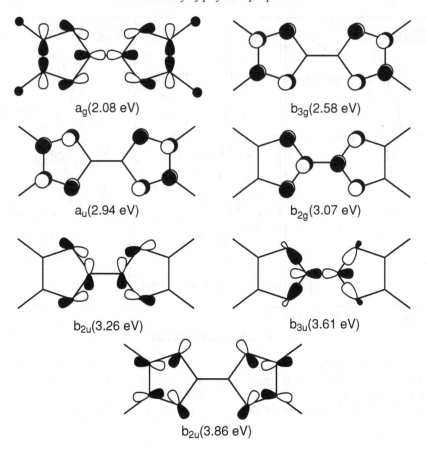

a_g(2.08 eV) b_{3g}(2.58 eV)

a_u(2.94 eV) b_{2g}(3.07 eV)

b_{2u}(3.26 eV) b_{3u}(3.61 eV)

b_{2u}(3.86 eV)

Figure 6.15. Schematic representation of the relevant TTF unoccupied MOs for the analysis of the NEXAFS spectra of TTF and TTF-TCNQ. The symmetry labels are those appropriate for the D_{2h} symmetry. The energy values given (in eV) are relative to those of the HOMO of TTF in the isolated TTF molecule. Reprinted with permission from J. Fraxedas, Y. J. Lee, I. Jiménez, R. Gago, R. M. Nieminen, P. Ordejón and E. Canadell, *Physical Review B*, **68**, 195115 (2003). Copyright (2003) by the American Physical Society.

In fact, the d-contributions extend over all the energy range of interest so that even if the p-contributions are dominant, there is always some d-contribution of the appropriate symmetry and, as a consequence, the levels are going to be visible to some degree in the S2p spectra.

The unoccupied TTF MOs are schematically shown in Fig. 6.15. The energy values associated with the orbitals are those calculated for a single molecule with the same geometry as for the neutral solid. The symmetry labels are those appropriate for idealized D_{2h} symmetry. For simplicity, the sulfur d orbital contributions have not been shown in these schematic drawings.

The first unoccupied peak in the total DOS of neutral TTF (see Fig. 6.5(b)) lies around 2 eV from the TTF HOMO peak. This peak originates from the a_g orbital (see Fig. 6.15), a σ-type orbital, which is essentially antibonding for all carbon–sulfur bonds and to a lesser extent between all carbon–hydrogen bonds. The reason why it is the lowest lying empty orbital is that it mixes in carbon–carbon bonding contributions thus making the level less antibonding. This feature will be labelled as $\sigma^*(a_g)$ owing to its antibonding character. This feature is not observed in the C1s NEXAFS spectrum of TTF (dashed line in Fig. 6.14 (b)) as if transitions from an initial s-state to a final $\sigma^*(a_g)$-state were forbidden. This is the first example of the breakdown of the atomic dipolar selection rules.

The next peak in the DOS of TTF (see Fig. 6.5(b)), which is the most prominent one, in fact consists of two contributions, a larger one at around 2.5 eV above the HOMO maximum and a smaller one, appearing as a shoulder slightly below. This is also the case for the C and S PDOS. The two contributions originate from π-type orbitals of TTF. The lower contribution originates from the b_{3g} orbital lying at 2.58 eV from the TTF HOMO (see Fig. 6.15). The large peak slightly above originates from the pair of a_u and b_{2g} orbitals in Fig. 6.15, which in the molecule are very close in energy, i.e., at 2.94 and 3.07 eV from the HOMO, and thus appear as a single peak in the DOS. The associated feature will be labelled as $\pi^*(b_{3g}, a_u, b_{2g})$, and it can be observed in the NEXAFS curves for both S2p (since it has a significant d-character in the sulfur atoms) and C1s ($s \rightarrow \pi^*$ transitions in carbon are allowed) spectra.

The next three peaks in the DOS (see Fig. 6.5(b)), seen as two peaks and a shoulder, are due to σ-type orbitals. The first one is due to the lower b_{2u} orbital of Fig. 6.15, which lies at 3.26 eV from the HOMO in the molecule and is another antibonding carbon–sulfur level which will be labelled $\sigma^*(b_{2u})$. The next two peaks originate from two orbitals that are quite similar in energy in the molecule: the b_{3u} and upper b_{2u} orbitals of Fig. 6.15, which lie at 3.61 and 3.86 eV from the HOMO in the molecule. These two orbitals are also carbon–sulfur antibonding but also include some carbon–carbon antibonding character and will be termed $\sigma^*(b_{3u}, b_{2u})$.

It is important to note that although not shown in Fig. 6.15, the lower b_{2u} orbital contains a sizeable contribution of the appropriate symmetry adapted combination of d_{z^2} orbitals of the sulfur atoms. This is not the case for any of the orbitals considered up to now. This observation will be important in order to understand the angular variation of the S2p spectra discussed later. At higher energies, the spectra are rather structureless because states lie in a quasi-continuum and electrons can be considered as nearly free. In fact the crystal field may have some influence even at such high lying states. In graphite, for instance, this influence remains even for surprisingly high energies, $c.$ 90 eV above the VBM (Bianconi *et al.*, 1977), while for III-V semiconductors it goes up to $c.$ 30 eV (Faul *et al.*, 1993).

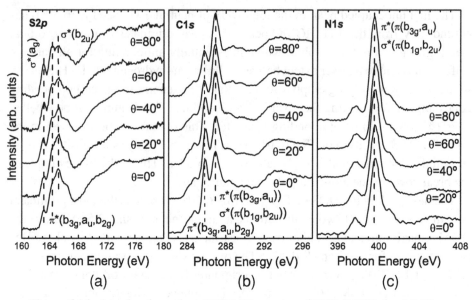

Figure 6.16. Angular dependent NEXAFS spectra of TTF-TCNQ for (a) S$2p$, (b) C1s and (c) N1s. The θ_E angle is defined as the angle between the light polarization vector and the surface plane. Reprinted with permission from J. Fraxedas, Y. J. Lee, I. Jiménez, R. Gago, R. M. Nieminen, P. Ordejón and E. Canadell, *Physical Review B*, **68**, 195115 (2003). Copyright (2003) by the American Physical Society.

The origin of the different peaks of the TTF-TCNQ S PDOS (Fig. 6.4(b)) can be understood exactly as for neutral TTF (Fig. 6.5(b)). This is not that surprising since the TTF 1D stacks in the two solids are very similar and TCNQ does not possess sulfur atoms. Summarizing, the first peaks above E_F in the S PDOS of Fig. 6.4(b) originate respectively from the $\sigma^*(a_g)$, the $\pi^*(b_{3g}, a_u, b_{2g})$ (the two peaks which were seen as a peak and a shoulder for TTF appear as a single peak now), the $\sigma^*(b_{2u})$ and the $\sigma^*(b_{3u}, b_{2u})$ orbitals of TTF (see Fig. 6.14(a)).

Figure 6.16 shows the NEXAFS spectra as functions of the incidence angle θ_E for the oriented TTF-TCNQ films. For $\theta_E = 0°$ and θ_E close to 90° E lies parallel and perpendicular to the molecular ab-plane, respectively. In this case the planar geometry of both TTF and TCNQ molecules is a clear advantage strongly simplifying the analysis.

The most relevant trends in Fig. 6.16(a), corresponding to the S$2p$ spectra, are the intensity reduction of $\sigma^*(b_{2u})$ and the increase of $\sigma^*(a_g)$ with increasing θ_E. According to the geometry of the TTF molecules and their molecular orbitals (see Figs. 1.15 and 6.15), the absorption intensity should exhibit a maximum when the light electric field coincides with the molecular plane, i.e. for large θ_E values. This is indeed observed for $\sigma^*(a_g)$, but not for $\sigma^*(b_{2u})$. The reason for this behaviour is that the $\sigma^*(b_{2u})$ feature exhibits a significant d_{z^2} contribution, an orbital pointing perpendicularly to the TTF molecular plane. Since the normal of this molecular plane forms an angle of about $60°$ with the direction perpendicular to the ab-plane (c^*-direction), the d_{z^2} contribution to the NEXAFS spectrum would be more evident for lower θ_E values and should be strongly reduced for large θ_E values, as experimentally observed. Similarly to the $\sigma^*(b_{2u})$ feature, $\pi^*(b_{3g}, a_u, b_{2g})$ should decrease at larger values of θ_E because of the orbital distribution. This behaviour is not evident from the spectra because of the overlap with $\sigma^*(b_{2u})$ but the intensity ratio between $\pi^*(b_{3g}, a_u, b_{2g})$ and $\sigma^*(a_g)$ decreases for increasing θ_E values, as expected.

TCNQ-derived spectral features

Let us now consider the case of neutral and charged TCNQ. Both the C1s and N1s spectra are quite well resolved (see Figs. 6.14(b) and (c)). The PDOS of both nitrogen and carbon in the region of interest are almost completely dominated by the p-type contributions. Looking at the TCNQ DOS in Fig. 6.5(d) we can see four clear peaks above that of the LUMO band. The first peak, appearing at around 2.8 eV above the LUMO minimum, originates from the lower a_u and b_{1u} orbitals of TCNQ, schematized in Fig. 6.17, which for the isolated molecule lie at 2.55 and 2.65 eV from the LUMO. These π-type orbitals, $\pi^*(a_u, b_{1u})$, are quite strongly concentrated on the benzenic region (small contributions are not shown in the schematic representations of Fig. 6.17). These orbitals originate from the degenerate pair of lowest empty π orbitals (e_{2u}) of C_6H_6, which only slightly delocalize towards the substituent in TCNQ.

The next two peaks are associated to the CN groups. The first one contains the contribution of the bands based on the a_g and b_{3u} orbitals of TCNQ lying at 3.51 and 3.63 eV from the LUMO, and will be labelled $\sigma^*(\pi(a_g, b_{3u}))$, while the second peak contains the contribution of bands based on four orbitals of TCNQ lying at 4.55 (b_{3g}) and 4.66 eV (a_u), labelled as $\pi^*(\pi(b_{3g}, a_u))$, and 4.77 (b_{1g}) and 4.86 eV (b_{2u}), labelled as $\sigma^*(\pi(b_{1g}, b_{2u}))$, from the LUMO (see Fig. 6.17).

Essentially, the $\pi^*(a_g, b_{3u})$ and $\pi^*(b_{1g}, b_{2u})$ orbitals are the four symmetry-adapted combinations of the in-plane π^* orbitals of the CN groups. It is important to distinguish between the two π^* orthogonal orbitals of the CN group. They are degenerate for CN itself because of the cylindrical symmetry but become non-degenerate in TCNQ. Because of the symmetry plane of the molecule, the two formally degenerate orbitals lead to one in-plane π^* orbital, denoted as $\sigma^*(\pi)$, which is symmetrical with respect to the molecular plane even if locally it is a π-type orbital, and one out-of-plane π^* orbital, denoted as $\pi^*(\pi)$, which is anti-symmetrical with respect to the molecular plane and is locally a π-type orbital.

The four $\sigma^*(\pi)$ orbitals lead to four symmetry-adapted combinations which are the main components of the four molecular orbitals. Two of them are lower in

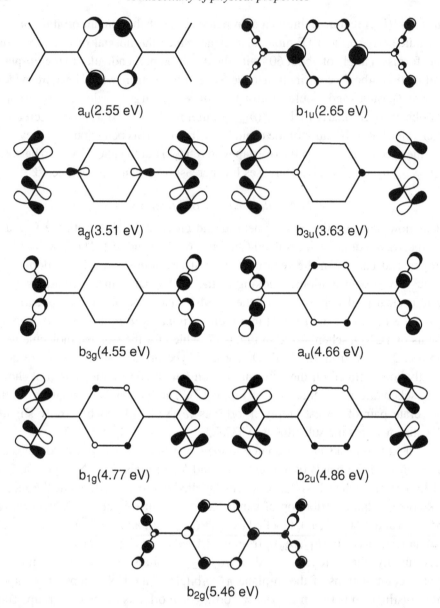

Figure 6.17. Schematic representation of the relevant TCNQ unoccupied MOs for the analysis of the NEXAFS spectra of TCNQ and TTF-TCNQ. The symmetry labels are those appropriate for D_{2h} symmetry. The energy values given (in eV) are relative to those of the LUMO of TCNQ in the isolated TCNQ molecule. Reprinted with permission from J. Fraxedas, Y. J. Lee, I. Jiménez, R. Gago, R. M. Nieminen, P. Ordejón and E. Canadell, *Physical Review B*, **68**, 195115 (2003). Copyright (2003) by the American Physical Society.

energy, $\sigma^*(\pi(a_g, b_{3u}))$, because they are symmetrical with respect to the C_2 axis along the long molecular axis and they exhibit positive overlap between the carbon components of the two adjacent $\sigma^*(\pi)$ orbitals. This lowers the energy considerably. The other two orbitals are based on the b_{1g} and b_{2u} orbitals, $\sigma^*(\pi(b_{1g}, b_{2u}))$. These orbitals are the two remaining symmetry-adapted combinations of the $\sigma^*(\pi)$ orbitals. By having negative overlap between the carbon components of the two adjacent $\sigma^*(\pi)$ orbitals they now lie around 1 eV higher in energy than their a_g and b_{3u} counterparts (see Fig. 6.17).

The $\pi^*(\pi(b_{3g}, a_u))$ orbitals are two of the symmetry-adapted combinations of the $\pi^*(\pi)$ orbitals. By being symmetrical with respect to the C_2 axis along the long molecular axis, the π orbital of the central carbon atom of the $C(CN)_2$ substituent cannot mix into these symmetry-adapted combinations, and thus, in practice, the $\pi^*(\pi)$ orbitals do not delocalize toward the C_6H_6 ring.

The next peak, at approximately 5.6 eV from the LUMO minimum, arises from a π-type orbital delocalized over all the molecule but originating from the highest π-type orbital of C_6H_6 (b_{2g}). Superposed to this contribution there is a second one, of σ-type and also localized on the C_6H_6 moiety, whose main character is σ_{C-H} antibonding. Thus, the first and fourth peaks are considerably more concentrated in the benzene ring whereas all other peaks are strongly based on π^*_{CN} orbitals. This is clear from the comparison of the C and N PDOS in Fig. 6.5(d).

The origin of the different peaks of the N PDOS of TTF-TCNQ (Fig. 6.4(b)) can be understood exactly as for the neutral TCNQ (Fig. 6.5(d)). C1s and N1s spectra of TCNQ are shown in Figs. 6.14(b) and (c), respectively. Note that the contribution of $\sigma^*(\pi(a_g, b_{3u}))$, which should appear (see the dotted line in Fig. 6.14(c)) in between those of $\pi^*(a_u, b_{1u})$ and $\pi^*(\pi(b_{3g}, a_u)) + \sigma^*(\pi(b_{1g}, b_{2u}))$, both clearly visible in the N1s spectra, does not appear at all. That the larger peak in the N1s spectra is partially associated with $\sigma^*(\pi(b_{1g}, b_{2u}))$ is clear not only from the energy separation from the first peak but, more importantly, from the angular dependence of the peaks, as discussed next.

Figures 6.16(b) and (c) show the angular dependence of the C1s and N1s NEXAFS spectra, respectively. The most salient feature of Fig. 6.16(c) is the intensity increase of the $\pi^*(\pi(b_{3g}, a_u)) + \sigma^*(\pi(b_{1g}, b_{2u}))$ peak for increasing θ_E values. However, the benzenic-type $\pi^*(a_u, b_{1u})$ and $\pi^*(b_{2g})$ peaks remain nearly unchanged. For TCNQ the C_2 molecular axis forms an angle of about 36° with the c^*-direction. Since CN bonds form an angle of 60° with this C_2 molecular axis within the molecular plane, the intensity associated to $\sigma^*(\pi(b_{3g}, a_u))$ should increase for larger θ_E. However, the intensity associated with $\pi^*(\pi)$ orbitals, the benzenic-like orbitals and $\pi^*(\pi(b_{3g}, a_u))$, should exhibit the opposite behaviour because they are perpendicular to the $\sigma^*(\pi)$ orbitals. From Fig. 6.16(c) it is clear that the $\pi^*(\pi(b_{3g}, a_u)) + \sigma^*(\pi(b_{1g}, b_{2u}))$ peak increases with regard to the benzenic-type orbitals for increasing θ_E values because of the increasing $\sigma^*(\pi)$ and decreasing $\pi^*(\pi)$ contributions. The same arguments apply for the $\pi^*(\pi(b_{3g}, a_u)) + \sigma^*(\pi(b_{1g}, b_{2u}))$ C1s peak (Fig. 6.16(b)), which indeed increases for increasing θ_E angles. In the case of TTF $\pi^*(b_{3g}, a_u, b_{2g})$ (Fig. 6.16 (a)), its intensity decreases for increasing θ_E angles, which is again ascribed to the π-character of the bonds as pointed out earlier.

Let us conclude this section with the intriguing observation of the absence of the $\sigma^*(\pi(a_g, b_{3u}))$ contribution in the N1s NEXAFS spectra of both neutral TCNQ and TTF-TCNQ. Let us recall the absence of precisely the a_g and b_{3u} contributions in the C1s spectra of TTF discussed above. However, the C1s NEXAFS spectrum of TCNQ shows some intensity in the $\sigma^*(\pi(a_g, b_{3u}))$ region [between $\pi^*(a_u, b_{1u})$ and $\pi^*(\pi(b_{3g}, a_u)) + \sigma^*(\pi(b_{1g}, b_{2u}))$]. This is due to the significant d-contribution from carbon for neutral TCNQ while nitrogen contributes negligibly. Thus, it seems that in addition to the intra-atomic selection rules there are additional restrictions apparently symmetry-related in MOMs like those discussed here. This unexplained phenomenon certainly calls for both theoretical and experimental future work.

CDW-induced surface modulation

As indicated above, the ARUPS results were all taken at cryogenic temperatures. In fact the TTF-TCNQ energy band dispersion cannot be observed at RT because of the considerable vibration of the surface molecules in UHV. These surface thermal vibrations, facilitated by the weak intermolecular interactions, should induce a reduction of ϱ at the surface. Indeed, if we model the 1D energy dispersion of the TTF HOMO and TCNQ LUMO bands using Eq. (1.33), then we obtain:

$$E_{TTF}(k) = E_{\alpha TTF} + 2E_{\beta TTF} \cos bk, \tag{6.1a}$$

$$E_{TCNQ}(k) = E_{\alpha TCNQ} + 2E_{\beta TCNQ} \cos bk, \tag{6.1b}$$

where $E_{\beta TTF} > 0$ and $E_{\beta TCNQ} < 0$ (compare to the band structures given in Figs. 1.28 and 6.3). ϱ can be calculated in the limit of negligible electron–electron interactions using Eq. (1.34), where k_F can be obtained from the condition $E_{TCNQ}(k) = E_{TTF}(k)$ at E_F. Under these assumptions ϱ can be expressed as:

$$\varrho = \frac{2}{\pi} \arccos \frac{E_{\alpha TCNQ} - E_{\alpha TTF}}{2(E_{\beta TTF} - E_{\beta TCNQ})}. \tag{6.2}$$

From this equation it turns out that ϱ decreases if $E_{\beta TTF} - E_{\beta TCNQ}$ decreases or, alternatively, if the bandwidth decreases. Hence surface thermal vibrations induce the reduction of ϱ because of the reduction of the $\pi - \pi$ intrastack overlap. Thus in order to keep the bulk value of ϱ at the surface in UHV it is mandatory to cool down the sample to cryogenic temperatures. However, when STM images are obtained at ambient conditions, submolecular resolution is indeed achieved because

Figure 6.18. (Top) STM image of the *ab*-plane of TTF-TCNQ taken at 63 K ($V_t = 50$ mV, $I_t = 1$ nA). The image area is 5.3 nm × 5.3 nm. Reprinted with permission from Z. Z. Wang, J. C. Girard, C. Pasquier, D. Jérome and K. Bechgaard, *Physical Review B*, **67**, 121401 (2003). Copyright (2003) by the American Physical Society. (Bottom) Simulation of the STM image of the *ab*-plane of TTF-TCNQ, obtained with DFT calculations in the GGA performed with the SIESTA code (Soler *et al.*, 2002) using the Tersoff–Hamann approximation (see Section 4.2). The value of the charge density is 2×10^{-5} electrons/a.u.^{3}, which is about 0.2 nm above the surface. Courtesy of Drs P. Ordejón and E. Canadell.

the oscillation amplitude is reduced as a consequence of the atmospheric pressure. The surface reconstruction example shown in Fig. 4.1, based on STM images taken at RT on BEDT-TTF-terminated *ac*-surfaces of β-(BEDT-TTF)$_2$PF$_6$, is not due to surface vibrations but to the breakdown of the 3D symmetry at the crystal surface, leading to differentiated surface and bulk CDWs (Ishida *et al.*, 2001).

Figure 6.18(top) displays a submolecular-resolution STM image of the *ab*-plane of an in-air cleaved TTF-TCNQ single crystal obtained in a constant current mode in

UHV at 63 K, thus above the Peierls transition (Wang *et al.*, 2003). This beautiful image reveals that the surface consists of two kinds of linear chains along the crystallographic *b*-direction. One type of chain, the most prominent, is characterized by a brighter central feature accompanied by two custodian weaker features. The second type of chain consists of two weak features. The image is compared to a simulated image obtained with DFT calculations as a height map of an isosurface of the LDOS at E_F, shown in Fig. 6.18(bottom).

The triplet and the single features of the simulated image are ascribed to the TCNQ and TTF chains, respectively. In the experimental STM image the distance between equivalent chains is 1.22 nm, corresponding to the *a* lattice constant ($a = 1.23$ nm), while the periodicity within the chains is 0.38 nm, coincident with the *b* lattice constant. The comparison of experimental and simulated images reveals two points of disagreement: (i) the single vs. double spots associated with TTF and (ii) the intensity and size of the outer spots of the triplet compared to the central one associated with TCNQ. The simulated images agree with simulations based on the extended Hückel TB method, also within the Tersoff–Hamann approximation, with a tip to surface distance of 0.05 nm (Magonov & Whangbo, 1996; p. 200). The simulated images simply reveal that according to the TTF-TCNQ bulk crystal structure, the hydrogen atoms of TTF and the nitrogen atoms of TCNQ protrude most on the *ab*-plane and that the nitrogen atoms have a higher electron density. However, experimental STM images taken in air show the TCNQ associated triplets with both intensity distributions, that is, chains with a more pronounced central feature together with chains with more pronounced external features (Ara *et al.*, 1995). These discrepancies suggest that perhaps the TTF and TCNQ molecules undergo a certain geometric deformation due to the intense applied electrical fields induced by the tip and/or that the Tersoff–Hamann approximation is a very good but insufficient starting point and therefore that the electronic structure of the tip should be included.

The real-space characterization of the CDW-induced modulation of a 2D surface lattice can be ideally performed with variable temperature STMs. The temperature-dependent modulation can be classified according to the HFW model introduced in Section 4.2 taking the ideal 1×1 surface structure as the reference lattice (a_s and b_s) and the projected CDW-modified structure as the overlayer system (a_o and b_o). In the case of TTF-TCNQ $a_s = a$ and $b_s = b$ and for the images taken at 63 K (see Fig. 6.18(top)) the transformation matrix corresponds to the identity. However, below 54 K a 2D superstructure appears in the image, as depicted in Fig. 6.19(a), which is characterized by the matrix $\begin{pmatrix} 2 & 0 \\ 0 & 3.3 \end{pmatrix}$.

The real-space modulation can be obtained by Fourier transforming the image, as shown in Fig. 6.19(b). Observe the $1/2$ periodicity along the a^*-direction and the

(a) (b)

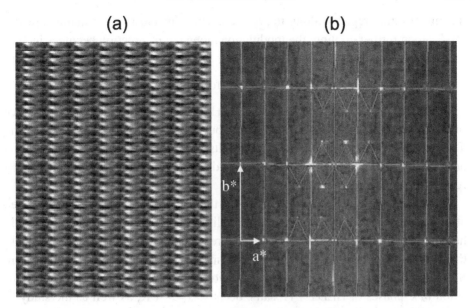

Figure 6.19. (a) STM image of the *ab*-plane of TTF-TCNQ taken at 49.2 K. The image area is 8.7 nm × 11.9 nm. (b) Fourier transformed pattern showing the 2*a* × 3.3*b* CDW ordering. Reprinted with permission from Z. Z. Wang, J. C. Girard, C. Pasquier, D. Jérome and K. Bechgaard, *Physical Review B*, **67**, 121401 (2003). Copyright (2003) by the American Physical Society.

Figure 6.20. STM image of the *ab*-plane of TTF-TCNQ taken at 33 K ($V_t =$ 100 mV, $I_t = 1$ nA). The image area is 9.0 nm × 9.0 nm. Courtesy of Dr Z. Z. Wang.

incommensurate registry along the b^*-axis. This superstructure remains down to 49 K. Upon further cooling, the modulation along the a-direction becomes incommensurate without noticeable change along b. However, a transverse commensurability arises abruptly at 38 K, with a transformation matrix $\begin{pmatrix} 4 & 0 \\ 0 & 3.3 \end{pmatrix}$.

A STM image taken at 33 K is presented in Fig. 6.20. Note the distinct appearance of this figure as compared to Fig. 6.18(top), but notice that their scales are different. These scanning tunnelling microscopy results on temperature-induced modulation are in very good agreement with previous detailed X-ray and neutron scattering reports (Pouget, 1988).

This work shows the exceptional physics that can be done with a STM operated at cryogenic temperatures and the availability of STMs working down to liquid helium temperature opens broad avenues of research in the coming years. No doubt that among the many future scientific experiments accessible with low temperature STMs, the real-space electronic characterization of the metal–superconductor transition in κ-phases of BEDT-TTF salts, because $T_c > 4$ K, as well as the study of magnetic ordering in MOMs, will certainly occupy a relevant position.

6.2 1D molecular magnets

After a detailed discussion on the electronic structure of 1D metals, in particular of TTF-TCNQ, we move now to molecular materials with magnetic properties confined to one dimension. We briefly analyse two examples. As already discussed in Section 1.7, one of the consequences of the Ising model is the absence of permanent magnetization in 1D magnetic systems since, in equilibrium, $\mathcal{M} = 0$ for $B = 0$ according to Eq. (1.49) (Ising, 1925). About forty years after Ising's work, R. J. Glauber predicted that for 1D magnetic systems $\mathcal{M}$ always decays exponentially with time, formulating the time-dependent Ising model (Glauber, 1963). The slow relaxation of $\mathcal{M}$ is a characteristic of 1D systems, but is also found for 0D clusters as in the case of Mn_{12} discussed in Section 1.5, and should be differentiated from paramagnetism, where the relaxation can be considered in a first approximation as instantaneous. The slow relaxation justifies using the term magnet.

The first experimental confirmation of Glauber's prediction was shown with the material $Co(hfac)_2(NITPhOMe)$, based on a nitronyl nitroxide radical (Caneschi *et al.*, 2001). $Co(hfac)_2(NITPhOMe)$ consists of alternating $Co(hfac)_2$ and radical moieties arranged in 1D arrays with a helical structure, and crystallizes in the $P3_1$ space group with $a = b = 1.129$ nm and $c = 2.057$ nm. A detail of the chain structure is depicted in Fig. 6.21.

The temperature dependence of magnetic susceptibility χ measured on single crystals reveals magnetic 1D behaviour. Below 50 K χ is strongly anisotropic with the trigonal axis corresponding to the easy axis of magnetization and therefore to

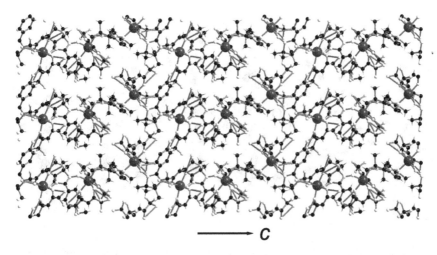

Figure 6.21. Chain structure of Co(hfac)$_2$(NITPhOMe) along the c-direction. Co and C atoms are represented by large dark grey and small black balls, respectively. The rest of the atoms are represented by white balls. Crystallographic data from Caneschi *et al.*, 2001.

Ising-type anisotropy. An estimation of the exchange coupling gives $J \simeq 19$ meV. At low temperature the spin of octahedral Co^{+2} centres can be handled as an effective $S = 1/2$ spin with anisotropic g values significantly different from 2. The chains behave as 1D ferrimagnets, because of the non-compensation of the magnetic moments of the Co^{+2} centre and the radical. In addition, the ac χ measured at $B = 0$ and with an oscillating field applied along the chain direction is strongly frequency dependent below 17 K, clearly indicating that the relaxation of $\mathcal{M}$ in this temperature range becomes slow on the time scale of the experiment (2 μs–1 s). This frequency dependence rules out any 3D transition.

We next turn our attention to the single-chain magnet formed by the heterometallic chain of Mn^{+3} and Ni^{+2} ions, [Mn$_2$(saltmen)$_2$Ni(pao)$_2$(C$_5$H$_5$N)$_2$](ClO$_4$)$_2$ (Clérac *et al.*, 2002). This compound ($C2/c$, $a = 2.114$ nm, $b = 1.597$ nm, $c = 1.862$ nm and $\beta = 98.059°$) consists of two fragments, the out-of-plane dimer [Mn$_2$(saltmen)$_2$]$^{2+}$ as a coordination acceptor building block and the neutral mononuclear unit [Ni(pao)$_2$(C$_5$H$_5$N)$_2$] as a coordination donor building block, forming an alternating chain having the repeating unit [-Mn-(O)$_2$-Mn-ON-Ni-NO-]$_n$. Figure 6.22 shows a detail of the complex chain structure revealing that the oximato group of the [Ni(pao)$_2$(C$_5$H$_5$N)$_2$] monomeric species bridges the Mn^{+3} dimers to form an alternating 1D chain. Each chain is contained in the ac-plane and is separated from the nearest chains with a minimum intermetallic distance between manganese and nickel of 1.039 nm, thus ensuring good magnetic isolation of the chains. No significant interchain π stacking is observed between the phenyl rings of the saltmen ligand.

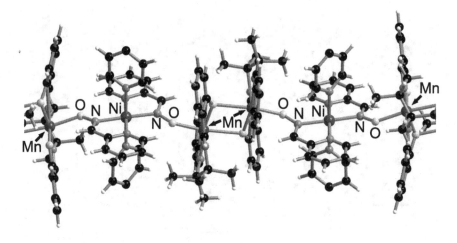

Figure 6.22. Chain structure of $[Mn_2(saltmen)_2Ni(pao)_2(C_5H_5N)_2](ClO_4)_2$. Crystallographic data from Clérac *et al.*, 2002.

Figure 6.23 shows magnetic measurements performed on oriented single crystals along the chains (empty circles), perpendicular to the chains in the *ac*-plane (full circles) and along the monoclinic *b*-axis (empty triangles). Below 60 K, χ becomes anisotropic and reveals above 20 K a quasi-uni-axial symmetry. This implies that the chain axis, where the maximum of the magnetic response is measured, is close to the easy magnetization axis. This anisotropy is confirmed by the field dependence of $\mathcal{M}$ (see inset in Fig. 6.23), which quickly saturates when the field is applied along the chain direction, at $\simeq 6\mu_B$. Using Eq. (1.49) to fit the data between 5 and 15 K (solid lines in Fig. 6.23 and its inset), the obtained effective parameters are $\mu_{m,eff}/\mu_B \simeq 5.4$ and $J_{eff} \simeq 1.0$ meV. The first one is consistent with the expected moment at saturation, $\mu_m/\mu_B = 6$, and the latter gives a value for the intrachain $Mn^{+3} \cdots Mn^{+3}$ interaction of 0.06 meV.

Below 6.5 K, the ac χ (χ' and χ'' are the real and imaginary components of the ac susceptibility, respectively, $\chi = \chi' + i\chi''$) are strongly frequency dependent (see Fig. 6.24). As pointed out above this result precludes any 3D ordering. When lowering the temperature, the decrease of χ' (Fig. 6.24(a)), with the appearance of χ'' (which becomes different from zero between 6.5 K at 1500 Hz and 4.5 K at 1 Hz, Fig. 6.24(b)), is directly associated to a blocking process of the magnetization. Indeed, the thermal energy is not sufficient to allow the magnetization to follow the ac field at a given frequency. At this frequency, the motion of the entire magnetization becomes frozen below the blocking temperature ($\sim$4 K). The estimated relaxation time of $\mathcal{M}$ is on the order of 5 months at 1.8 K, hence deserving to be considered as a magnet.

An important characteristic of this material is the high spin value of the magnetic unit inside the chain ($S = 3$). It is remarkable that with the small

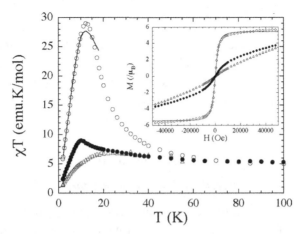

Figure 6.23. Magnetic measurements performed on oriented single crystals of [Mn$_2$(saltmen)$_2$Ni(pao)$_2$(C$_5$H$_5$N)$_2$](ClO$_4$)$_2$ along the chains (empty circles), perpendicular to the chains in the ac-plane (full circles) and along the monoclinic b-axis (empty triangles). Temperature dependence of the χT product at 10 kOe below 100 K. The solid line corresponds to the best fits obtained with the 1D Ising model (see text). Inset: Field dependence of $\mathcal{M}$ at 10 K. M and H stand for $\mathcal{M}$ and B, respectively. The solid line corresponds to the best fit obtained with the 1D Ising model. Reprinted with permission from Clérac *et al.*, 2002. Copyright (2002) American Chemical Society.

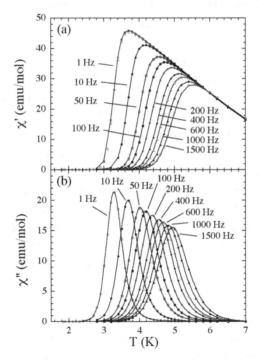

Figure 6.24. Temperature and frequency dependence of the (a) real (χ') and (b) imaginary (χ'') part of the ac χ. The solid lines are guides to the eye. Reprinted with permission from Clérac *et al.*, 2002. Copyright (2002) American Chemical Society.

interunit couplings the same order of magnitude for the blocking temperature as for $Co(hfac)_2(NITPhOMe)$ is obtained. Therefore, the use of high-spin magnetic units combined with strong interunit magnetic 1D interactions appears to be an attractive route to design metastable magnets at high temperature.

6.3 Interfacial doping

Organic field-effect transistors

In Section 1.5 we referred to the different ways of doping (disordered, ordered, internal and interfacial) leading to conducting (metallic, semiconducting and superconducting) materials. Here we will concentrate on interfacial doping, which corresponds to the direct injection of electrons or holes at an organic/insulator interface, generating a confined 2D charge carrier gas. The insulator can be of either inorganic or organic (polymer) nature. Charge injection, induced by the application of an external electrical field, represents a versatile way of reversible doping, since it can be continuously varied and the polarity can be selected. However, disordered, ordered and internal doping give fixed doping levels since once the material is prepared such levels can be only irreversibly modified. On the other hand, the effective injection of carriers at the organic/metal contact interfaces depends not only on the quality of the interface but also on the injection barriers, since many MOMs exhibit dipole barriers with most common metals.

The artificial structure that permits such continuous control is the FET that here will be termed the organic FET, abbreviated as OFET. Nowadays, the silicon-based metal-oxide-semiconductor FET, the MOSFET, is the most prominent constituent of modern microelectronics, both as discrete devices and in integrated circuits. MOSFETs are mainly fabricated with single-crystalline silicon, because of the excellent quality of the silicon/silicon oxide interface. Let us briefly recall the working principle of a FET applied to organic insulating materials. Basically, a FET operates as a capacitor where one plate is a conducting channel between two ohmic contacts, the source and drain electrodes. The density of charge carriers in the channel is modulated by the voltage applied to the second plate of the capacitor, the gate electrode. A scheme of an OFET based on an organic thin film was given in Fig. 4.31.

As in traditional inorganic semiconductors, organic materials can function either as p-type or n-type semiconductors. In p-type semiconductors the majority carriers are holes, while in n-type the majority carriers are electrons. The most widely studied organic semiconductors are of p-type. For a p-type insulator when the gate electrode is biased positively with respect to the grounded source electrode, OFETs operate in the depletion mode, and the channel region is depleted of carriers resulting in a high channel resistance (off state). When the gate electrode is biased negatively, OFETs operate in the accumulation mode and a large concentration of carriers is accumulated in the transistor channel, resulting in a low channel resistance (on state). For n-type OFET operation, the electrode polarity is reversed and the majority carriers are electrons instead of holes.

For an OFET, the dependence of the drain current I_D on the source–drain voltage V_D and the source–gate voltage V_G is described by the classical equations derived from inorganic-based thin film transistors (Horowitz, 1998):

$$I_D = \frac{W_{ch}}{L_{ch}} \mu_{e,h} C_i \left(V_G - V_T - \frac{V_D}{2} \right) V_D, \tag{6.3a}$$

$$I_D = \frac{W_{ch}}{2L_{ch}} \mu_{e,h} C_i (V_G - V_T)^2. \tag{6.3b}$$

Equations (6.3a) and (6.3b) correspond to the linear and saturation regimes, respectively. Here, W_{ch} and L_{ch} are the channel width and length, respectively, C_i the insulator capacitance per unit area and V_T the threshold voltage. V_T is the gate voltage for which the channel conductance (at low drain voltages) is equal to that of the whole semiconducting layer. $\mu_{e,h}$ is thus evaluated by using either Eq. (6.3a) or (6.3b) from the experimental I_D–V_D curves. When using these expressions one has to be aware that they are valid in the accumulation mode, under the condition of constant mobility (not always fulfilled in organic semiconductors) and where L_{ch} is much larger than the insulator thickness.

The experimental determination of μ_e and μ_h strongly relies on the quality of both the material grown on the insulator and the interface. OFETs can be prepared either by deposition of a thin film on a lithographically obtained surface or by growth of a suitable dielectric on top of organic single crystals. In addition, the metallic contacts for the source, drain and gate have to be prepared. OFETs are easier to process as thin films but then the defect density (e.g., grain boundaries) is higher than for single crystals. However, sputtering dielectrics such as Al_2O_3 onto the surface of organic molecular crystals unavoidably results in a very high density of traps that may completely suppress the field effect.

The breakthrough in the direct fabrication of free-standing single-crystal OFETs came with using thin polymer films of parylene, the generic name for poly-*p*-xylylene, as a gate-dielectric material (Podzorov *et al.*, 2003). Parylene, not to be confused with perylene, forms transparent pinhole-free conformal coatings as thin as 0.1 μm with excellent dielectric and mechanical properties. The deposition is obtained by vaporization of di-*p*-xylylene dimers, which transform into the monomeric form (*p*-xylylene) by pyrolysis (typically below 900 K) and polymerize in the deposition zone at RT. In addition, parylene is chemically inert and does not react with organic molecular crystals. Parylene may also be used in flexible thin films. Single crystals of rubrene, with a parylene polymer film as the gate insulator, exhibit p-type conductivity with $\mu_h \sim 0.1$–1 $cm^2 V^{-1} s^{-1}$ (Podzorov *et al.*, 2003).

High-performance OFETs have also been fabricated on the surface of freestanding organic single crystals with the lamination technique. This conceptually simple method consists of laminating a monolithic elastomeric transistor stamp against the surface of a crystal and has enabled the fabrication of rubrene transistors with charge carrier mobilities as high as ~15 $cm^2 V^{-1} s^{-1}$ (Sundar *et al.*, 2004). The fabrication process uses a flexible elastomer as a substrate (e.g., PDMS) on which the transistor stamp is constructed. The gate and source–drain electrodes are evaporated through shadow masks and are separated by an additively transferred thin film (2–4 μm) of PDMS. The flexibility of both the dielectric and the substrate enables assembly of devices through simple lamination of the organic crystal and the elastomeric stamp surface. Slight pressure applied to one edge of the crystal initiates contact with the stamp. The ease of this assembly process together with its inherently non-invasive nature enable the systematic analysis of the semiconducting properties of pristine organic crystals.

The simple drop casting crystallization method, discussed in Section 3.1, has also shown its validity in the determination of hole mobilities at least for the TTF-derivatives DTTTF (Mas-Torrent *et al.*, 2004), EDT-TTF-(CONHMe)$_2$ (Colin *et al.*, 2004) and DBTTF (Mas-Torrent *et al.*, 2005). In all cases, $\mu_h \sim 1$ $cm^2 V^{-1} s^{-1}$. In this case the formation of high-quality organic/insulating interfaces is a matter of serendipity (see the discussion on the formation of nano-volcanoes at the surfaces of drop-cast single crystals in Section 3.1).

Table 6.1 lists the experimental values of the mobilities obtained for some selected MOMs. Such values vary depending on the preparation of the materials as thin films or single crystals as well as on the preparation of the interfaces and metallic contacts. OFETs obtained from rubrene single crystals exhibit mobility values between 0.1 and 15 $cm^2 V^{-1} s^{-1}$, depending on the quality of the crystal/dielectric interface. Once the preparation of the interface has been mastered, the maximum RT $\mu_{e,h}$ values should keep increasing with time with the aim of soon obtaining their intrinsic values. For pentacene, mobilities as high as 3 $cm^2 V^{-1} s^{-1}$ have been claimed (Dimitrakopoulos & Malenfant, 2002; de Boer *et al.*, 2004). In the case of thin films, orientation is also a major issue, in particular for anisotropic molecules. When planar molecules stand perpendicularly to the insulator surface, the mobilities are higher because of the more effective π–π interactions, in clear contrast to when the molecules lie flat on the surface.

Charge mobilities determined by the time of flight (TOF) method show that values of 1–10 $cm^2 V^{-1} s^{-1}$ are not rare for MOMs (Karl *et al.*, 1999). In TOF experiments the transit time τ for a sheet of photoinjected carriers to move across a sample of thickness L is monitored. The sheet of carriers is usually generated

Table 6.1. *Values of* $\mu_{e,h}$ *for selected organic semiconductors and insulators as active components of OFETs*

Material	$\mu_{e,h}$ [cm^2 V^{-1} s^{-1}]	Form	Reference
p-type			
rubrene	15	sc	Sundar *et al.*, 2004
rubrene	1	sc	Podzorov *et al.*, 2003
DTTTF	1.4	sc	Mas-Torrent *et al.*, 2004
DBTTF	1	sc	Mas-Torrent *et al.*, 2005
EDT-TTF-(CONHMe)$_2$	1	sc	Colin *et al.*, 2004
pentacene	1.5	tf	Lin *et al.*, 1997
pentacene	0.5	sc	Takeya *et al.*, 2003
tetracene	0.4	sc	de Boer *et al.*, 2003
tetramethyl-pentacene	0.3	tf	Meng *et al.*, 2003
TMTSF	0.2	sc	Nam *et al.*, 2003
CuPc/CoPc	0.1	tf	Zhang *et al.*, 2004
α-6T	0.03	tf	Dodabalapur *et al.*, 1995
n-type			
C$_{60}$	0.08	tf	Haddon *et al.*, 1995
F$_{16}$CuPc	0.03	tf	Bao *et al.*, 1998
PTCDA	10^{-4}	tf	Ostrick *et al.*, 1997
TCNQ	3×10^{-5}	tf	Brown *et al.*, 1994

Note: sc and tf stand for single crystal and thin film. The mobilities given here correspond to the maximum reported values of the corresponding references.

by a short pulse of strongly absorbed light of sufficiently high photon energy. The sample is sandwiched between two electrodes (one semitransparent) in order to apply an external voltage V. The carrier mobility is related to τ through the expression $\mu_{e,h} \simeq L^2/V\tau$. Perylene (not parylene!) exhibits $\mu_e \sim 3$ cm^2 V^{-1} s^{-1} at RT while for biphenyl $\mu_e \sim 1$ cm^2 V^{-1} s^{-1}, also at RT. At low temperature, carrier mobilities increase reaching values of more than 100 cm^2 V^{-1} s^{-1}.

An important parameter, from the application point of view, in OFETs is the on–off current ratio. For pentacene this ratio can be as high as 10^8 near $V_T = 0$ (Lin *et al.*, 1997), comparable to values typically obtained for hydrogenated amorphous silicon, making pentacene suitable for display and other low-voltage applications.

Doping-induced superconductivity?

Doping-induced superconductivity in MOMs has been demonstrated, e.g., for C$_{60}$, where Cs$_x$Rb$_y$C$_{60}$ reaches T_c values as high as 33 K (Tanigaki *et al.*, 1991).

The neutral insulator TMTSF, which shows field-effect conduction with $\mu_h \simeq 0.2$ cm^2 V^{-1} s^{-1} (Nam et al., 2003), when transformed into a Bechgaard salt also becomes superconducting, but at lower temperatures. In this case the perfect segregation of organic and inorganic molecular planes leads to confined electronic systems, which in the normal state are quasi 1D. Organic superconductors based on the BEDT-TTF molecule represent the case of pure 2D electronic systems.

The fundamental question that many researchers have formulated is whether a 2D electron gas can exhibit zero resistance below a given temperature, that is, become superconducting. The answer seems to be positive if one thinks of the BEDT-TTF salts or of the inorganic material MgB$_2$ but is not yet solved for artificial structures such as OFETs and inorganic superlattices, e.g., those involving III-V semiconductors. This extremely interesting subject was, however, surrounded by an unprecedented scandal in the year 2002, when it was proven that some articles published in highly rated journals showing the discovery of superconductivity in OFETs were false. The internal investigation conducted at Bell Laboratories, Lucent Technologies proved scientific misconduct by the falsification or fabrication of data.[3] For instance, anthracene and C$_{60}$ were claimed to become superconductors below 4 and 52 K, respectively, upon charge injection. Such articles were retracted. This situation was very prejudicial for the scientific community and calls for a profound reflection on the way some scientists are working or forced to work (quick spectacular results and many publications seem to guarantee glory and funding!).[4]

At this point it should be made clear that the falsified data mainly concerned superconductivity of organic semiconductors or insulators but not OFETs. It suffices to read articles published before 2002, such as Garnier (1996). However, because of the charge injection procedure involved, suspicion unfairly fell over the OFETs community, a community that was rapidly incremented after the false claims. Fortunately, it is rewarding to see that from about 2003 onwards, many research groups have published several works on OFETs (some of them summarized in Table 6.1), showing that they pursued their research on this subject, and showing that the quality of the semiconductor/insulator interfaces is improving. This is clearly the way forward. However, since 2D electron systems can in principle become

---

[3] The full report of the investigation committee has been available at http://www.lucent.com/news_events/pdf/researchreview.pdf

[4] D' être plus qu' un homme, dans un monde d' hommes. Echapper à la condition humaine, vous disais-je. Non pas puissant: tout-puissant. La maladie chimérique, donc la volonté de puissance n' est que la justification intellectuelle, c' est la volonté de déité: tout homme rêve d' être dieu. To be more than a man, in a world of men. To escape the human condition, I was saying to you. Not powerful: all-powerful. The visionary disease, of which the desire for power is only the intellectual justification, is the desire for deity: every man dreams of being god. André Malraux, La condition humaine.

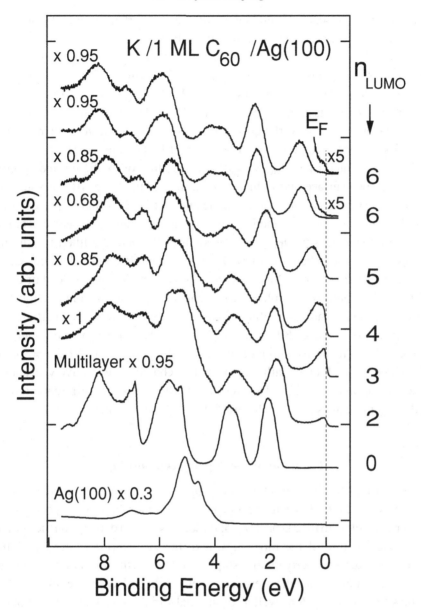

Figure 6.25. Valence band photoemission spectra of 1 ML C_{60} on a Ag(100) surface as a function of potassium doping. Also shown are the spectra of the clean Ag(100) surface and of a C_{60} multilayer (bottom). All binding energies are referred to the E_F of polycrystalline silver. Reprinted from *Surface Science*, Vols. 454–456, C. Cepek, M. Sancrotti, T. Greber and J. Osterwalder, *Electronic structure of K doped C_{60} monolayers on Ag(001)*, 467–471, Copyright (2000), with permission from Elsevier.

superconducting, research should be continued and perhaps in a few years super-conductivity will be conclusively found in OFETs.

Let us end this section with the interesting case of the controlled doping of thin C_{60} films with potassium (Cepek *et al.*, 2000). Figure 6.25 shows the evolution measured *in situ* of the valence band spectra of 1 ML C_{60} grown on a Ag(100) surface as a function of potassium exposure. The C_{60} molecules form a closed-packed compact layer, with an overlayer–substrate commensurable register represented by a $c(6 \times 4)$ reconstruction, as evidenced by LEED patterns. The LUMO band filling is represented by $n_{LUMO} = N_{at}$. At the bottom of the figure, the spectra of a clean Ag(100) surface and a thick C_{60} multilayer film (>5 ML) are shown for comparison. The C_{60} HOMO has a binding energy of *c.* 2 eV and disperses towards higher binding energies as a function of potassium coverage. The peak that emerges close to E_F for increasing n_{LUMO} corresponds to the partially filled LUMO up to the $n_{LUMO} \simeq 6$ case, where the LUMO becomes completely filled. In this case we can no longer talk about LUMO, since it is fully occupied, but instead of the new HOMO. This beautiful example complements the discussion from Section 1.7 on electronic structure and band filling, because it clearly shows that a partially filled band, e.g., $n_{LUMO} \simeq 3$, leads to metallic character ($N(E_F) \neq 0$) while a completely filled band leads to an insulating or semiconducting character ($N(E_F) = 0$). Solid K_3C_{60} and K_6C_{60} exhibit metallic and insulating character, respectively. In addition, the fact that for $n_{LUMO} \leq 3$ the HOMO and LUMO bands shift differently indicates that the C_{60} molecules become distorted in this doping regime.

6.4 Structure–property relationships

The relationships between structures and properties can be classified as intrinsic and extrinsic owing to the molecular arrangement and morphology, respectively. The term intrinsic refers to the 3D packing of molecules, which depends on the geometry and chemical nature of the molecules, and thus on intermolecular interactions. Extrinsic structure–property relationships are related to the formation of interfaces, e.g., grain boundaries, and the presence of defects. In both cases the role of external variables such as T, P, B as well as of internal variables such as the type of guest molecules is essential.

Intrinsic structure–property relationships

The role of solvents

Organic solvents not only participate in the synthesis and crystallization of MOMs but they can also form part of new crystallographic phases as guest molecules. In general these new structures can be regarded as perturbations of the solvent-free phases. Changing the solvent molecule in a given structure may induce dramatic differences in the observed physical properties. A remarkable example is the ternary hybrid RCS β''-(BEDT-TTF)$_4$·(guest)·[(H$_3$O)Fe(C$_2$O$_4$)$_3$] (Turner *et al.*, 1999). If the solvent molecule is C$_6$H$_5$CN the material is a superconductor with $T_c \simeq 8.6$ K, while when the guest molecule is C$_5$H$_5$N the resulting material exhibits a metal–insulator transition at 116 K, associated with an imperfect ordering of the terminal ethylenedithio groups. Both structures consist of 2D layers of BEDT-TTF with +0.5 formal charge per molecule and layers of approximately hexagonal symmetry containing H$_3$O$^+$ and [Fe(C$_2$O$_4$)$_3$]$^{-3}$. The solvent molecules occupy hexagonal cavities formed by the anionic layer. Thus, the physical properties can be modified with a certain degree of freedom by this discrete variable of the parameter hyperspace.

Another relevant example is the series of pseudopolymorphs β''-(BEDT-TTF)$_4$·(guest)$_n$·[Re$_6$X$_6$Cl$_8$], where X = S or Se (Deluzet *et al.*, 2002a). Among the considerable number of synthesized materials we focus on two of them: β''-(BEDT-TTF)$_4$·(H$_2$O)$_2$·[Re$_6$S$_6$Cl$_8$] and β''-(BEDT-TTF)$_4$·(C$_4$H$_8$O$_2$)·[Re$_6$S$_6$Cl$_8$]. Views of their crystallographic structures along the *a*-axis are given in Fig. 6.26. Both phases adopt the same layered hybrid architecture in which single slabs of BEDT-TTF and inorganic cluster dianions alternate along the *b*-axis direction. The BEDT-TTF slabs have the β'' topology (see Fig. 1.20). The cluster dianions are arranged within the inorganic slab so as to leave enough room to incorporate neutral molecules within the hybrid host structure.

Both phases exhibit metallic character at RT with $\sigma_{RT} \simeq 90$ and 20–70 Ω^{-1} cm^{-1} for the H$_2$O- and C$_4$H$_8$O$_2$-derived materials, respectively. The water pseudoisomorph remains metallic down to 4.2 K while the C$_4$H$_8$O$_2$ pseudoisomorph shows a rather sharp metal–insulator transition below *c*. 100 K. In this case the metallic state is restored down to 4.2 K by application of hydrostatic pressures of 15 kbar.

The metallic behaviour can be quite simply understood with the band structure arguments given in Section 1.7, although the chemical formula may frighten at first sight. Since there are four donors per repeat unit of the layer, then the band structure close to E_F consists of four BEDT-TTF-induced bands. Since the average charge per molecule of the donors is +0.5 there should be $4 \times 0.5 = 2$ holes in the bands. If the topmost band is empty (we have 2 electrons per band), the salt can be either a semiconductor, in the case that the band does not overlap with the neighbouring band (*weak* interaction) where a band gap would form, or a metal if the two upper bands overlap (*strong* interaction). Since in this case the overlap is important, they exhibit metallic character. Note that the introduction of guest molecules in the structure permits a control on the balance of interactions, in particular with hydrogen bonding.

(a)   **(b)**

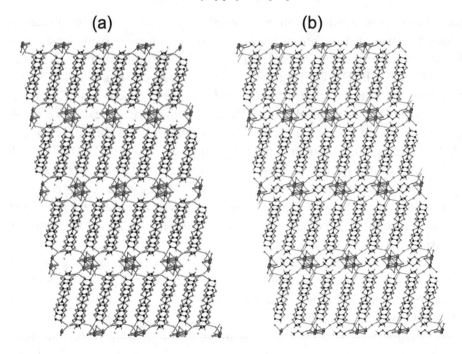

Figure 6.26. Projected view along the a-direction of (a) β''-(BEDT-TTF)$_4$·
(H$_2$O)$_2$·[Re$_6$S$_6$Cl$_8$], $P\bar{1}$, $a = 0.910$ nm, $b = 2.006$ nm, $c = 1.157$ nm, $\alpha = 84.380°$, $\beta = 101.405°$, $\gamma = 79.318°$ and (b) β''-(BEDT-TTF)$_4$·(C$_4$H$_8$O$_2$)·
[Re$_6$S$_6$Cl$_8$], $P\bar{1}$, $a = 0.910$ nm, $b = 2.003$ nm, $c = 1.170$ nm, $\alpha = 84.85°$, $\beta = 101.32°$, $\gamma = 78.495°$. Crystallographic data from Deluzet *et al.*, 2002a.

In the C$_4$H$_8$O$_2$ case the metal–insulator phase transition seems to originate from structural modifications as a function of temperature. Dimerization would explain such a transition because of the induced opening of the gap.

Polymorphism of the Bechgaard–Fabre salts

As we saw in Section 3.1 the BFS are usually grown electrochemically on a platinum wire as high-quality single-crystalline needles by constant low dc oxidation of an organic solution of the corresponding neutral π-donor molecule and the TBA salt of the anion as electrolyte, an experiment which, repeated countless times in many laboratories, has consistently delivered the triclinic $P\bar{1}$ phase. However, when synthesized by EC in a physically confined environment polymorphs of the BFS have been recently obtained. We discuss here the crystal structure and the corresponding transport properties of μ'-(TMTTF)$_2$ReO$_4$, μ'-(TMTTF)$_2$IO$_4$ and μ''-(TMTSF)$_2$ClO$_4$, a series of monoclinic polymorphs (Perruchas *et al.*, 2005). The stacks are significantly dimerized in the first two resulting in activated conductivity (see discussion on page 67). On the contrary, the quasi-regular zig-zag stacking pattern and electronic structure of the third are essentially similar to those of the prototypical triclinic metal and superconductor $(TMTSF)_2ClO_4$.

The CEC was carried out between two glass substrates held close together. The working electrode was a gold deposit on one of the glass slides. As discussed in Chapter 3, the CEC technique is well suited for the synthesis of single crystals of known compounds already obtained by classical EC, as well as for the preparation of new materials such as $(TMTSF)_2[W_6O_{19}]$ (Deluzet *et al.*, 2002b).

Let us start first with thin single crystals of $(TMTTF)_2ReO_4$. When prepared in the confined geometry, single crystals with two different shapes are obtained: rectangular and square platelets. The rectangular crystals are identified as the classical $P\bar{1}$ phase with cell parameters identical to those reported in the literature for $(TMTTF)_2ReO_4$ (Kobayashi *et al.*, 1984). The crystal structure of the square-shaped crystals is different, as determined by XRD. In Fig. 6.27 the structure for both the classical, $(TMTTF)_2ReO_4$, and the new phase, μ'-$(TMTTF)_2ReO_4$, are reported.

Both salts have the same composition with a stoichiometry of two TMTTF molecules per one ReO_4 anion. In each structure the organic molecules form slabs but in the new one each molecular plane is rotated with respect to the next due to a glide plane c in the crystal symmetry $C2/c$. The anion is located on a C_2 axis and not on an inversion centre. The new phase μ'-$(TMTTF)_2ReO_4$ is thus a polymorph of the $(TMTTF)_2ReO_4$ salt. The molecular slabs are quite different between the two structures. The TMTTF columns are more dimerized with a larger longitudinal displacement of the molecule in μ'-$(TMTTF)_2ReO_4$ as shown in Fig. 6.28.

Overlap between sulfur atoms is less favourable, so interactions between molecules are weaker in μ'-$(TMTTF)_2ReO_4$ than in $(TMTTF)_2ReO_4$. The electronic structure of μ'-$(TMTTF)_2ReO_4$ is reported in Fig. 6.29. With the cell axis chosen, a' and b' are the intra- and interstack directions, respectively. The band structure of $(TMTTF)_2ReO_4$ has been calculated using the same cell axis system in order to facilitate the comparison (see Fig. 1.30 and discussion about the origin of the bands). The main differences between the two band structures are the larger dimerization gap and thus the smaller dispersion of the upper band for μ'-$(TMTTF)_2ReO_4$. These are ascribed to the decrease of the interdimer interaction in μ'-$(TMTTF)_2ReO_4$ as a result of the considerably larger displacement along the long molecular axis (see Fig. 6.28), whereas the intradimer interaction remains almost identical in both polymorphs. Consequently, electronic repulsion can take over and lead to a localized system with activated conductivity.

In agreement with this analysis, single-crystal conductivity measurements using the four-probe technique reveals semiconducting behaviour for μ'-$(TMTTF)_2ReO_4$, as shown in Fig. 6.30. In this case $\sigma_{RT} \simeq 0.011$ Ω^{-1} cm^{-1} and $E_a \simeq 0.17$ eV.

(a)

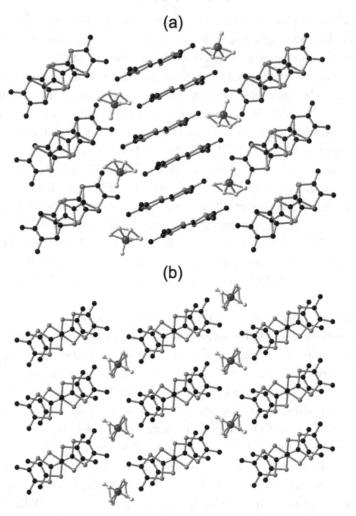

(b)

Figure 6.27. (a) View of the crystal structure of μ'-(TMTTF)$_2$ReO$_4$ along the [110] direction ($C2/c$, $a = 1.303$ nm, $b = 0.879$ nm, $c = 2.386$ nm, $\beta = 97.94°$). Crystallographic data from Perruchas *et al.*, 2005. (b) Crystal structure of (TMTTF)$_2$ReO$_4$ at 293 K. Crystallographic data from Kobayashi *et al.*, 1984. C, Re, S and O atoms are represented by black, dark grey, medium grey and light grey balls, respectively. H atoms are omitted for clarity. The two disordered positions of the anions are represented.

For comparison, the classical triclinic phase (TMTTF)$_2$ReO$_4$ is metallic down to 230 K with $\sigma_{RT} \simeq 33$ Ω^{-1} cm^{-1} (Coulon *et al.*, 1982).

Remarkably, when electrocrystallizing TMTTF in a TBA·IO$_4$ solution in a confined environment, only rectangular-shaped thin crystals are obtained. The crystal structure gives a 2:1 stoichiometry salt with the formula μ'-(TMTTF)$_2$IO$_4$,

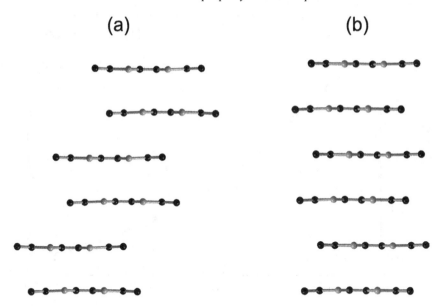

Figure 6.28. Stack of molecules in (a) μ'-(TMTTF)$_2$ReO$_4$ and (b) in triclinic (TMTTF)$_2$ReO$_4$. C and S atoms are represented by black and medium grey balls, respectively. H atoms are omitted for clarity.

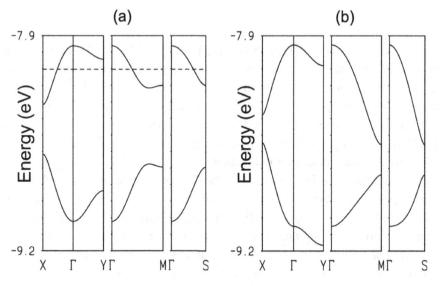

Figure 6.29. (a) Band structure of μ'-(TMTTF)$_2$ReO$_4$. $\Gamma = (0, 0)$, X $= (a'^*/2, 0)$, Y $= (0, b'^*/2)$, M $= (a'^*/2, b'^*/2)$, S $= (-a'^*/2, b'^*/2)$ with $a' = \frac{1}{2}(b - a)$ and $b' = \frac{1}{2}(a + b)$. (b) Band structure of (TMTTF)$_2$ReO$_4$. Reprinted with permission from Perruchas *et al.*, 2005.

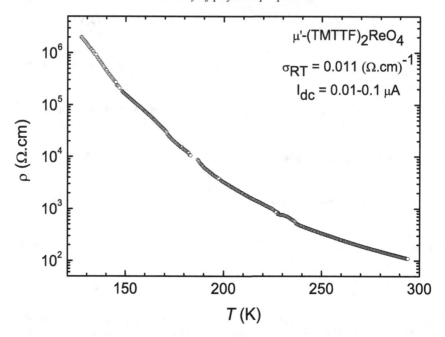

Figure 6.30. Resistivity measurements using the four-probe technique performed on a μ'-(TMTTF)$_2$ReO$_4$ thin single crystal. Reprinted with permission from Perruchas *et al.*, 2005.

presenting the same polymorphism observed for μ'-(TMTTF)$_2$ReO$_4$ with dimerized columns of TMTTF molecules. However, the anion IO$_4^-$ is not disordered ($C2/c$, $a = 1.304$ nm, $b = 0.879$ nm, $c = 2.399$ nm, $\beta = 98.03°$). Conductivity measurements give $\sigma_{RT} \simeq 0.008$ Ω^{-1} cm^{-1}. The classical phase (TMTTF)$_2$IO$_4$ is a semiconductor even at RT (Yakushi *et al.*, 1986). As for the previous μ'-(TMTTF)$_2$ReO$_4$ polymorph, the new structural arrangement of organic molecules involves weaker intermolecular interactions.

When preparing (TMTSF)$_2$ClO$_4$ with the CEC technique, crystals with two different shapes are collected between the substrates: rectangles and squares. The rectangular platelets correspond to the original triclinic (TMTSF)$_2$ClO$_4$ salt. The crystal structure of the square platelets is shown in Fig. 6.31(a) with that of triclinic (TMTSF)$_2$ClO$_4$ (Fig. 6.31(b)) for comparison. With the same formula and stoichiometry, a new phase is obtained which corresponds to a polymorph of (TMTSF)$_2$ClO$_4$ called μ''-(TMTSF)$_2$ClO$_4$. Organic molecules form slabs in both structures but, due to the glide plane c in the new salt, the slabs present different orientations along c. μ''-(TMTSF)$_2$ClO$_4$ shows a slightly different type of polymorphism from the one observed in μ'-(TMTSF)$_2$ReO$_4$ and μ'-(TMTSF)$_2$IO$_4$, hence using μ'' instead of μ'. The TMTSF slabs in μ''-(TMTSF)$_2$ClO$_4$ are structurally

(a)

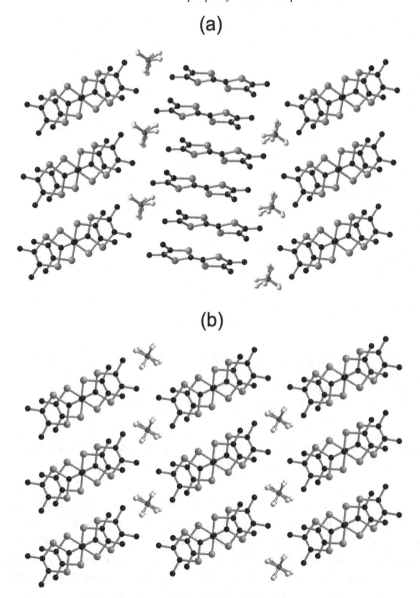

(b)

Figure 6.31. (a) View along the $[1\bar{1}0]$ direction of μ''-(TMTSF)$_2$ClO$_4$ ($C2/c$, $a = 1.206$ nm, $b = 0.879$ nm, $c = 2.639$ nm, $\beta = 95.30°$). Crystal structure from Perruchas *et al.*, 2005. (b) View along the a-axis of triclinic (TMTSF)$_2$ClO$_4$ ($P\bar{1}$, $a = 0.723$ nm, $b = 0.768$ nm, $c = 1.327$ nm, $\alpha = 84.58°$, $\beta = 86.73°$, $\gamma = 70.43°$). Crystal structure from Rindorf *et al.*, 1982. C, Cl, S and O atoms are represented by black, dark grey, medium grey and light grey balls, respectively. H atoms are omitted for clarity. The two disordered positions of the anions are represented.

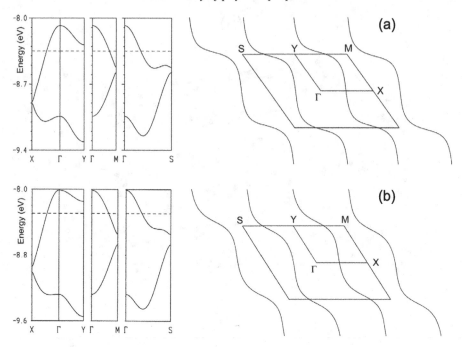

Figure 6.32. (a) Band structure and FS of μ''-(TMTSF)$_2$ClO$_4$. $\Gamma = (0, 0)$, X $=$ $(a'^*/2, 0)$, Y $= (0, b'^*/2)$, M $= (a'^*/2, b'^*/2)$, S $= (-a'^*/2, b'^*/2)$ with $a' = \frac{1}{2}(b +$ $a)$ and $b' = b$. (b) Band structure and FS of triclinic (TMTSF)$_2$ClO$_4$. Reprinted with permission from Perruchas *et al.*, 2005.

identical to those in (TMTSF)$_2$ClO$_4$, the overlap between all TMTSF molecules being the same. The electronic structure of μ''-(TMTSF)$_2$ClO$_4$ is reported in Fig. 6.32(a).

The cell axes are chosen so that a' is the interstack direction as for μ'-(TMTSF)$_2$ReO$_4$. For comparison, the band structure and the FS of (TMTSF)$_2$ClO$_4$ have been calculated using the same cell axes and are also reported in Fig. 6.32(b). Both are very similar but the band dispersions in μ''-(TMTSF)$_2$ClO$_4$ are approximately 20% weaker than in (TMTSF)$_2$ClO$_4$. Then interactions between TMTSF molecules in the organic slabs are slightly weaker in the new phase favouring electronic localization. Moreover, the FSs calculated for the two salts, although similar, exhibit slightly different nesting vectors, which means that the interstack interactions are also somewhat different in the two polymorphs. Thus, despite the structural similarity, the physical behaviour of the two polymorphs could well be different. Clearly, conductivity measurements are needed, but for the time being μ''-(TMTSF)$_2$ClO$_4$ is a disappearing polymorph (Dunitz & Bernstein, 1995). See page 49 for the definition of a disappearing polymorph.

(a)   (b)

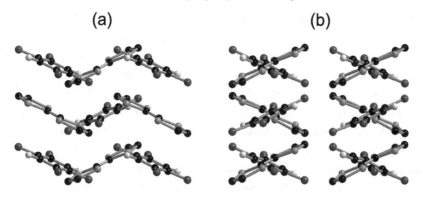

Figure 6.33. View along the c-axis of the crystal structure of TTF-CA: (a) N-phase ($P2_1/n$, $a = 0.740$ nm, $b = 0.762$ nm, $c = 1.459$ nm, $\beta = 99.10°$) and (b) I-phase (Pn, $a = 0.719$ nm, $b = 0.754$ nm, $c = 1.444$ nm, $\beta = 98.60°$). C, Cl, S and O atoms are represented by black, dark grey, medium grey and light grey balls, respectively. H atoms are omitted for clarity. Crystallographic data from Le Cointe *et al.*, 1995.

It is worth noting that when trying to obtain other polymorphs of Bechgaard salts incorporating octahedral, fluorine-containing anions as in $(TMTSF)_2PF_6$ or $(TMTSF)_2SbF_6$ only the triclinic phase is obtained.

Phase transitions

Some semiconducting organic CT complexes of mixed-stack architecture exhibit the rather unusual neutral-to-ionic (N–I) phase transition upon variation of an external variable of parameter hyperspace, such as P or T. The transition manifests itself by a change of ϱ and a dimerization distortion with the formation of donor–acceptor dimers along the stacking axis in the I-phase.

This situation is different from what we previously saw for the BFS polymorphs and, e.g., for TMTSF-TCNQ, because in those cases the physical properties are associated with different structures in a *discrete way*, that is the synthetic routes lead to the different polymorphs and each polymorph exhibits its specific properties. However, what we encounter here is the *continuous* transition between two polymorphs, with the resulting change of properties. TTF-CA is a prototype compound undergoing such a N–I transition (Torrance *et al.*, 1981). At atmospheric pressure and RT, TTF-CA crystallizes in the N-phase (see Table 1.9). The space group is $P2_1/n$ with two symmetry related undimerized donor–acceptor pairs in the unit cell (see Fig. 6.33(a)). TTF and CA molecules are located on sites with a centre of inversion i and form alternating stacks along the crystallographic a-axis. For the N-phase $\varrho \simeq 0.3$. Below $T_{N-I} \simeq 81$ K the system undergoes a first order transition, making

TTF-CA more ionic, increasing ϱ up to 0.7. The space group of the I-phase is *Pn* with two equivalent donor–acceptor dimers related by a glide plane with a ferroelectric arrangement (see Fig. 6.33(b)). Further examples of mixed-stack organic CT materials exhibiting N–I transitions are tetramethylbenzidine-TCNQ ($T_{N-I} \simeq$ 205 K) (Iwasa *et al.*, 1990) and DMTTF-CA ($T_{N-I} \simeq 65$ K) (Aoki *et al.*, 1993).

In addition to the *P*- and *T*-induced N–I transition, such a transformation can also be induced by laser irradiation (Collet *et al.*, 2003). With the development of ultrafast lasers and reliable optics, it is now possible to induce a phase transition by a light pulse and then to change the macroscopic state and the physical properties of a material. An optical 300 fs laser pulse effectively switches the material from the N to the I state on a 500 ps time scale, generating long-range structural order. This kind of study has become possible thanks to time-resolved XRD possibilities offered by latest generation synchrotron sources. The experiments referred to here were performed with laser pulses of about 300 fs at 1.55 eV, with the light polarization parallel to the CT stack. After laser irradiation, drastic changes are observed in the intensity of some of the Bragg reflections. These intensity changes are a direct signature of a strong structural reorganization in the photoinduced state. The equilibrium state recovers in much less than 1 ms.

The case of TTTA is also extremely interesting since it shows RT magnetic bistability (Fujita & Awaga, 1999). As TTTA samples are cooled from RT, the paramagnetic χ shows a slight decrease, which becomes rather abrupt at 230 K, and below 170 K $\chi \simeq 0$. TTTA is thus diamagnetic below this temperature (LT-phase from Table 1.11). When heated from below this temperature (150 K), $\chi \simeq 0$ up to 230 K and slightly increases above this temperature but a sudden increase is observed at 305 K, recovering the RT value (HT-phase from Table 1.11). This paramagnetic–diamagnetic transition is of first order and exhibits a surprisingly wide hysteresis loop (230–305 K). The major structural difference between both phases is the strong dimerization found in the LT-phase. For the HT-phase the intra- and intercolumn exchange coupling constants are -27.6 and -5.2 meV, respectively, while for the LT-phase the intradimer constant is much larger, -112.2 meV.

Extrinsic structure–property relationships

The role of grain boundaries

In this section we illustrate the effect of grain boundaries on the conductivity of thin films of the TTF-based molecular metals TTF-TCNQ and TTF[Ni(dmit)$_2$]$_2$. In general the existence of grain boundaries hinders the metallic behaviour, inducing semiconducting-like activated conduction ($\sigma_{RT} < 10\ \Omega^{-1}\ cm^{-1}$) because at the intergrain interfaces stoichiometry is not necessarily preserved and because

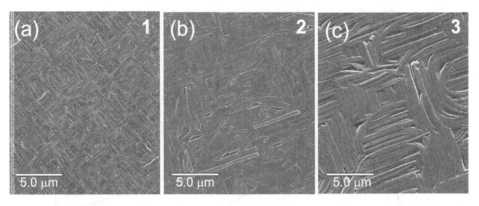

Figure 6.34. SEM images of TTF-TCNQ/KCl(100) films obtained at (a) **1**, T_{sub} = RT and $T_{ann} \simeq 350$ K, (b) **2**, $T_{sub} \simeq 310$ K and $T_{ann} \simeq 360$ K and (c) **3**, $T_{sub} \simeq 325$ K and $T_{ann} \simeq 350$ K. Reprinted from *Journal of Solid State Chemistry*, Vol. 168, J. Fraxedas, S. Molas, A. Figueras, I. Jiménez, R. Gago, P. Auban-Senzier and M. Goffman, *Thin films of molecular metals: TTF-TCNQ*, 384-389, Copyright (2002), with permission from Elsevier.

of the uncontrolled nature of intergrain contacts. However, we will see that under certain conditions the metallic character can be maintained down to cryogenic temperatures.

Let us start with thin TTF-TCNQ films grown by PVD in HV (Fraxedas *et al.*, 2002b). Figure 6.34 shows SEM images of films **1**, **2** and **3** prepared on KCl(100) substrates. The thickness of these films is ~1 μm. Films **1**, **2** and **3** exhibit an increase of microcrystal size upon increasing T_{sub}. When the films are grown at T_{sub} = RT and $T_{ann} \simeq 350$ K (**1**) the mean length of the microcrystals along the long crystallographic axis (*b*-axis) is *c.* 1.5 μm, increasing to *c.* 2.5 μm when $T_{sub} \simeq 310$ K and $T_{ann} \simeq 360$ K (**2**) and up to about 5.0 μm when obtained at $T_{sub} \simeq 325$ K and $T_{ann} \simeq 350$ K (**3**). Note the relevant increase of size for a moderate increase in temperature (25 K). This is a characteristic of MOMs.

When the KCl(100) substrates are held at RT we can observe that the microcrystals are equally distributed in perpendicular directions. In addition to the increase in size with increasing T_{sub}, the microcrystals in the same direction tend to agglomerate leaving, however, the final microcrystal distribution unaltered: no temperature-induced in-plane preferential orientation is observed. In case **3** a considerable number of microcrystals become curved. This is due to the triggering of in-plane growth directions other than the equivalent [110] directions caused by the contributions of the electrostatic and van der Waals interactions to the total energy at the interface (Caro *et al.*, 1997).

Figure 6.35 shows ρ measurements as a function of temperature performed on films grown at T_{sub} = RT and $T_{ann} \simeq 350$ K (**1a** and **1b**) and $T_{sub} \simeq 325$ K and

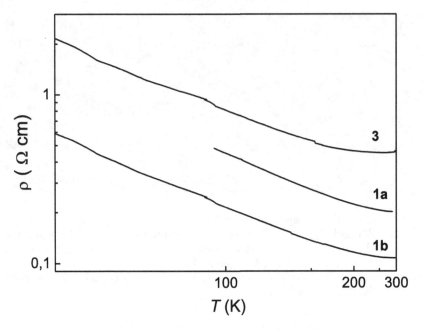

Figure 6.35. $\rho(T)$ curves of thin TTF-TCNQ films grown at (**1a** and **1b**) $T_{sub} = $ RT and $T_{ann} \simeq 350$ K and (**3**) $T_{sub} \simeq 325$ K and $T_{ann} \simeq 350$ K. The measurements were performed using the four contacts low-frequency lock-in technique with 10 µA for **1a** and 0.1 µA for **1b** and **3**. Reprinted from *Journal of Solid State Chemistry*, Vol. 168, J. Fraxedas, S. Molas, A. Figueras, I. Jiménez, R. Gago, P. Auban-Senzier and M. Goffman, *Thin films of molecular metals: TTF-TCNQ*, 384-389, Copyright (2002), with permission from Elsevier.

$T_{ann} \simeq 350$ K (**3**). The conduction barrier energy E_a can be obtained from the Arrhenius plot: $E_a \simeq 296.8$ and 295.4 K for the nominally identical samples **1a** and **1b** (grown in different experiments) and 273.7 K for sample **3**. Earlier reported E_a values for oriented thin TTF-TCNQ films lie between 170 and 580 K (Reinhardt *et al.*, 1980; de Caro *et al.*, 2000a). The decrease of E_a upon increase of T_{sub} is in line with the increase in size of the microcrystals, which reduces the number of grain boundaries. The obtained σ_{RT} values ($2 < \sigma_{RT} < 10 \ \Omega^{-1} \ cm^{-1}$) are comparable with earlier determinations on thin TTF-TCNQ films ($1 < \sigma_{RT} < 30 \ \Omega^{-1} \ cm^{-1}$) (Reinhardt *et al.*, 1980; Sumimoto *et al.*, 1995; Figueras *et al.*, 1999; de Caro *et al.*, 2000a), the conductivity of the films resulting from a random contribution of σ_a and σ_b values.

A detail of the $\rho(T)$ curve is displayed for sample **3** in Fig. 6.36. The non-linear increase of resistivity below *c.* 50 K corresponds to the metal-insulator Peierls transition. The Peierls transition is more readily observed for the $E_a \simeq 273.7$ K sample because of its lower activation energy as compared to the *c.* $E_a \simeq 296$ K samples.

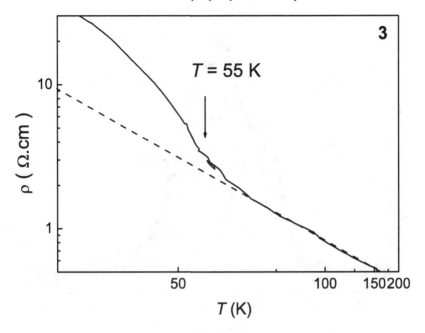

Figure 6.36. Detail of the $\rho(T)$ curve for **3** showing the Peierls transition. The dashed line corresponds to a linear interpolation of the higher temperature region. Reprinted from *Journal of Solid State Chemistry*, Vol. 168, J. Fraxedas, S. Molas, A. Figueras, I. Jiménez, R. Gago, P. Auban-Senzier and M. Goffman, *Thin films of molecular metals: TTF-TCNQ*, 384–389, Copyright (2002), with permission from Elsevier.

Our next example deals with thin films of TTF[Ni(dmit)$_2$]$_2$ grown by electrode-position on silicon wafers (de Caro *et al.*, 2004). We recall here that TTF[Ni(dmit)$_2$]$_2$ single crystals exhibit metallic behaviour down to 3 K, with $\sigma_{RT} \simeq 300 \ \Omega^{-1} \ cm^{-1}$ and superconductivity is observed at $T_c \simeq 1.6$ K under application of a hydrostatic pressure of 7 kbar (Brossard *et al.*, 1986). The electrodeposited films exhibit metal-lic character down to *c.* 12 K in spite of their polycrystalline morphology, as will be shown next. The advantage of the EC technique over vapour-phase deposition methods is that vapour-phase deposition is limited to sublimation/evaporation of neutral species. The EC-grown films consist of crystals with sizes ranging from 0.6 to 1 μm and the estimated thickness of the films is *c.* 1 μm.

Figure 6.37 shows the XPS S$2p$ line measured *ex situ* for a TTF[Ni(dmit)$_2$]$_2$ thin film. The experimental lineshape can be satisfactorily decomposed into three contributions. The most intense line, with a binding energy of 163.5 eV, corre-sponds to C–S–C bonds, and the 161.8 and 165.3 eV lines to C–S–Ni and C=S bonds, respectively. In the nominal formula TTF[Ni(dmit)$_2$]$_2$ there are 12 C–S–C, 8 C–S–Ni and 4 C=S bonds, which results in a ratio 3:2:1. From the fit, the intensity

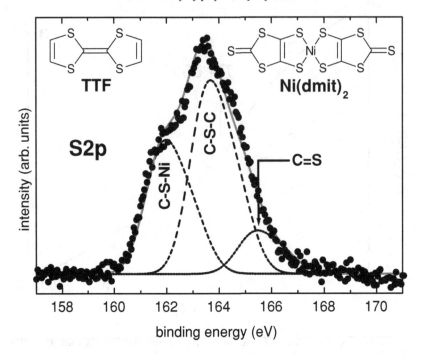

Figure 6.37. XPS S2*p* line (black dots) measured *ex situ* at RT for a TTF[Ni(dmit)₂]₂ thin film. Also shown is a least-squares fit (Gaussians), after a Shirley-type background subtraction, where each line (dashed) is composed of a spin-orbit split doublet (compare to Fig. 1.31). The addition of the three lines results in the continuous grey line. Reprinted with permission from de Caro *et al.*, 2004.

ratio between the C–S–C and C–S–Ni contributions is 1.4, in agreement with the nominal 1.5 value.

Finally, Fig. 6.38 shows the electrical behaviour of the electrodeposited material. The ratio between the resistance at a given temperature T over the RT resistance is reported as a function of the temperature. The curve clearly indicates metallic behaviour down to c. 12 K. Below this temperature the resistance slightly increases. Such behaviour is totally reversible when warming back to RT (see inset in Fig. 6.38). The conductivity values are $\sigma_{RT} \simeq 12~\Omega^{-1}~cm^{-1}$ and $\sigma > 50~\Omega^{-1}~cm^{-1}$ at 12 K. As expected because of grain boundary effects, these values are much lower than the conductivity values of single crystals (300 $\Omega^{-1}~cm^{-1}$ at 300 K, and $1.5 \times 10^5~\Omega^{-1}~cm^{-1}$ at 2 K). The metallic behaviour shows that low intergrain conduction energy barriers are achieved, in contrast with most thin films of highly conducting materials reported to date.

Another example of thin films showing metallic transport properties down to liquid helium temperature is θ-(BET-TTF)₂Br.3H₂O, where bilayers are grown on transparent polycarbonate substrates (Mas-Torrent *et al.*, 2001). The bilayers are

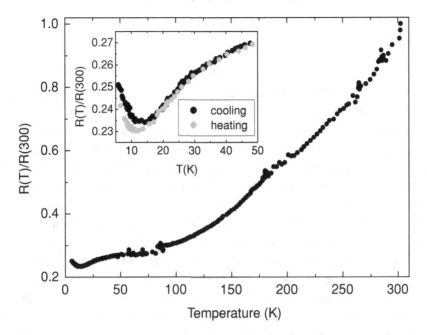

Figure 6.38. Resistance normalized to RT resistance as a function of temperature for electrodeposited TTF[Ni(dmit)$_2$]$_2$ thin films (cooling cycle). The inset shows a detail of the low-temperature region for both cooling (black dots) and heating (grey dots) cycles. Reprinted with permission from de Caro *et al.*, 2004.

prepared by dispersing BET-TTF molecules into the polymeric substrate and by exposure to bromine vapours. The detailed procedure for the preparation of these films is discussed on page 136. σ_{RT} values as high as $\sim 120 \ \Omega^{-1} \ cm^{-1}$ have been measured for such films.

Grain boundaries have an influence not only on the transport properties but also on the optical properties of thin films. This has been illustrated for thin α-4T and α-6T films grown by vapour deposition. In the case of α-6T films, they exhibit a morphology transition from grains to lamellae depending on the growing conditions, which is ascribed to the change from diffusion-limited growth to a strong adsorption regime (Biscarini *et al.*, 1997). This point was discussed in Section 5.2. As the film morphology evolves from 2D to 3D structures with film thickness, the change from exciton to aggregate emission is mainly determined by the increase of the number of aggregate states (coalescence of domains) with the number of layers. Such changes in morphology induce specific photoluminescence and polarized electroluminescence spectral features (Muccini *et al.*, 2001). On the other hand the island growth mode of α-4T and α-6T films deposited on silica and KAP generates photoluminescence features not found in pristine α-4T and

α-6T, and are thus identified as arising from the boundaries between grains (Tavazzi *et al.*, 2002).

The role of defects

In Sections 5.5 and 5.6 we analysed the nanometre-scale morphology of thin films of the remarkable material *p*-NPNN and we saw that at ambient conditions the films transform from the initially formed kinetic α-phase to the thermodynamically stable β-phase. It is interesting to note that this transformation is hindered when the thickness of the films is below 1 μm due to residual stress after growth. The defect density and thus stress can be tuned with the film thickness, decreasing with increasing film thickness. The accumulated stress field after growth induces an increase in the activation energy of formation of critical nuclei of the β-phase in the as-grown α-phase matrix (Fraxedas *et al.*, 1999). Thus, under these conditions the metastable α-phase becomes stabilized. Since the α-phase needs to incorporate defects in order to become stable we should perhaps no longer talk about the α-phase but of a different structure. However, because the XRD patterns of the films can be precisely identified with those of α-phase single crystals it seems reasonable to maintain the name α-phase. The stability of α-*p*-NPNN in the form of thin films allows the determination of the intrinsic physical properties of this elusive phase and, as an example, the determination of the magnetic properties is illustrated next (Molas *et al.*, 2003).

Figure 6.39(a) shows the χ^{-1} vs. T curve, normalized to the RT value, for a 100 nm thick α-*p*-NPNN/glass film obtained from electron paramagnetic resonance (EPR) measurements with the static magnetic field applied perpendicular to the substrate plane. As previously shown in Fig. 3.19, the molecular *ab*-planes are parallel to the substrate's surface. The data points closely follow the Curie–Weiss law $\chi^{-1} = (T - \Theta_W)/C$, where C stands for the Curie constant. In this case $\Theta_W \simeq -0.3$ K, indicating that the net intermolecular interactions are weakly antiferromagnetic. No hint of a transition at low temperature is observed. These results coincide with those derived from SQUID measurements on a single α-*p*-NPNN crystal (Tamura *et al.*, 2003), where $0.5 < \Theta_W < 0$, which are displayed in Fig. 6.40.

However, the χ^{-1} vs. T curve for completely transformed samples shows the beginning of a magnetic transition below 5 K (see Fig. 6.39(b)), more clearly observed in the $\chi^{-1}T$ vs. T curve in the inset of Fig. 6.39(b). As discussed in Section 1.5 the β-phase becomes ferromagnetic below 0.6 K.

The dependence of the *g*-factor on temperature for both stabilized (α) and transformed (β) phases with the static magnetic field applied perpendicularly to the substrate plane is shown in Fig. 6.41. At low temperature the *g*-factor decreases for

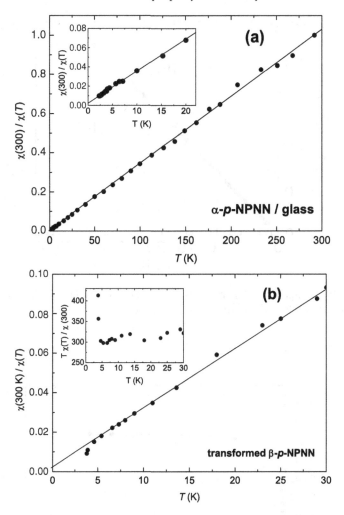

Figure 6.39. χ^{-1} as a function of T normalized at RT for (a) a stabilized α-p-NPNN thin film grown on a glass substrate and (b) a fully transformed β-p-NPNN thin film. The inset in (a) shows a detail of the low-temperature region while in (b) the low-temperature part of the χT vs. T curve is displayed. Molas *et al.*, 2003. Reproduced by permission of the Royal Society of Chemistry.

both phases suggesting the initiation of a bulk magnetic transition. This is in line with the results of the β-phase but raises some questions for the α-phase. A partial transformation of the films to the β-phase can be ruled out since the films are stable upon temperature cycles. Note that at RT the g-factors are noticeably different for the two phases: $g(\beta) = 2.0101$ and 2.0057 for the magnetic field perpendicular and parallel to the substrate plane, respectively, and $g(\alpha) = 2.0068$ and 2.0070 for the

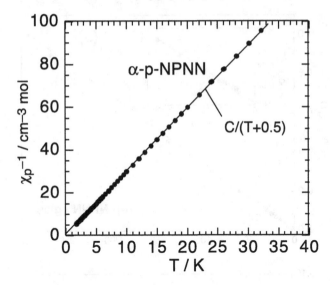

Figure 6.40. Temperature dependence of χ^{-1}. The line denotes the fit to the Curie–Weiss law. Reprinted with permission from Tamura *et al.*, 2003.

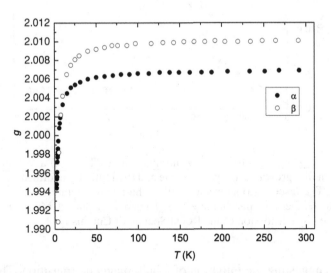

Figure 6.41. Dependence of the *g*-factor on temperature for both phases, stabilized (α) and transformed (β). The static magnetic field is applied perpendicularly to the substrate surface. Molas *et al.*, 2003. Reproduced by permission of the Royal Society of Chemistry.

magnetic field perpendicular and parallel to the molecular *ab*-plane, respectively. Note that the *g*-values are almost isotropic in the *α*-phase. For the *α*-phase the maximum *g* value, 2.0092, is obtained when the applied field is parallel to the *b*-axis (along the molecular ONCNO units), while the minimum value, 2.0030, corresponds to the field rotated 30° from the *a*-axis to the *c**-direction (*c** stands for the projection of the *c*-axis along the normal to the *ab*-plane), as determined for an *α*-*p*-NPNN single crystal (Tamura *et al.*, 2003). In spite of the fact that the *α*-phase films exhibit no in-plane texture due to their spherulitic morphology the larger *g*-values are obtained when the magnetic field is applied parallel to the substrate surface (*ab*-plane), because this plane contains the *b*-axis. The *g*-values along the *a*-, *b*- and *c*-directions for *β*-*p*-NPNN are 2.0070, 2.0030 and 2.0106, respectively (Tamura *et al.*, 1991).

In summary, the stabilization of metastable phases by means of defects is indeed not restricted to *p*-NPNN and should allow the determination of physical properties of elusive short-lived phases.

6.5 Multiproperty materials

Transition metals such as iron, cobalt and nickel undergo spontaneous ferromagnetic order below T_C, hence belonging to the bifunctional class. Both physical properties, metallicity and ferromagnetism, are interdependent and cannot be decoupled, and are ascribed to the partial occupancy of the *d* electrons shell. As far as superconductivity is concerned, its coexistence with magnetic order apparently seems to be hindered because both properties are considered incompatible, i.e., Cooper pairs are disrupted by externally applied magnetic fields. However, there are a few examples of coexistence of superconductivity and long-range magnetic order in the Chevrel phases $REMo_6S_8$ (Lynn *et al.*, 1981) and in $RERh_4B_4$ (Fertig *et al.*, 1977), where RE stands for rare-earth elements. Both materials are known as reentrant superconductors since they become superconductors below a given transition temperature T_c but reenter the normal state at a lower temperature. In some of these reentrant superconductors there may be a tendency for physical separation between the superconducting electrons and the magnetic electrons. For example, in $RERh_4B_4$ $4d$ Rh electrons are most probably responsible for superconductivity while magnetic order arises from $4f$ RE electrons.

This decoupling may be achieved by segregation of the electronic sublattices and MOMs are perhaps the best suited materials concerning lattice decoupling because of their intrinsic segregated weakly interacting 2D structure. When combining electrical conductivity with magnetism the key point is the interplay between π electrons from the donor and the *d* electrons from the transition metal of the inorganic or organometallic part. Let us explore some examples.

Coexistence of magnetism and superconductivity

The first selected material is the *paramagnetic superconductor* β''-(BEDT-TTF)$_4$(H$_2$O)Fe(C$_2$O$_4$)$_3$·C$_6$H$_5$CN. This CTS was the first example of a molecular superconductor containing magnetic ions (Kurmoo *et al.*, 1995). Its structure consists of successive layers of BEDT-TTF in the β''-arrangement (see Fig. 1.20) and layers with approximately hexagonal geometry containing alternating H$_2$O and Fe(C$_2$O$_4$)$_3^{-3}$, with C$_6$H$_5$CN lying within the hexagonal cavities. β''-(BEDT-TTF)$_4$(H$_2$O)Fe(C$_2$O$_4$)$_3$·C$_6$H$_5$CN crystallizes in the $C2/c$ space group with lattice parameters $a = 1.023$ nm, $b = 2.004$ nm, $c = 3.497$ nm and $\beta = 93.25°$. This salt is a metal with a resistivity of $\sim 10^{-2}$ Ω cm at 200 K, decreasing monotonically by a factor of about 8 down to just below 7 K, where superconductivity sets in. χ obeys the Curie–Weiss law from 300 K down to about 8 K. The resulting Weiss temperature, $\Theta_W \simeq -0.2$ K, implies very weak antiferromagnetic exchange between the iron magnetic moments.

κ-(BEDT-TSF)$_2$FeBr$_4$ represents the case of an *antiferromagnetic superconductor* (Ojima *et al.*, 1999). Within its crystal structure ($Pnma$, $a = 1.179$ nm, $b = 3.661$ nm, $c = 0.850$ nm) the conduction layers composed of BEDT-TSF dimers and the layers of tetrahedral anions with localized magnetic Fe^{+3} moments are arranged alternately along the b-direction. A fit of χ to the Curie–Weiss law gives a Curie constant of 4.70 emu K mol^{-1} and a Weiss temperature of $\Theta_W \simeq -5.5$ K. The 4.70 emu K mol^{-1} value is close to 4.38 emu K mol^{-1}, the expected value for the $S = 5/2$ localized spin system with $g = 2.0$. Therefore, the FeBr$_4^-$ anion is considered to be in a high-spin state. The onset of antiferromagnetic ordering of Fe^{+3} spins is found at 2.5 K ($T_N \simeq 2.5$ K). The superconducting transition is observed at $T_c \simeq 1.0$ K and no anomaly has been detected at this temperature, showing that the antiferromagnetic order of Fe^{+3} spin system is not destroyed by the onset of superconductivity.

The closely related material λ-(BEDT-TSF)$_2$FeCl$_4$ is not a multiproperty material but is unique in the sense that it becomes a superconductor when elevated external magnetic fields are applied, in this case above 17 T (Uji *et al.*, 2001).

Coexistence of ferromagnetism and metallic conductivity

[BEDT-TTF]$_3$[MnCr(C$_2$O$_4$)$_3$] exemplifies the case of a material where ferromagnetism and metallic conductivity coexist (Coronado *et al.*, 2000). The structure consists of organic layers of BEDT-TTF cations alternating with honeycomb-ordered layers of the bimetallic oxalato complex [MnCr(C$_2$O$_4$)$_3$]$^-$. [BEDT-TTF]$_3$[MnCr(C$_2$O$_4$)$_3$] crystallizes in the $P\bar{1}$ space group, with lattice parameters $a = 0.517$ nm, $b = 0.638$ nm, $c = 1.661$ nm, $\alpha = 87.15°$, $\beta = 85.12°$ and

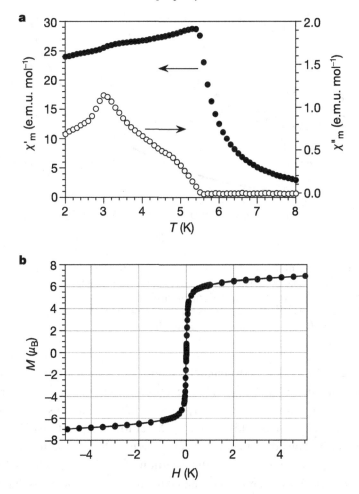

Figure 6.42. Magnetic characterization of the hybrid material [BEDT-TTF]$_3$[MnCr(C$_2$O$_4$)$_3$]: (a) χ' (filled circles) and χ'' (open circles) measured at 110 Hz. (b) Plot of $\mathscr{M}$ versus magnetic field at 2 K. M and H represent $\mathscr{M}$ and $\boldsymbol{B}$, respectively. Reprinted with permission from Coronado *et al.*, 2000. Copyright (2000) by the Nature Publishing Group.

$\gamma = 66.60°$. The BEDT-TTF cations are tilted with respect to the inorganic layer by an angle of 45°. The layers of BEDT-TTF are constructed by only one crystallographically independent BEDT-TTF molecule located with the central carbon–carbon double bond on an inversion centre. In this model the donor molecules adopt the β-type packing while the inorganic oxalate-based bimetallic layer is crystallographically disordered.

The magnetic properties of this compound show that it is a ferromagnet below $T_C \simeq 5.5$ K. In particular, ac magnetic measurements exhibit a sharp peak in the in-phase signal (χ'), accompanied by an out-of-phase signal (χ'') below T_C as

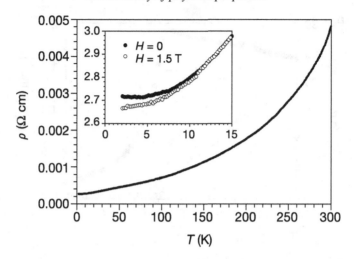

Figure 6.43. Thermal variation of ρ in the hybrid material [BEDT-TTF]$_3$ [MnCr(C$_2$O$_4$)$_3$]. The inset shows the low-temperature magnetoresistance with **B** applied perpendicular to the layers (open circles). The ρ values have been multiplied by a factor 10^4. Reprinted with permission from Coronado *et al.*, 2000. Copyright (2000) by the Nature Publishing Group.

shown in Fig. 6.42(a). The ferromagnetic nature of the transition is confirmed by the field dependence of the isothermal magnetization performed at 2 K, which shows a rapid saturation at $\mathcal{M} \simeq 7.1\mu_B$ (Fig. 6.42(b)). The hysteretic behaviour shows quite small coercive fields ($\simeq 5-10$ Oe) at low temperatures, therefore this compound can be considered as a soft ferromagnet. The properties indicate that the interlayer separation imposed by the insertion of BEDT-TTF into the bimetallic oxalato layers (1.661 nm), large as compared to other layered magnets, has no effect on the magnetic properties. This establishes that the magnetic ordering in these 2D phases occurs within the bimetallic layers and is triggered by the small magnetic anisotropy of the metallic ions.

The transport properties of this material determined using the standard four-contact dc method are depicted in Fig. 6.43. The current is set to flow in the plane of the layers. σ_{RT} reaches values as high as 250 Ω^{-1} cm^{-1}. The temperature dependence of the resistance exhibits metallic behaviour down to 2 K. Application of a magnetic field (up to 5 T) perpendicular to the layers leads to the appearance of a negative magnetoresistance at temperatures below about 10 K (inset of Fig. 6.43), whereas the resistance is practically field independent when the field is applied parallel to the layers. This is the only evident interplay between the two kinds of sublattices, ferromagnetic and conducting. No other influence of the conducting sublattice on the magnetic properties has been encountered and, as a consequence, the two sublattices can be considered as quasi-independent from the electronic point of view.

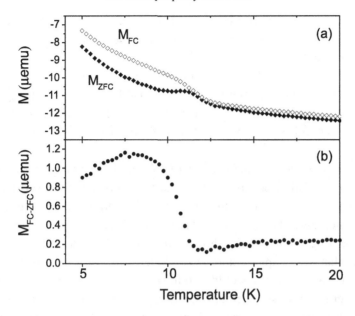

Figure 6.44. Magnetization vs. temperature for a 100-bilayer LB film transferred with Mn^{+2} with the measuring field applied parallel to the film plane. (a) Comparison of the data taken upon warming the film after cooling in zero applied field (ZFC) and cooling in a field of 1000 G (FC). For both cases, the measuring field is 100 G. (b) The difference in the FC and ZFC data showing a net magnetization below 11.5 ± 0.5 K. M represents $\mathcal{M}$. Reprinted with permission from Petruska *et al.*, 2002. Copyright (2002) American Chemical Society.

Organic/inorganic Langmuir–Blodgett films

The intrinsic segregated architecture of LB films makes them ideal systems for combining more than one targeted physical property into a single material. The example discussed here corresponds to a mixed organic/inorganic dual-network LB film in which the polar network is a magnetic manganese phosphonate lattice and the organic network contains the bis(phosphonic acid) amphiphile TTF-LB1 (Petruska *et al.*, 2002). OPA forms LB films with a variety of divalent, trivalent or tetravalent metal ions resulting in layers of the metal ions sandwiched within bilayers of the organophosphonate. Solid-state metal phosphonates already form layered continuous lattice structures.

The most extensive magnetic studies have been on manganese phosphonate LB films. Sheets of manganese ions, in a distorted square array, are bonded on top and bottom by layers of the organophosphonate. Within a layer, each phosphonate group bridges four metal ions, and each metal ion is coordinated by five oxygen atoms from four different phosphonate groups. The distorted octahedral coordination of the metal is completed by a water molecule. Magnetic exchange within the layers is antiferromagnetic, and the solid-state alkylphosphonates $(C_nH_{2n+1}PO_3)Mn.H_2O$

($n = 2-6$) each order antiferromagnetically between 13 K and 15 K. A weak moment develops below the ordering temperature and the solid-state phosphonates are known to be canted antiferromagnets.

LB deposition of the TTF-LB1 amphiphile from an aqueous Mn^{+2} subphase forms Y-type films with stoichiometry $Mn_2(TTF-LB1)(H_2O)_2$. Static χ measurements show that the films become magnetic near 11.5 K when the manganese phosphonate network orders as a canted antiferromagnet. This is illustrated in Fig. 6.44.

Attempts to subsequently oxidize the TTF network have given unstable phases. Iodine vapour oxidation occurs initially, but the films quickly revert back to the neutral donor form. In spite of the not-yet proven bifunctionality, this example represents a step towards the controlled preparation of a dual-network LB film that is both conducting and magnetic, showing the way to be followed.

It is clear that this chapter could be continued with more examples of physical properties of MOMs, but it would certainly become very long and difficult to digest. I rather prefer to stop here, aware, as mentioned at the very beginning, that many interesting examples are missing, making this chapter inevitably incomplete. However, other examples have already been discussed in the previous chapters, so a broad range of examples has been covered.

Afterword

... que tot està per fer i tot és possible.

Miquel Martí i Pol, *Ara mateix*

We discussed in Section 1.1 the complexity that a rather simple molecule such as N_2 can manifest when it is made to form 2D and 3D periodic and non-periodic structures. This system was used to introduce the myriad molecules that we have been studying throughout this book and now it will help us to conclude. Given that solid nitrogen becomes semiconducting at the elevated pressure of 240 GPa, we can ask ourselves if, from the practical point of view, solid nitrogen is of any use. It is clear that it is hard to find technological applications in these extreme conditions, unless somebody dreams of making business on distant planets or stars. However, the important point is the scientific knowledge that is acquired working with almost any system, no matter how difficult it is. Unfortunately, many scientists fancy themselves as pseudo-technologists mainly because of the desperate search for financial funding. This means that the fabulous science behind the studied systems is not fully explored and that the claims for incredible applications usually remain as claims. Science and technology should go hand in hand, but scientists should have enough free rein to explore the unknown. The results always come.

Sometimes, among the MOMs community one gets the feeling that this field is losing strength, with many scientists getting bored of obtaining and characterizing similar materials with similar properties and hurrying to publish the results. I do not share this opinion. In fact, and I hope I have made it clear in this book, only a few systems out of an incredibly large number are well known, though not completely characterized. There are still many open questions of fundamental interest. Therefore, the incorporation of new scientific communities, as in the case of the surface science community, is always good news, and others should also be welcome.

As made evident in this book, one of the leading and thus most studied systems is derivatives of the very special TTF molecule. Although TTF is rather peculiar because of its electronic structure, why should chemists and physicists not give birth in the coming years to a new branch of materials based on a different molecule? Nothing prevents this future discovery, except perhaps that effort and investment are necessary. It is not always desirable to wait for visionary people who define the way to be followed. Scientific progress can be viewed as a tree (the Tree of Science), where the time axis goes from the roots to the top. With TTF a new branch was born and many other subbranches have seen the light from the mother TTF branch. However, as in any regular multibranched tree, other branches exist (C_{60}, perylenes, etc.) and new ones are waiting. Think of carbon nanotubes (this is perhaps the best example of a subbranch because it is made exclusively out of carbon!). But the tree needs time to grow and water to survive and it is not always advisable to wait for water fallen from heaven in the form of rain. Research has to continue on a systematic basis.

Another field where, in my opinion, effort should be placed concerns relating small molecules with biomolecules, thus knocking on the door of biochemists and biologists. As already discussed, one of the nice things when one works with small molecules is that they tend to aggregate in ordered solids, a property that makes possible the characterization of the correlation between structure and physical properties. Studying intermolecular interactions and their balance is much easier with small molecules and the information that is obtained is valid and valuable for larger molecules. For polymers and for biomolecules this is extremely difficult, and in fact the structures of only a few *small* biomolecules such as proteins are known. This kind of investigation should include surfaces and interfaces and should lead to intelligent surfaces, which should be capable of recognizing and inducing order in biomolecules. This is an open and fascinating field.

It is evident that a multidisciplinary academic formation is needed for young scientists. This is valid not only for the field of organic materials but for many others. Universities should worry about it and adapt their academic structures to this new situation. I consider this a very important point.

To conclude let me cite the great Catalan poet Miquel Martí i Pol, who wrote that everything is still to be done and everything is possible. This is also my feeling.

Appendix A

Symmetry operations of relevant point groups

Table A.1 shows the symmetry operations of the point groups more commonly used in the book. The notation for the symmetry operations is the following:

E identity transformation

C_n proper rotation of $2\pi/n$ radians ($n \in \mathbb{N}$): the axis for which n is greatest is termed the principal axis

S_n improper rotation of $2\pi/n$ radians ($n \in \mathbb{N}$): improper rotations are regular rotations followed by a reflection in the plane perpendicular to the axis of rotation

i inversion operator (equivalent to S_2)

σ mirror plane

σ_h horizontal reflection plane: perpendicular to the principal rotation axis

σ_v vertical reflection plane: contains the principal rotation axis

σ_d diagonal or dihedral reflection plane: bisects two C_2 axes

Table A.1. *Symmetry operations of relevant point groups*

Point group	Symmetry operations
C_{2v}	E, C_2, $\sigma_v(xz)$, $\sigma_v'(yz)$
C_{2h}	E, C_2, i, σ_h
C_{3v}	E, $2C_3$, $3\sigma_v$
D_{2h}	E, $C_2(x)$, $C_2(y)$, $C_2(z)$, σ_{xy}, σ_{xz}, σ_{yz}, i
D_{4h}	E, $2C_4$, C_2, $2C_2'$, $2C_2''$, i, $2S_4$, σ_h, $2\sigma_v$, $2\sigma_d$
D_{6h}	E, $2C_6$, $2C_3$, C_2, $3C_2'$, $3C_2''$, i, $2S_3$, $2S_6$, σ_h, $3\sigma_d$, $3\sigma_v$

Appendix B

DFT approximation

The fundamental quantity in DFT is the local electronic charge density of the solid $\rho(r)$. The total energy E of the full many-body problem of interacting quantum mechanical particles is expressed as a functional of this density:

$$E\left[\rho(r)\right] = K\left[\rho(r)\right] + \int V_{\text{eff}}\rho(r)\,d^3r$$

$$+ \frac{1}{2} \int \frac{\rho(r)\rho(r')}{|r - r'|}\,d^3r\,d^3r' + E_{\text{xc}}\left[\rho(r)\right]. \tag{B.1}$$

Equation B.1 contains three parts:

(i) the kinetic energy $K\left[\rho(r)\right]$ of a non-interacting system,
(ii) the effective potential energy of the crystal, V_{eff}, which includes the Hartree contribution to the Coulomb interaction between the charges and
(iii) the exchange and correlation energy E_{xc}.

Minimizing the functional results in the Kohn–Sham equation:

$$\left[-\frac{\hbar^2}{2m_e}\nabla^2 + V_{\text{KS}}\right]\Psi_i = E_i\Psi_i, \tag{B.2}$$

which has the form of the one-electron Schrödinger equation with the Hamiltonian given by $H_{\text{KS}} = -\hbar^2/2m_e\nabla^2 + V_{\text{KS}}$, where V_{KS} stands for the Kohn–Sham potential. V_{KS} represents a static mean field of the electrons and has to be determined from the self-consistency condition:

$$V_{\text{KS}}\left[\rho(r)\right] = V_{\text{eff}}(r) + \int \frac{\rho(r')}{|r - r'|}\,d^3r' + \frac{\delta E_{\text{xc}}\left[\rho(r)\right]}{\delta\rho(r)}. \tag{B.3}$$

The Kohn–Sham equation becomes a reference system for DFT because it yields the correct ground state via:

$$\rho(r) = \sum_i f(E_i)|\Psi_i(r)|^2, \tag{B.4}$$

where $f(E_i)$ is the Fermi function described on page 59. Although the Kohn–Sham equation describes a non-interacting single-particle system, it gives the correct density of the many-body interacting system.

References

Acker, D. S. and Hertler, W. R. (1962). *J. Am. Chem. Soc.* **84**, 3370–3374.

Adachi, T., Ojima, E., Kato, K. *et al.* (2000). *J. Am. Chem. Soc.* **122**, 3238–3239.

Ajami, D., Oeckler, O., Simon, A. and Herges, R. (2003). *Nature* **426**, 819–821.

Akamatu, H., Inokuchi, H. and Matsunaga, Y. (1954). *Nature* **173**, 168–169.

Allemand, P. M., Khemani, K. C., Koch, A. *et al.* (1991). *Science* **253**, 301–303.

Almeida, M. and Henriques, R. T. (1997). In *Handbook of Organic Conductive Molecules and Polymers*, ed. H. S. Nalwa, Vol. 1, Chapter 2, pp. 87–149. Chichester: John Wiley & Sons.

Alonso, M. I., Garriga, M., Ossó, J. O. *et al.* (2003). *J. Chem. Phys.* **119**, 6335–6340.

Alvarado, S. F., Rossi, L., Müller, P. and Riess, W. (2001a). *Synth. Met.* **122**, 73–77.

Alvarado, S. F., Rossi, L., Müller, P., Seidler, P. F. and Riess, W. (2001b). *IBM J. Res. & Dev.* **45**, 89–100.

Al-Jishi, R. (1983). *Phys. Rev. B* **28**, 112–116.

An, J. M. and Pickett, W. E. (2001). *Phys. Rev. Lett.* **86**, 4366–4369.

Ando, T., Fowler, A. B. and Stern, F. (1982). *Rev. Mod. Phys.* **54**, 437–672.

Angelova, A., Moradpour, A., Auban-Senzier, P., Akaaboune, N.-E. and Pasquier, C. (2000). *Chem. Mater.* **12**, 2306–2310.

Aoki, S., Nakayama, T. and Miura, A. (1993). *Phys. Rev. B* **48**, 626–629.

Ara, N., Kawazu, A., Shigekawa, H., Yase, K. and Yoshimura, M. (1995). *Appl. Phys. Lett.* **66**, 3278–3280.

Ashcroft, N. W. (1968). *Phys. Rev. Lett.* **21**, 1748–1749.

Ashcroft, N. W. and Mermin, N. D. (1976). In *Solid State Physics*. New York: Holt, Rinehart and Winston.

Ashida, M., Uyeda, N. and Suito, E. (1966). *Bull. Chem. Soc. Jpn.* **39**, 2616–2624.

Auban-Senzier, P. and Jérome, D. (2003). *Synth. Met.* **133–134**, 1–5.

Auban-Senzier, P., Pasquier, C., Jérome, D., Carcel, C. and Fabre, J. M. (2003). *Synth. Met.* **133–134**, 11–14.

Awaga, K., Inabe, T., Nagashima, U. and Maruyama, Y. (1989). *J. Chem. Soc., Chem. Commun.*, 1617–1618.

    (1990) *J. Chem. Soc., Chem. Commun.*, 520.

Bacon, G. E., Curry, N. A. and Wilson, S. A. (1964). *Proc. R. Soc. London Ser. A* **279**, 98–110.

Bain, C. D., Troughton, E. B., Tao, Y.-T. *et al.* (1989). *J. Am. Chem. Soc.* **111**, 321–335.

Baker, K. N., Fratini, A. V., Resch, T. *et al.* (1993). *Polymer* **34**, 1571–1587.

Balicas, L., Behnia, K., Kang, W. *et al.* (1994). *Adv. Mater.* **6**, 762–765.

Balicas, L., Brooks, J. S., Storr, K. *et al.* (2001). *Phys. Rev. Lett.* **87**, 067002.

Banister, A. J., Bricklebank, N., Clegg, W. *et al.* (1995). *J. Chem. Soc., Chem. Commun.*, 679–680.

Banister, A. J., Bricklebank, N., Lavender, I. *et al.* (1996). *Angew. Chem. Int. Ed. Engl.* **35**, 2533–2535.

Bao, Z., Lovinger, A. J. and Brown, J. (1998). *J. Am. Chem. Soc.* **120**, 207–208.

Barabási, A.-L. and Stanley, H. E. (1995). In *Fractal Concepts in Surface Growth*. Cambridge: Cambridge University Press.

Barrena, E., Palacios-Lidón, E., Munuera, C. *et al.* (2004). *J. Am. Chem. Soc.* **126**, 385–395.

Barth, J. V., Weckesser, J., Cai, C. *et al.* (2000). *Angew. Chem. Int. Ed. Engl.* **39**, 1230–1234.

Batail, P., Boubekeur, K., Fourmigué, M. and Gabriel, J.-C. P. (1998). *Chem. Mater.* **10**, 3005–3015.

Bauer, E. (1994). *Rep. Prog. Phys.* **57**, 895–938.

Bechgaard, K., Kistenmacher, T. J., Bloch, A. N. and Cowan, D. O. (1977). *Acta Cryst. B* **33**, 417–422.

Bechgaard, K., Jacobsen, C. S. and Andersen, N. H. (1978). *Solid State Commun.* **25**, 875–879.

Bechgaard, K., Jacobsen, C. S., Mortensen, K., Pedersen, H. J. and Thorup, N. (1980). *Solid State Commun.* **33**, 1119–1125.

Bechgaard, K., Carneiro, K., Olsen, M., Rasmussen, F. B. and Jacobsen, C. S. (1981). *Phys. Rev. Lett.* **46**, 852–855.

Bender, K., Hennig, I., Schweitzer, D. *et al.* (1984). *Mol. Cryst. Liq. Cryst.* **108**, 359–371.

Bernstein, J. (2002). In *Polymorphism in Molecular Crystals*. Oxford: Clarendon Press.

Bernstein, J. and Hagler, A. T. (1978). *J. Am. Chem. Soc.* **100**, 673–681.

Bernstein, M. P., Sandford, S. A., Allamandola, L. J. *et al.* (1999). *Science* **283**, 1135–1138.

Bethe, H. (1928). *Annalen der Physik* **87**, 55–129.
  (1931). *Z. Phys.* **71**, 205–226.

Bethune, D. S., Kiang, C. H., de Vries, M. S. *et al.* (1993). *Nature* **363**, 605–607.

Bianconi, A., Hagstrom, S. B. M. and Bachrach, R. Z. (1977). *Phys. Rev. B* **16**, 5543–5548.

Bini, R., Ulivi, L., Kreutz, J. and Jodl, H. J. (2000). *J. Chem. Phys.* **112**, 8522–8529.

Binnig, G., Rohrer, H., Gerber, Ch. and Weibel, E. (1982). *Phys. Rev. Lett.* **49**, 57–60.

Binnig, G., Quate, C. F. and Gerber, Ch. (1986). *Phys. Rev. Lett.* **56**, 930–933.

Biscarini, F., Samorí, P., Greco, O. and Zamboni, R. (1997). *Phys. Rev. Lett.* **78**, 2389–2392.

Bjørnholm, T., Hassenkam, T., Greve, D. R. *et al.* (1999). *Adv. Mater.* **11**, 1218–1221.

Bloch, A. N. (1977). *Lecture Notes in Physics* **65**, 317–348.

Bloch, A. N., Cowan, D. O., Bechgaard, K. *et al.* (1975). *Phys. Rev. Lett.* **34**, 1561–1564.

Blodgett, K. B. (1935). *J. Am. Chem. Soc.* **57**, 1007–1022.

Blundell, S. J. and Pratt, F. L. (2004). *J. Phys.: Condens. Matter.* **16**, R771–R828.

Bode, M., Getzlaff, M. and Wiesendanger, R. (1998). *Phys. Rev. Lett.* **81**, 4256–4259.

Böhringer, M., Morgenstern, K. and Schneider, W.-D. (1997). *Phys. Rev. B* **55**, 1384–1387.

Böhringer, M., Berndt, R., Mauri, F., De Vita, A. and Car, R. (1999). *Phys. Rev. Lett.* **83**, 324–327.

Bouchoms, I. P. M., Schoonveld, W. A., Vrijmoeth, J. and Klapwijk, T. M. (1999). *Synth. Met.* **104**, 175–178.

Boukherroub, R., Morin, S., Bensebaa, F. and Wayner, D. D. M. (1999). *Langmuir* **15**, 3831–3835.

Bousseau, M., Valade, L., Legros, J. P. *et al.* (1986). *J. Am. Chem. Soc.* **108**, 1908–1916.

Bradshaw, A. M. and Richardson, N. V. (1996). *Pure & Appl. Chem.* **68**, 457–467.

Braslau, A., Deutsch, M., Pershan, P. S. *et al.* (1985). *Phys. Rev. Lett.* **54**, 114–117.

Brinkmann, M., Gadret, G., Muccini, M. *et al.* (2000). *J. Am. Chem. Soc.* **122**, 5147–5157.

Brinkmann, M., Wittmann, J.-C., Barthel, M., Hanack, M. and Chaumont, C. (2002a). *Chem. Mater.* **14**, 904–914.

Brinkmann, M., Graff, S. and Biscarini, F. (2002b). *Phys. Rev. B* **66**, 165430.

Brinkmann, M., Graff, S., Straupé, C. *et al.* (2003). *J. Phys. Chem. B* **107**, 10531–10539.

Brossard, L., Ribault, M., Valade, L. and Cassoux, P. (1986). *Physica* **143B**, 378–380.

Brossard, L., Hurdequint, H., Ribault, M. *et al.* (1988). *Synth. Met.* **27**, B157–B162.

Brown, C. J. (1968). *J. Chem. Soc. A*, 2488–2493.

Brown, A. R., de Leeuw, D. M., Lous, E. J. and Havinga, E. E. (1994). *Synth. Met.* **66**, 257–261.

Bryce, M. R. and Petty, M. C. (1995). *Nature* **374**, 771–776.

Buchholz, J. C. and Somorjai, G. A. (1977). *J. Chem. Phys.* **66**, 573–580.

Buravov, L. I., Zvereva, G. I., Kaminskii, V. F. *et al.* (1976). *J. Chem. Soc., Chem. Comm.*, 720–721.

Burns, G. (1993). In *High-Temperature Superconductivity*. Boston: Academic Press.

Burton, W. K., Cabrera, N. and Frank, F. C. (1951). *Philos. Trans. R. Soc. London A* **243**, 299–358.

Calcott, W. S., Tinker, J. M. and Weinmayr, V. (1939). *J. Am. Chem. Soc.* **61**, 949–951.

Campbell, R. B., Roberston, J. M. and Trotter, J. (1961). *Acta Cryst.* **14**, 705–711.

Canadell, E. (1997). *New. J. Chem.* **21**, 1147–1159.

Canadell, E. and Whangbo, M.-H. (1991). *Chem. Rev.* **91**, 965–1034.

Candela, G. A., Swartzendruber, L. J., Miller, J. S. and Rice, M. J. (1979). *J. Am. Chem. Soc.* **101**, 2755–2756.

Caneschi, A., Gatteschi, D., Lalioti, N. *et al.* (2001). *Angew. Chem. Int. Ed. Engl.* **40**, 1760–1763.

Caro, J., Fraxedas, J. and Figueras, A. (1997). *Chem. Vap. Deposition* **3**, 263–269.

Caro, J., Fraxedas, J., Jürgens, O. *et al.* (1998). *Adv. Mater.* **10**, 608–610.

Caro, J., Fraxedas, J., Santiso, J. *et al.* (1999). *Synth. Met.* **102**, 1607–1608.

Caro, J., Fraxedas, J. and Figueras, A. (2000). *J. Cryst. Growth* **209**, 146–158.

Carswell, W. E., Zugrav, M. I., Wessling, F. C. and Haulenbeek, G. (2000). *J. Cryst. Growth* **211**, 428–433.

Casalis, L., Danisman, M. F., Nickel, B. *et al.* (2003). *Phys. Rev. Lett.* **90**, 206101.

Cava, R. J., Batlogg, B., Krajewski, J. J. *et al.* (1988). *Nature* **332**, 814–816.

Cavallini, M. and Biscarini, F. (2003). *Nano Lett.* **3**, 1269–1271.

Cavallini, M., Murgia, M. and Biscarini, F. (2001). *Nano Lett.* **1**, 193–195.

Cavallini, M., Biscarini, F., Gómez-Segura, J., Ruiz, D. and Veciana, J. (2003). *Nano Lett.* **3**, 1527–1530.

Cepek, C., Sancrotti, M., Greber, T. and Osterwalder, J. (2000). *Surf. Sci.* **454–456**, 467–471.

Chappell, J. S., Bloch, A. N., Bryden, W. A. *et al.* (1981). *J. Am. Chem. Soc.* **103**, 2442–2443.

Chassé, T., Wu, C.-I., Hill, I. G. and Kahn, A. (1999). *J. Appl. Phys.* **85**, 6589–6592.

Chen, Q., Rada, T., McDowall, A. and Richardson, N. V. (2002). *Chem. Mater.* **14**, 743–749.

Chernov, A. A. (1984). In *Modern Crystallography III: Crystal Growth*. Berlin: Springer.

Chiang, C. K., Fincher Jr., C. R., Park, Y. W. *et al.* (1977). *Phys. Rev. Lett.* **39**, 1098–1101.
Chiarelli, R., Novak, M. A., Rassat, A. and Tholence, J. L. (1993). *Nature* **363**, 147–149.
Chkoda, L., Schneider, M., Shklover, V. *et al.* (2003). *Chem. Phys. Lett.* **371**, 548–552.
Chou, S. Y., Krauss, P. and Renstrom, P. J. (1996). *Science* **272**, 85–87.
Chow, D. S., Zamborszky, F., Alavi, B. *et al.* (2000). *Phys. Rev. Lett.* **85**, 1698–1701.
Chrisey, D. B., Piqué, A., McGill, R. A. *et al.* (2003). *Chem. Rev.* **103**, 553–576.
Christensen, C. A., Goldenberg, L. M., Bryce, M. B. and Becher, J. (1998). *Chem. Commun.*, 509–510.
Claessen, R., Sing, M., Schwingenschlögl, U. *et al.* (2002). *Phys. Rev. Lett.* **88**, 096402.
Clemente, D. A. and Marzotto, A. (1996). *J. Mater. Chem.* **6**, 941–946.
Clemente-León, M., Mingotaud, C., Agricole, B. *et al.* (1997). *Angew. Chem. Int. Ed. Engl.* **36**, 1114–1116.
Clemente-León, M., Soyer, H., Coronado, E. *et al.* (1998). *Angew. Chem. Int. Ed. Engl.* **37**, 2842–2845.
Clérac, R., Miyasaka, H., Yamashita, M. and Coulon, C. (2002). *J. Am. Chem. Soc.* **124**, 12837–12844.
Cohen, M. J., Coleman, L. B., Garito, A. F. and Heeger, A. J. (1974). *Phys. Rev. B* **10**, 1298–1307.
Coleman, L. B., Cohen, M. J., Sandman, D. J. *et al.* (1973a). *Solid State Commun.* **12**, 1125–1132.
Coleman, L. B., Cohen, J. A., Garito, A. F. and Heeger, A. J. (1973b). *Phys. Rev. B* **7**, 2122–2128.
Colin, C., Pasquier, C. R., Auban-Senzier, P. *et al.* (2004). *Synth. Met.* **146**, 273–277.
Cölle, M., Dinnebier, R. E. and Brütting, W. (2002). *Chem. Comm.*, 2908–2909.
Collet, E., Lemée-Cailleau, M.-H., Buron-Le Cointe, M. *et al.* (2003). *Science* **300**, 612–615.
Coluzza, C., Berger, H., Alméras, P. *et al.* (1993). *Phys. Rev. B* **47**, 6625–6629.
Conwell, E. (1988). In *Semiconductors and Semimetals: Highly Conducting Quasi-One-Dimensional Organic Crystals*, ed. E. Conwell, Vol. 27. Boston: Academic Press.
Cooper, W. F., Kenney, N. C., Edmonds, J. W. *et al.* (1971). *Chem. Commun.*, 889–890.
Coronado, E., Galán-Mascarós, J. R., Gómez-García, C. J. and Laukhin, V. (2000). *Nature* **408**, 447–449.
Cotton, F. A. (1971). In *Chemical Applications of Group Theory*. New York: Wiley-Interscience.
Coulon, C., Delhaes, P., Flandrois, S. *et al.* (1982). *J. Physique* **43**, 1059–1067.
Cox, J. J. and Jones, T. S. (2000). *Surf. Sci.* **457**, 311–318.
Cox, J. J., Bayliss, S. M. and Jones, T. S. (1999). *Surf. Sci.* **433–435**, 152–156.
Creuzet, F., Creuzet, G., Jérome, D., Schweitzer, D. and Keller, H. J. (1985). *J. Physique Lett.* **46**, L1079–L1085.
Dahlen, M. A. (1939). *Ind. Eng. Chem.* **31**, 839–847.
Dähne, L. (1995). *J. Am. Chem. Soc.* **117**, 12855–12860.
Dardel, B., Malterre, D., Grioni, M. *et al.* (1991). *Phys. Rev. Lett.* **67**, 3144–3147.
Davey, R. J., Maginn, S. J., Andrews, S. J. *et al.* (1994). *J. Chem. Soc. Faraday Trans.* **90**, 1003–1009.
Day, P. (1993). *Phys. Scr.* **T49**, 726–730.
(2002). *Notes Rec. R. Soc. Lond.* **56**, 95–103.
Debe, M. K. and Kam, K. K. (1990). *Thin Solid Films* **186**, 289–325.
de Boer, R. W. I., Klapwijk, T. M. and Morpurgo, A. F. (2003). *Appl. Phys. Lett.* **83**, 4345–4347.

de Boer, R. W. I., Gershenson, M. E., Morpurgo, A. F. and Podzorov, V. (2004). *phys. stat. sol. (a)* **201**, 1302–1331.

de Caro, D., Sakah, J., Basso-Bert, M. *et al.* (2000a). *C. R. Acad. Sci. Paris, Série IIC, Chimie* **3**, 675–680.

de Caro, D., Basso-Bert, M., Sakah, J. *et al.* (2000b). *Chem. Mater.* **12**, 587–589.

de Caro, D., Fraxedas, J., Faulmann, C. *et al.* (2004). *Adv. Mater.* **16**, 835–838.

de Diesbach, H. and von der Weid, E. (1927). *Helv. Chim. Acta* **10**, 886–888.

Deegan, R. D., Bakajin, O., Dupont, T. F. *et al.* (1997). *Nature* **389**, 827–829.

de Kruif, C. G. and Govers, H. A. J. (1980). *J. Chem. Phys.* **73**, 553–555.

Delhaes, P., Dupart, E., Manceau, J. P. *et al.* (1985). *Mol. Cryst. Liq. Cryst.* **119**, 269–276.

Deluzet, A., Rousseau, R., Guilbaud, C. *et al.* (2002a). *Chem. Eur. J.* **8**, 3884–3900.

Deluzet, A., Perruchas, S., Bengel, H. *et al.* (2002b). *Adv. Funct. Mater.* **12**, 123–128.

Desiraju, G. R. (1989). In *Crystal Engineering: The Design of Organic Solids*. Amsterdam: Elsevier.

(1995). *Angew. Chem. Int. Ed. Engl.* **34**, 2311–2327.

Dhindsa, A. S., Song, Y.-P., Badyal, J. P. *et al.* (1992). *Chem. Mater.* **4**, 724–728.

Dimitrakopoulos, C. D. and Malenfant, P. R. L. (2002). *Adv. Mater.* **14**, 99–117.

Dimitrakopoulos, C. D., Brown, A. R. and Pomp, A. (1996). *J. Appl. Phys.* **80**, 2501–2508.

Dodabalapur, A., Torsi, L. and Katz, H. E. (1995). *Science* **268**, 270–271.

Donaldson, D. M., Robertson, J. M. and White, J. G. (1953). *Proc. R. Soc. London A* **220**, 311–321.

Doublet, M. L., Canadell, E. and Shibaeva, R. P. (1994). *J. Phys. I France* **4**, 1479–1490.

Dourthe, C., Izumi, M., Garrigou-Lagrange, C. *et al.* (1992). *J. Phys. Chem.* **96**, 2812–2820.

Dresselhaus, M. S., Dresselhaus, G. and Eklund, P. (1996). In *The Science of Fullerenes and Carbon Nanotubes*. New York: Academic Press.

Dromzee, Y., Chiarelli, R., Gambardelli, S. and Rassat, A. (1996). *Acta Cryst. C* **52**, 474–477.

Dunitz, J. D. and Bernstein, J. (1995). *Acc. Chem. Res.* **28**, 193–200.

Dunphy J. C., Rose M., Behler S. *et al.* (1998). *Phys. Rev. B* **57**, R12705–R12708.

Dürr, A. C., Schreiber, F., Münch, M. *et al.* (2002a). *Appl. Phys. Lett.* **81**, 2276–2278.

Dürr, A. C., Schreiber, F., Kelsch, M., Carstanjen, H. D. and Dosch, H. (2002b). *Adv. Mater.* **14**, 961–963.

Dürr, A. C., Koch, N., Kelsch, M. *et al.* (2003a). *Phys. Rev. B* **68**, 115428.

Dürr, A. C., Schreiber, F., Ritley, K. A. *et al.* (2003b). *Phys. Rev. Lett.* **90**, 016104.

Einstein, A. (1905). *Ann. Physik* **17**, 132–148.

Eland, J. H. D. (1984). In *Photoelectron Spectroscopy*, 2nd edn. London: Butterworths.

Ellern, A., Bernstein, J., Becker, J. Y. *et al.* (1994). *Chem. Mater.* **6**, 1378–1385.

Enkelmann, V., Morra, B. S., Kröhnke, Ch., Wegner, G. and Heinze, J. (1982). *Chem. Phys.* **66**, 303–313.

Eremets, M. I., Hemley, R. J., Mao, Ho-kwang and Gregoryanz, E. (2001). *Nature* **411**, 170–174.

Eremtchenko, M., Schaefer, J. A. and Tautz, F. S. (2003). *Nature* **425**, 602–605.

Etter, M. C. (1990). *Acc. Chem. Res.* **23**, 120–126.

Family, F. (1990). *Physica A* **168**, 561–580.

Faraday, M. (1825). *Philos. Trans. R. Soc. London*, 440–466.

(1859). In *Faraday, Michael, 1791–1867. Experimental Researches in Chemistry and Physics*, 1991 edn. London: Taylor & Francis.

Fasel, R., Cossy, A., Ernst, K.-H. *et al.* (2001). *J. Chem. Phys.* **115**, 1020–1027.

Fasel, R., Parschau, M. and Ernst, K.-H. (2003). *Angew. Chem. Int. Ed. Engl.* **42**, 5178–5181.

Faul, J., Neuhold, G., Ley, L. *et al.* (1993). *Phys. Rev. B* **47**, 12625–12635.

Faulmann, Ch., Rivière, E., Dorbes, S. *et al.* (2003). *Eur. J. Inorg. Chem.*, 2880–2888.

Fenter, P., Schreiber, F., Zhou, L., Eisenberger, P. and Forrest, S. R. (1977). *Phys. Rev. B* **56**, 3046–3053.

Ferraris, J., Cowan, D. O., Walatka, V. and Perlstein, J. H. (1973). *J. Am. Chem. Soc.* **95**, 948–949.

Fertig, W. A., Johnston, D. C., DeLong, L. E. *et al.* (1977). *Phys. Rev. Lett.* **38**, 987–990.

Figueras, A., Caro, J., Fraxedas, J. and Laukhin, V. (1999). *Synth. Met.* **102**, 1611–1612.

Fischer, G. and Dormann, E. (1998). *Phys. Rev. B* **58**, 7792–7794.

Forrest, S. R. (1997). *Chem. Rev.* **97**, 1793–1896.

Fourmigué, M., Krebs, F. C. and Larsen, J. (1993). *Synthesis*, 509–512.

Frank, F. C. (1949). *Disc. Faraday Soc.* **5**, 48–54.

  (1951). *Acta Cryst.* **4**, 497–501.

Frank, F. C. and Read, W. T. (1950). *Phys. Rev.* **79**, 722–723.

Fraxedas, J. (2002). *Adv. Mater.* **14**, 1603–1614.

Fraxedas, J., Caro, J., Figueras, A., Gorostiza, P. and Sanz, F. (1998). *Surf. Sci.* **415**, 241–250.

Fraxedas, J., Caro, J., Santiso, J. *et al.* (1999). *Europhys. Lett.* **48**, 461–467.

Fraxedas, J., Garcia-Manyes, S., Gorostiza, P. and Sanz, F. (2002a). *Proc. Natl. Acad. Sci. USA* **99**, 5228–5232.

Fraxedas, J., Molas, S., Figueras, A. *et al.* (2002b). *J. Solid State Chem.* **168**, 384–389.

Fraxedas, J., Lee, Y. J., Jiménez, I. *et al.* (2003). *Phys. Rev. B* **68**, 195115.

Fraxedas, J., Verdaguer, A., Sanz, F., Baudron, S. and Batail, P. (2005). *Surf. Sci.* **588**, 41–48.

Fujita, W. and Awaga, K. (1999). *Science* **286**, 261–262.

Gador, D., Buchberger, C., Fink, R. and Umbach, E. (1998). *Europhys. Lett.* **41**, 231–236.

Gallani, J. L., Le Moigne, J., Oswald, L., Bernard, M. and Turek, P. (2001). *Langmuir* **17**, 1104–1109.

Gao, W. and Kahn, A. (2001). *Appl. Phys. Lett.* **79**, 4040–4042.

Garnier, F. (1996). *Pure & Appl. Chem.* **68**, 1455–1462.

Garoche, P., Brusetti, R., Jérome, D. and Bechgaard, K. (1982). *J. Physique* **43**, L147–L152.

Garreau, B., de Montauzon, D., Cassoux, P. *et al.* (1995). *New J. Chem.* **19**, 161–171.

Gates, J. A. and Kesmodel, L. L. (1982). *J. Chem. Phys.* **76**, 4281–4286.

Gatteschi, D. (1994). *Adv. Mater.* **6**, 635–645.

Gavezzotti, A. and Desiraju, G. R. (1988). *Acta Cryst. B* **44**, 427–434.

Geiser, U., Kini, A. M., Wang, H. H., Beno, M. A. and Williams, J. M. (1991). *Acta Cryst. C* **47**, 190–192.

Giamarchi, T. (2004). In *Quantum Physics in One Dimension*. Oxford: Oxford University Press.

Gibaud, A., Cowlam, N., Vignaud, G. and Richardson, T. (1995). *Phys. Rev. Lett.* **74**, 3205–3208.

Glauber, R. J. (1963). *J. Math. Phys.* **4**, 294–307.

Gomar-Nadal, E., Abdel-Mottaleb, M. M. S., De Feyter, S. *et al.* (2003). *Chem. Comm.*, 906–907.

Gómez-Navarro, C., de Pablo, P. J. and Gómez-Herrero, J. (2004). *Adv. Mater.* **16**, 549–552.

Goodings, E. P., Mitchard, D. A. and Owen, G. (1972). *J. Chem. Soc., Perkin Trans. 1* **11**, 1310–1314.

Greene, R. L., Street, G. B. and Suter, L. J. (1975). *Phys. Rev. Lett.* **34**, 577–579.

Gregoryanz, E., Goncharov, A. F., Hemley, R. J. *et al.* (2002). *Phys. Rev. B* **66**, 224108.

Grobman, W. D. and Koch, E. E. (1979). In *Photoemission in Solids II*, eds. M. Cardona and L. Ley, Vol. 2, Chapter 5, pp. 261–298. Berlin: Springer-Verlag.

Gubser, D. U., Fuller, W. W., Poehler, T. O. *et al.* (1982). *Mol. Cryst. Liq. Cryst.* **79**, 225–234.

Gundlach, D. J., Lin, Y. Y., Jackson, T. N., Nelson, S. F. and Schlom, D. G. (1997). *IEEE Electron Device Lett.* **18**, 87–89.

Gutierrez-Llorente, A., Pérez-Casero, R., Pajot, B. *et al.* (2003). *Appl. Phys. A* **77**, 785–788.

Gütlich, P., Hauser, A. and Spiering, H. (1994). *Angew. Chem. Int. Ed. Engl.* **33**, 2024–2054.

Haddon, R. C., Perel, A. S., Morris, R. C. *et al.* (1995). *Appl. Phys. Lett.* **67**, 121–123.

Hammond, R. B., Roberts, K. J., Docherty, R., Edmonson, M. and Gairns, R. (1996). *J. Chem. Soc., Perkin Trans. 2*, 1527–1528.

Hark, Th. E. M. van den and Beurskens, P. T. (1976). *Cryst. Struct. Comm.* **5**, 247–252.

Harvey, R. G. (2004). *Curr. Org. Chem.* **8**, 303–323.

Hashizume, T., Motai, K., Wang, X. D. *et al.* (1993). *Phys. Rev. Lett.* **71**, 2959–2962.

Hauschild, A., Karki, K., Cowie, B. C. C. *et al.* (2005). *Phys. Rev. Lett.* **94**, 036106.

He, J., Patitsas, S. N., Preston, K. F., Wolkow, R. A. and Wayner, D. D. M. (1998). *Chem. Phys. Lett.* **286**, 508–514.

Hebard, A. F., Rosseinsky, M. J., Haddon, R. C. *et al.* (1991). *Nature* **350**, 600–601.

Heeger, A. J. (2001). *Rev. Mod. Phys.* **3**, 681–700.

Heilbronner, E. (1964). *Tetrahedron Lett.* **29**, 1923–1928.

Heisenberg, W. (1928). *Z. Phys.* **49**, 619–636.

Hennig, I., Bender, K., Schweitzer, D. *et al.* (1985). *Mol. Cryst. Liq. Cryst.* **119**, 337–341.

Heutz, S. and Jones, T. S. (2002). *J. Appl. Phys.* **92**, 3039–3046.

Heuzé, K., Fourmigué, M. and Batail, P. (1999). *J. Mater. Chem.* **9**, 2373–2379.

Hill, I. G., Rajagopal, A., Kahn, A. and Hu, Y. (1998). *Appl. Phys. Lett.* **73**, 662–664.

Hill, I. G., Kahn, A., Soos, Z. G. and Pascal, R. A. (2000). *Chem. Phys. Lett.* **327**, 181–188.

Hiller, W., Strähle, J., Kobel, W. and Hanack, M. (1982). *Z. Kristallogr.* **159**, 173–183.

Hillier, A. C. and Ward, M. D. (1994). *Science* **263**, 1261–1264.

Hillier, A. C., Maxson, J. B. and Ward, M. D. (1994). *Chem. Mater.* **6**, 2222–2226.

Hilti, B., Mayer, C. W. and Rihs, G. (1981). *Solid State Commun.* **38**, 1129–1134.

Hiraki, K. and Kanoda, K. (1998). *Phys. Rev. Lett.* **80**, 4737–4740.

Hla, S.-W., Bartels, L., Meyer, G. and Rieder, K.-H. (2000). *Phys. Rev. Lett.* **85**, 2777–2780.

Hoffmann, R. (1963). *J. Chem. Phys.* **39**, 1397–1412.

  (1987). *Angew. Chem. Int. Ed. Engl.* **26**, 846–878.

  (1988). *Rev. Mod. Phys.* **60**, 601–628.

Holden, A. N., Matthias, B. T., Anderson, P. W. and Lewis, H. W. (1956). *Phys. Rev.* **102**, 1463.

Hooks, D. E., Yip, C. M. and Ward, M. D. (1998). *J. Phys. Chem. B* **102**, 9958–9965.

Hooks, D. E., Fritz, T. and Ward, M. D. (2001). *Adv. Mater.* **13**, 227–241.

Horowitz, G. (1998). *Adv. Mater.* **10**, 365–377.

Horowitz, G., Bachet, B., Yassar, A. *et al.* (1995). *Chem. Mater.* **7**, 1337–1341.

Hoshino, A., Takenaka, Y. and Miyaji, H. (2003). *Acta Cryst. B* **59**, 393–403.

Hurtley, W. R. H. and Smiles, S. (1926). *J. Chem. Soc.*, 2263–2270.

Iijima, S. (1991). *Nature* **354**, 56–58.

Ilakovac, V., Ravy, S., Pouget, J. P. *et al.* (1993). *J. Physique IV* **3**, 137–140.

Imakubo, T., Tajima, N., Tamura, M. *et al.* (2002). *J. Mater. Chem.* **12**, 159–161.

Imer, J. M., Patthey, F., Dardel, B. *et al.* (1989). *Phys. Rev. Lett.* **62**, 336–339.

Inokuchi, M., Tajima, H., Ohta, T. *et al.* (1996). *J. Phys. Soc. Jpn.* **65**, 538–544.

Inoue, I. H., Watanabe, M., Kinoshita, T. *et al.* (1993). *Phys. Rev. B* **47**, 12917–12920.

Ishibashi, S. and Kohyama, M. (2000). *Phys. Rev. B* **62**, 7839–7844.

Ishida, M., Mori, T. and Shigekawa, H. (1999). *Phys. Rev. Lett.* **83**, 596–599.

Ishida, M., Takeuchi, O., Mori, T. and Shigekawa, H. (2001). *Phys. Rev. B* **64**, 153405.

Ishiguro, T., Yamaji, K. and Saito, G. (1998). In *Organic Superconductors*. Berlin: Springer.

Ishii, H., Sugiyama, K., Yoshimura, D. *et al.* (1998). *IEEE J. Sel. Top. Quant. Elec.* **4**, 24–33.

Ising, E. (1925). *Z. Physik* **31**, 253–258.

Iung, Ch. and Canadell, E. (1997). In *Description orbitalaire de la structure électronique des solides. De la molécule aux composés 1D*. Paris: Ediscience International.

Iwasa, Y., Koda, T., Tokura, Y. *et al.* (1990). *Phys. Rev. B* **42**, 2374–2377.

Iyoda, M., Ogura, E., Hara, K. *et al.* (1999). *J. Mater. Chem.* **9**, 335–337.

Iyoda, M., Hasegawa, M. and Miyake, Y. (2004). *Chem. Rev.* **104**, 5085–5113.

Jacobsen, C. S., Mortensen, K., Andersen, J. R. and Bechgaard, K. (1978). *Phys. Rev. B* **18**, 905–921.

Jacobsen, C. S., Pedersen, H. J., Mortensen, K. *et al.* (1982). *J. Phys. C: Solid State Phys.* **15**, 2651–2663.

Jeong, H.-C. and Williams, E. D. (1999). *Surf. Sci. Rep.* **34**, 171–294.

Jeppesen, J. O., Brønsted Nielsen, M. and Becher, J. (2004). *Chem. Rev.* **104**, 5115–5131.

Jérome, D., Mazaud, A., Ribault, M. and Bechgaard, K. (1980). *J. Physique* **41**, L95–L98.

Kagan, J. and Arora, S. K. (1983). *Heterocycles* **20**, 1937–1940.

Kahn, O. (1993). In *Molecular Magnetism*. New York: VCH Publishers, Inc.

Kaiser, R. I. and Roessler, K. (1997). *Astrophys. J.* **475**, 144–154.

Kang, W., Montambaux, G., Cooper, J. R. *et al.* (1989). *Phys. Rev. Lett.* **62**, 2559–2562.

Karl, N. and Günther, Ch. (1999). *Cryst. Res. Technol.* **34**, 243–254.

Karl, N., Kraft, K.-H., Marktanner, J. *et al.* (1999). *J. Vac. Sci. Technol. A* **17**, 2318–2328.

Kartsovnik, M. V. (2004). *Chem. Rev.* **104**, 5737–5781.

Katnani, A. D. and Margaritondo, G. (1983). *Phys. Rev. B* **28**, 1944–1956.

Kawamoto, T., Mori, T., Takimiya, K. *et al.* (2002). *Phys. Rev. B* **65**, 140508(R).

Kelty, S. P., Lu, Z. and Lieber, C. M. (1991). *Phys. Rev. B* **44**, 4064–4067.

Kikuchi, K., Murata, K., Honda, Y. *et al.* (1987). *J. Phys. Soc. Jpn.* **56**, 4241–4244.

Kini, A. M., Geiser, U., Wang, H. H. *et al.* (1990). *Inorg. Chem.* **29**, 2555–2557.

Kinoshita, M. (1994). *J. Phys. Soc. Jpn.* **33**, 5718–5733.

Kistenmacher, T. J., Phillips, T. E. and Cowan, D. O. (1974). *Acta Cryst. B* **30**, 763–768.

Kistenmacher, T. J., Emge, T. J., Shu, P. and Cowan, D. O. (1979), *Acta Cryst. B* **35**, 772–775.

Kistenmacher, T. J., Emge, T. J., Bloch, A. N. and Cowan, D. O. (1982), *Acta Cryst. B* **38**, 1193–1199.

Kitaigorodskii, A. J. (1961). In *Organic Chemical Crystallography*. New York: Consultants Bureau.

Klemenz, C. (1998). *J. Cryst. Growth* **187**, 221–227.

Kloc, Ch., Simpkins, P. G., Siegrist, T. and Laudise, R. A. (1997). *J. Cryst. Growth* **182**, 416–427.

Klots, C. E., Compton, R. N. and Raaen, V. F. (1974). *J. Chem. Phys.* **60**, 1177–1178.

Knotek, M. L. and Feibelman, P. J. (1978). *Phys. Rev. Lett.* **40**, 964–967.

Knupfer, M. and Peisert, H. (2004). *phys. stat. sol. (a)* **201**, 1055–1074.

Knupfer, M., Peisert, H. and Schwieger, T. (2001). *Phys. Rev. B* **65**, 033204.

Kobayashi, H., Mori, T., Kato, R. *et al.* (1983). *Chem. Lett.*, 581–584.

Kobayashi, H., Kobayashi, A., Sasaki, Y., Saito, G. and Inokuchi, H. (1984). *Bull. Chem. Soc. Jpn.* **57**, 2025–2026.

(1986a). *Bull. Chem. Soc. Jpn.* **59**, 301–302.

Kobayashi, H., Kato, R., Kobayashi, A. *et al.* (1986b). *Chem. Lett.*, 789–792.

Kobayashi, A., Kato, R., Kobayashi, H. *et al.* (1986c). *Chem. Lett.*, 2017–2020.

(1987). *Chem. Lett.*, 459–462.

Kobayashi, H., Akutsu, H., Arai, E., Tanaka, H. and Kobayashi, A. (1997). *Phys. Rev. B* **56**, R8526–R8529.

Koch, N., Ghijsen, J., Johnson, R. L. *et al.* (2002). *J. Phys. Chem. B* **106**, 4192–4196.

Koch, N., Chan, C., Kahn, A. and Schwartz, J. (2003). *Phys. Rev. B* **67**, 195330.

Krätschmer, W., Lamb, L. D., Fostiropoulos, K. and Huffman, D. R. (1990). *Nature* **347**, 354–358.

Krause, B., Dürr, A. C., Schreiber, F., Dosch, H. and Seeck, O. H. (2003). *J. Chem. Phys.* **119**, 3429–3435.

Kroemer, H. (2001). *Rev. Mod. Phys.* **73**, 783–793.

Krömer, J., Rios-Carreras, I., Fuhrmann, G. *et al.* (2000). *Angew. Chem. Int. Ed. Engl.* **39**, 3481–3486.

Kroto, H. W., Heath, J. R., O' Brien, S. C., Curl, R. F. and Smalley, R. E. (1985). *Nature* **318**, 162–163.

Krug, J. (1997). *Adv. Phys.* **46**, 139–282.

Krummacher, S., Schmidt, V. and Wuilleumier, F. (1980). *J. Phys. B: Atom. Molec. Phys.* **13**, 3993–4005.

Kühnle, A., Linderoth, T. R., Hammer, B. and Besenbacher, F. (2002). *Nature* **415**, 891–893.

Kurmoo, M., Graham, A. W., Day, P. *et al.* (1995). *J. Am. Chem. Soc.* **117**, 12209–12217.

Kushmerick, J. G., Holt, D. B., Yang, J. C. *et al.* (2002). *Phys. Rev. Lett.* **89**, 086802.

Lackinger, M., Griessl, S., Heckl, W. M. and Hietschold, M. (2002). *J. Phys. Chem. B* **106**, 4482–4485.

Lang, H. P., Wiesendanger, R., Thommen-Geiser, V. and Güntherodt, H.-J. (1992). *Phys. Rev. B* **45**, 1829–1837.

Lang, H. P., Erler, B., Rossberg, A. *et al.* (1996). *J. Vac. Sci. Technol. B* **14**, 970–973.

Langmuir, I. (1917). *J. Am. Chem. Soc.* **39**, 1848–1906.

(1918). *J. Am. Chem. Soc.* **40**, 1361–1403.

Laudise, R. A., Kloc, Ch., Simpkins, P. G. and Siegrist, T. (1998). *J. Cryst. Growth* **187**, 449–454.

Laukhina, E. E., Merzhanov, V. A., Pesotskii, S. I. *et al.* (1995). *Synth. Met.* **70**, 797–800.

Laukhina, E., Ribera, E., Vidal-Gancedo, J. *et al.* (2000) *Adv. Mater.* **12**, 54–58.

Law, K.-Y. (1993). *Chem. Rev.* **93**, 449–486.

Le Cointe, M., Lemée-Cailleau, M. H., Cailleau, H. *et al.* (1995). *Phys. Rev. B* **51**, 3374–3386.

Lee, S. T., Wang, Y. M., Hou, X. Y. and Tang, C. W. (1999). *Appl. Phys. Lett.* **74**, 670–672.

Lee, T.-W., Zaumseil, J., Bao, Z., Hsu, J. W. P. and Rogers, J. A. (2004). *Proc. Natl. Acad. Sci. USA* **101**, 429–433.

Lin, Y.-Y., Gundlach, D. J., Nelson, S. F. and Jackson, T. N. (1997). *IEEE Electron Device Lett.* **18**, 606–608.

Linford, M. R. and Chidsey, C. E. D. (1993). *J. Am. Chem. Soc.* **115**, 12631–12632.

Little, W. A. (1964). *Phys. Rev.* **134**, A1416-A1424.

Llusar, R., Uriel, S., Vicent, C. *et al.* (2004). *J. Am. Chem. Soc.* **126**, 12076–12083.

Lockhart, T. P., Comita, P. B. and Bergman, R. G. (1981). *J. Am. Chem. Soc.* **103**, 4082–4090.

Lof, R. W., van Veendaal, M. A., Koopmans, B., Jonkman, H. T. and Sawatzky, G. A. (1992). *Phys. Rev. Lett.* **68**, 3924–3937.

Long, R. E., Sparks, R. A. and Trueblood, K. N. (1965). *Acta Cryst.* **18**, 932–939.

Lorenz, T., Hofmann, M., Grüninger, M., Freimuth, A. *et al.* (2002). *Nature* **418**, 614–617.

Loveland, R. J., LeComber, P. G. and Spear, W. E. (1972). *Phys. Rev. B* **6**, 3121–3127.

Luttinger, J. M. (1963). *J. Math. Phys.* **4**, 1154–1162.

Lynn, J. W., Shirane, G., Thomlinson, W. and Shelton, R. N. (1981). *Phys. Rev. Lett.* **46**, 368–371.

Madelung, O. (1978). In *Introduction to Solid-State Theory*, eds. M. Cardona, P. Fulde and H. J. Queisser. Berlin: Springer-Verlag.

Magonov, S. N. and Whangbo, M.-H. (1996). In *Surface Analysis with STM and AFM*. Weinheim: Wiley VCH.

Mallah, T., Thiébaut, S., Verdaguer, M. and Veillet, P. (1993). *Science* **262**, 1554–1557.

Manriquez, J. M., Yee, G. T., McLean, R. S., Epstein, A. J. and Miller, J. S. (1991). *Science* **252**, 1415–1417.

Mao, Ho-kwang and Hemley, R. J. (1992). *Am. Sci.* **80**, 234–246.

Martin, R. L., Kress, J. D., Campbell, I. H. and Smith, D. L. (2000). *Phys. Rev. B* **61**, 15804–15811.

Martinsen, J., Palmer, S. M., Tanaka, J., Greene, R. C. and Hoffman, B. M. (1984). *Phys. Rev. B* **30**, 6269–6276.

Marx, D. and Wiechert, H. (1996). In *Advances in Chemical Physics*, eds. I. Prigogine and S. A. Rice, Vol. XCV, pp. 213–395. John Wiley & Sons, Inc.

Mas-Torrent, M., Laukhina, E., Rovira, C. *et al.* (2001). *Adv. Funct. Mater.* **11**, 299–303.

Mas-Torrent, M., Durkut, M., Hadley, P., Ribas, X. and Rovira, C. (2004). *J. Am. Chem. Soc.* **126**, 984–985.

Mas-Torrent, M., Hadley, P., Bromley, S. T. *et al.* (2005). *Appl. Phys. Lett.* **86**, 012110.

Matsumoto, N., Shima, H., Fujii, T. and Kannari, F. (1997). *Appl. Phys. Lett.* **71**, 2469–2471.

Mattheus, C. C., Dros, A. B., Baas, J. *et al.* (2001). *Acta Cryst. C* **57**, 939–941.

McCoy, H. N. and Moore, W. C. (1911). *J. Am. Chem. Soc.* **33**, 273–292.

McKweon, N. B. (1998). In *Phthalocyanine Materials: Synthesis, Structure and Function*. Cambridge: Cambridge University Press.

McPherson, A. (1999). In *Crystallization of Biological Macromolecules*. New York: Cold Spring Harbor Laboratory Press.

Mena-Osteritz, E. and Bäuerle, P. (2001). *Adv. Mater.* **13**, 243–246.

Meng, H., Bendikov, M., Mitchell, G. *et al.* (2003). *Adv. Mater.* **15**, 1090–1093.

Metzger, R. M. (1977). *J. Chem. Phys.* **66**, 2525–2533.

Meyer zu Heringdorf, F.-J., Reuter, M. C. and Tromp, R. M. (2001). *Nature* **412**, 517–520.

Miller, J. S. and Drillon, M. (2001). In *Magnetism: Molecules to Materials*, eds. J. S. Miller and M. Drillon. Weinheim: Wiley-VCH.

Miller, J. S. and Epstein, A. J. (1994). *Angew. Chem. Int. Ed. Engl.* **33**, 385–415.

Miller, J. S., Epstein, A. J. and Reiff, W. M. (1988). *Chem. Rev.* **88**, 201–220.

Misaki, Y., Higuchi, N., Fujiwara, H. *et al.* (1995). *Angew. Chem. Int. Ed. Engl.* **34**, 1222–1225.

Mitsui, T., Rose, M. K., Fomin, E., Ogletree, D. F. and Salmeron, M. (2002). *Science* **297**, 1850–1852.

Miura, Y. F., Ohnishi, S., Hara, M., Sasabe, H. and Knoll, W. (1996). *Appl. Phys. Lett.* **68**, 2447–2449.

Molas, S., Caro, J., Santiso, J. *et al.* (2000). *J. Cryst. Growth* **218**, 399–409.

Molas, S., Coulon, C. and Fraxedas, J. (2003). *Cryst. Eng. Comm.* **5**, 310–312.

Moret, R. and Pouget, J. P. (1986). In *Crystal Chemistry and Properties of Materials with Quasi-One-Dimensional Structures*, ed. J. Rouxel. Dordrecht: D. Reidel Publishing Company.

Moret, R., Pouget, J. P., Comès, R. and Bechgaard, K. (1982). *Phys. Rev. Lett.* **49**, 1008–1012.

(1983). *J. Physique* **44**, C3-957–C3-962.

Moulin, J.-F., Brinkmann, M., Thierry, A. and Wittmann, J.-C. (2002). *Adv. Mater.* **14**, 436–439.

Moulton, B. and Zaworotko, M. J. (2001). *Chem. Rev.* **101**, 1629–1658.

Moulton, B., Lu, J., Hajndl, R., Hariharan, S. and Zaworotko, M. J. (2002). *Angew. Chem. Int. Ed. Engl.* **41**, 2821–2824.

Muccini, M., Murgia, M., Biscarini, F. and Taliani, C. (2001). *Adv. Mater.* **13**, 355–358.

Müller, H., Petersen, J., Strohmaier, R. *et al.* (1996). *Adv. Mater.* **8**, 733–737.

Müller, A., Peters, F., Pope, M. T. and Gatteschi, D. (1998). *Chem. Rev.* **98**, 239–271.

Müller, P., Alvarado, S. F., Rossi, L. and Reiss, W. (2001). *Mater. Phys. Mech.* **4**, 76–80.

Müller-Wiegand, M., Georgiev, G., Oesterschulze, E., Fuhrmann, T. and Salbeck, J. (2002). *Appl. Phys. Lett.* **81**, 4940–4942.

Nagamatsu, J., Nakagawa, N., Muranaka, T., Zenitani, Y. and Akimitsu, J. (2001). *Nature* **410**, 63–64.

Nakamura, T., Kojima, K., Matsumoto, M. *et al.* (1989). *Chem. Lett.*, 367–368.

Nakamura, T., Yunome, G., Azumi, R. *et al.* (1994). *J. Phys. Chem.* **98**, 1882–1887.

Nakazawa, Y., Tamura, M., Shirakawa, N. *et al.* (1992). *Phys. Rev. B* **46**, 8906–8914.

Nalwa, H. S. (1997). In *Handbook of Organic Conductive Molecules and Polymers*, ed. H. S. Nalwa. Chichester: John Wiley & Sons.

Nam, M. S., Ardavan, A., Cava, R. J. and Chaikin, P. M. (2003). *Appl. Phys. Lett.* **83**, 4782–4784.

Narloch, B. and Menzel, D. (1997). *Chem. Phys. Lett.* **270**, 163–168.

Nickel, B., Barabash, R., Ruiz, R. *et al.* (2004). *Phys. Rev. B* **70**, 125401.

Niles, D. W. and Margaritondo, G. (1986). *Phys. Rev. B* **34**, 2923–2925.

Nilsson, A., Weinelt, M., Wiell, T. *et al.* (1997). *Phys. Rev. Lett.* **78**, 2847–2850.

Noh, Y.-Y., Kim, J.-J., Yoshida, Y. and Yase, K. (2003). *Adv. Mater.* **15**, 699–702.

Northrup, J. E., Tiago, M. L. and Louie, S. G. (2002). *Phys. Rev. B* **66**, 121404(R).

Nowakowski, R., Seidel, C. and Fuchs, H. (2001). *Phys. Rev. B* **63**, 195418.

Nuzzo, R. G. and Allara, D. L. (1983). *J. Am. Chem. Soc.* **105**, 4481–4483.

Ogawa, T., Kuwamoto, K., Isoda, S., Kobayashi, T. and Karl, N. (1999). *Acta Cryst. B* **55**, 123–130.

Ojima, E., Fujiwara, H., Kato, K. and Kobayashi, H. (1999). *J. Am. Chem. Soc.* **121**, 5581–5582.

Osiecki, J. H. and Ullman, E. F. (1968). *J. Am. Chem. Soc.* **90**, 1078–1079.

Ostrick, J. R., Dodabalapur, A., Torsi, L. *et al.* (1997). *J. Appl. Phys.* **81**, 6804–6808.

Papageorgiou, N., Giovanelli, L., Faure, J. B. *et al.* (2001). *Surf. Sci.* **482–485**, 1199–1204.

Papavassiliou, G. C., Monsdis, G. A., Zambounis, J. S. *et al.* (1988). *Synth. Met.* **27**, B379–B383.

Park, R. L. and Madden Jr., H. H. (1968). *Surf. Sci.* **11**, 188–202.

Parkin, S. S. P., Engler, E. M., Schumaker, R. R. *et al.* (1983a). *Phys. Rev. Lett.* **50**, 270–273.

Parkin, S. S. P., Mayerle, J. J. and Engler, E. M. (1983b). *J. Physique* **44**, C3-1105–C3-1109.

Pascual, J. I., Gómez-Herrero, J., Rogero, C. *et al.* (2000). *Chem. Phys. Lett.* **321**, 78–82.

Pascual, J. I., Jackiw, J. J., Song Z. *et al.* (2001). *Phys. Rev. Lett.* **86**, 1050–1053.

Pascual, J. I., Gómez-Herrero, J., Sánchez-Portal, D. and Rust, H.-P. (2002). *J. Chem. Phys.* **117**, 9531–9534.

Pascual J. I., Lorente N., Song Z., Conrad, H. and Rust, H.-P. (2003). *Nature* **423**, 525–528.

Pedio, M., Felici, R., Torrelles, X. *et al.* (2000). *Phys. Rev. Lett.* **85**, 1040–1043.

Pérez-Murano, F., Abadal, G., Barniol, N. *et al.* (1995). *J. Appl. Phys.* **78**, 6797–6801.

Perruchas, S., Fraxedas, J., Canadell, E., Auban-Senzier, P. and Batail, P. (2005). *Adv. Mater.* **17**, 209–212.

Petruska, M. A., Watson, B. C., Meisel, M. W. and Talham, D. R. (2002). *Chem. Mater.* **14**, 2011–2019.

Petty, M. C. (1996). In *Langmuir-Blodgett films*, 1st edn. Cambridge: Cambridge University Press.

Pilia, L., Malfant, I., de Caro, D. *et al.* (2004). *New. J. Chem.* **28**, 52–55.

Piner, R. D., Zhu, J., Xu, F., Hong, S. and Mirkin, C. A. (1999). *Science* **283**, 661–663.

Pisignano, D., Raganato, M. F., Persano, L. *et al.* (2004). *Nanotechnology* **15**, 953–957.

Podzorov, V., Pudalov, V. M. and Gershenson, M. E. (2003). *Appl. Phys. Lett.* **82**, 1739–1741.

Poirier, G. E., Fitts, W. P. and White, J. M. (2001). *Langmuir* **17**, 1176–1183.

Pomerantz, M. (1978). *Solid State Comm.* **27**, 1413–1416.

Pomerantz, M., Dacol, F. H. and Segmüller, A. (1978). *Phys. Rev. Lett.* **40**, 246–249.

Pope, M. and Swenberg, C. E. (1999). In *Electronic Processes in Organic Crystals and Polymers*, 2nd edn. New York: Oxford University Press.

Poppensieker, J., Röthig, Ch. and Fuchs, H. (2001). *Adv. Funct. Mater.* **11**, 188–192.

Postma, H. W. Ch., Teepen, T., Yao, Z., Grifoni, M. and Dekker, C. (2001). *Science* **293**, 76–79.

Pouget, J. P. (1988). In *Semiconductors and Semimetals*, ed. E. Conwell, Vol. 27, pp. 88–214. San Diego: Academic Press.

Pouget, J. P., Moret, R., Comes, R. *et al.* (1982). *Mol. Cryst. Liq. Cryst.* **79**, 129–143.

Pouget, J. P., Shirane, G., Bechgaard, K. and Fabre, J. M. (1983). *Phys. Rev. B* **27**, 5203–5206.

Qiu, X. H., Nazin, G. V. and Ho, W. (2003). *Science* **299**, 542–546.

Rajagopal, A., Wu, C. I. and Kahn, A. (1998). *J. Appl. Phys.* **83**, 2649–2655.

Redhead, P. A., Hobson, J. P. and Kornelsen, E. V. (1993). In *The Physical Basis of Ultra High Vacuum*. New York: American Institute of Physics.

Reinhardt, C., Vollmann, W., Hamann, C., Libera, L. and Trompler, S. (1980). *Krist. Tech.* **13**, 243–251.

Repp, J., Meyer, G., Stojković, S. M., Gourdon, A. and Joachim, C. (2005). *Phys. Rev. Lett.* **94**, 026803.

Resel, R. (2003). *Thin Solid Films* **433**, 1–11.

Resel, R., Ottmar, M., Hanack, M., Keckes, J. and Leising, G. (2000). *J. Mater. Res.* **15**, 934–939.

Rindorf, G., Soling, H. and Thorup, N. (1982). *Acta Cryst. B* **38**, 2805–2808.

Robertson, J. M. (1936). *J. Chem. Soc.*, 1195–1209.

Robertson, J. M. and White, J. G. (1945). *J. Chem. Soc.*, 607–617.

Rojas, C., Caro, J., Grioni, M. and Fraxedas, J. (2001). *Surf. Sci.* **482–485**, 546–551.

Rosei, F., Schunack, M., Jiang, P. *et al.* (2002). *Science* **296**, 328–331.

Rovira, C., Tarrés, J., Llorca, J. *et al.* (1995). *Phys. Rev. B* **52**, 8747–8758.

Rovira, C., Veciana, J., Ribera, E. *et al.* (1997). *Angew. Chem. Int. Ed. Engl.* **36**, 2324–2326.

Ruiz, R., Nickel, B., Koch, N. *et al.* (2003). *Phys. Rev. Lett.* **91**, 136102.

Ruocco, A., Donzello, M. P., Evangelista, F. and Stefani, G. (2003). *Phys. Rev. B* **67**, 155408.

Saito, K., Yamamura, Y., Akutsu, H. *et al.* (1999). *J. Phys. Soc. Japan* **68**, 1277–1285.

Salih, A. J., Lau, S. P., Marshall, J. M. *et al.* (1996). *Appl. Phys. Lett.* **71**, 2469–2471.

Salmerón-Valverde, A., Bernés, S. and Robles-Martínez, J. G. (2003). *Acta Cryst. B* **59**, 505–511.

Sato, N., Inokuchi, H. and Slinish, E. (1987). *Chem. Phys.* **115**, 269–277.

Sautet, P. and Joachim, C. (1988). *Phys. Rev. B* **38**, 12238–12247.

(1991). *Chem. Phys. Lett.* **185**, 23–30.

Schaffer, A. M., Gouterman, M. and Davidson, E. R. (1973). *Theor. Chim. Acta* **30**, 9–30.

Schlaf, R., Merritt, C. D., Crisafulli, L. A. and Kafafi, Z. H. (1999). *J. Appl. Phys.* **86**, 5678–5686.

Schlettwein, D., Hesse, K., Tada, H. *et al.* (2000). *Chem. Mater.* **12**, 989–995.

Schneegans, O., Moradpour, A., Houzé, F. *et al.* (2001). *J. Am. Chem. Soc.* **123**, 11486–11487.

Schreiber, F. (2004). *J. Phys.: Condens. Matter* **16**, R881–R900.

Schreiber, F., Eberhardt, A., Leung, T. Y. B. *et al.* (1998). *Phys. Rev. B* **57**, 12476–12481.

Schroeder, P. G., France, C. B., Park, J. B. and Parkinson, B. A. (2002). *J. Appl. Phys.* **91**, 3010–3014.

Schukat, G., Richter, A. M. and Fanghänel, E. (1987). *Sulfur Rep.* **7**, 155–240.

Schunack, M., Petersen, L., Kühnle, A. *et al.* (2001). *Phys. Rev. Lett.* **86**, 456–459.

Schwartz, A., Dressel, M., Gruner, G. *et al.* (1998). *Phys. Rev. B* **58**, 1261–1271.

Schwieger, T., Peisert, H., Knupfer, M., Golden, M. S. and Fink, J. (2001). *Phys. Rev. B* **63**, 165104.

Segovia, P., Purdie, D., Hengsberger, M. and Baer, Y. (1999). *Nature* **402**, 504–507.

Seguin, J. L., Suzanne, J., Bienfait, M., Dash, J. G. and Venables, J. A. (1983). *Phys. Rev. Lett.* **51**, 122–125.

Seidel, C., Kopf, H. and Fuchs, H. (1999). *Phys. Rev. B* **60**, 14341–14347.

Seip, C. T., Granroth, G. E., Meisel, M. W. and Talham, D. R. (1997). *J. Am. Chem. Soc.* **119**, 7084–7094.

Seki, K., Karlsson, U. O., Engelhardt, R., Koch, E. E. and Schmidt, W. (1984). *Chem. Phys.* **91**, 459–470.

Sellam, F., Schmitz-Hübsch, T., Toerker, M. *et al.* (2001). *Surf. Sci.* **478**, 113–121.

Servet, B., Horowitz, G., Ries, S. *et al.* (1994). *Chem. Mater.* **6**, 1809–1815.

Sessoli, R., Gatteschi, D., Caneschi, A. and Novak, M. A. (1993). *Nature* **365**, 141–143.

Shen, Ch. and Kahn, A. (2001). *J. Appl. Phys.* **90**, 4549–4554.

Shibaeva, R. P., Kaminskii, V. F., Yagubskii, E. B. (1985). *Mol. Cryst. Liq. Cryst.* **119**, 361–367.

Shtein, M., Peumans, P., Benziger, J. B. and Forrest, S. R. (2003). *J. Appl. Phys.* **93**, 4005–4016.

Siegrist, T., Fleming, R. M., Haddon, R. C. *et al.* (1995). *J. Mater. Res.* **10**, 2170–2173.

Siegrist, T., Kloc, C., Laudise, R. A., Katz, H. E. and Haddon, R. C. (1998). *Adv. Mater.* **10**, 379–382.

Siegrist, T., Kloc, Ch., Schön, J. H. *et al.* (2001). *Angew. Chem. Int. Ed. Engl.* **40**, 1732–1736.

Sing, M., Schwingenschlögl, U., Claessen, R., Dressel, M. and Jacobsen, C. S. (2003a). *Phys. Rev. B* **67**, 125402.

Sing, M., Schwingenschlögl, U., Claessen, R. *et al.* (2003b). *Phys. Rev. B* **68**, 125111.

Skulason, H. and Frisbie, C. D. (2002). *J. Am. Chem. Soc.* **124**, 15125–15133.

Smith, K. E. (1993). *Ann. Rep. Prog. Chem.* **90**, 115–154.

Smith, C. G. (1996). *Rep. Prog. Phys.* **59**, 235–282.

Smolenyak, P., Peterson, R., Nebesny, K. *et al.* (1999). *J. Am. Chem. Soc.* **121**, 8628–8636.

Soler, J. M., Artacho, E., Gale, J. D. *et al.* (2002). *J. Phys.: Condens. Matter* **14**, 2745–2779.

Soukopp, A., Glöckler, K., Kraft, P. *et al.* (1998). *Phys. Rev. B* **58**, 13882–13894.

Stipe, B. C., Rezaei, M. A. and Ho, W. (1998). *Science* **280**, 1732–1735.

Sumimoto, T., Kudo, K., Nagashima, T., Kuniyoshi, S. and Tanaka, K. (1995). *Synth. Met.* **70**, 1251–1252.

Sundar, V. C., Zaumseil, J., Podzorov, V. *et al.* (2004). *Science* **303**, 1644–1646.

Sutton, A. P. and Balluffi, R. W. (1995). In *Interfaces in Crystalline Materials*. Oxford: Clarendon Press.

Taborski, J., Väterlein, P., Dietz, H., Zimmermann, U. and Umbach, E. (1995). *J. Electron Spectrosc. Relat. Phenom.* **75**, 129–147.

Takeya, J., Goldmann, C., Haas, S. *et al.* (2003). *J. Appl. Phys.* **94**, 5800–5804.

Talham, D. R. (2004). *Chem. Rev.* **104**, 5479–5501.

Tamura, M., Nakazawa, Y., Shiomi, D. *et al.* (1991). *Chem. Phys. Lett.* **186**, 401–404.

Tamura, M., Hosokoshi, Y., Shiomi, D. *et al.* (2003). *J. Phys. Soc. Jpn.* **72**, 1735–1744.

Tanaka, J. (1963). *Bull. Chem. Soc. Jpn.* **36**, 1237–1249.

Tanaka, H., Okano, Y., Kobayashi, H., Suzuki, W. and Kobayashi, A. (2001). *Science* **291**, 285–287.

Tanda, S., Tsuneta, T., Okajima, Y. *et al.* (2002). *Nature* **417**, 397–398.

Tang, C. W. and VanSlyke, S. A. (1987). *Appl. Phys. Lett.* **51**, 913–915.

Tanigaki, K., Ebbesen, T. W., Saito, S. *et al.* (1991). *Nature* **352**, 222–223.

Tavazzi, S., Besana, D., Borghesi, A. *et al.* (2002). *Phys. Rev. B* **65**, 205403.

Tejedor, C., Flores, F. and Louis, E. (1977). *J. Phys. C* **10**, 2163–2177.

Tersoff, J. (1984). *Phys. Rev. B* **30**, 4874–4877.

Tersoff, J. and Hamann, D. R. (1985). *Phys. Rev. B* **31**, 805–813.

Thakur, M., Haddon, R. C. and Glarum, S. H. (1990). *J. Cryst. Growth* **106**, 724–727.

Thomas, R. and Coppens, P. (1972). *Acta Cryst. B* **28**, 1800–1806.

Thorup, N., Ringdorf, G., Soling, H. and Bechgaard, K. (1981). *Acta Cryst. B* **37**, 1236–1240.

Tomonaga, S. (1950). *Prog. Theor. Phys.* **5**, 544–569.

Torrance, J. B., Scott, B. A., Welber, B., Kaufman, F. B. and Seiden, P. E. (1979). *Phys. Rev. B* **19**, 730–741.

Torrance, J. B., Girlando, A., Mayerle, J. J. *et al.* (1981). *Phys. Rev. Lett.* **47**, 1747–1750.

Tromp, R. M., Mankos, M., Reuter, M. C., Ellis, A. W. and Copel, M. (1998). *Surf. Rev. Lett.* **5**, 1189–1197.

Turek, P., Nozawa, K., Shiomi, D. *et al.* (1991). *Chem. Phys. Lett.* **180**, 327–331.

Turner, S. S., Day, P., Abdul Malik, K. M. *et al.* (1999). *Inorg. Chem.* **38**, 3543–3549.

Ugarte, D. (1992). *Nature* **359**, 707–709.

Uji, S., Shinagawa, H., Terashima, T. *et al.* (2001). *Nature* **410**, 908–910.

Ulman, A. (1996). *Chem. Rev.* **96**, 1533–1554.

Ullmann, F., Meyer, G. M., Loewenthal, O. and Gilli, E. (1904). *Justus Liebigs Annalen der Chemie* **331**, 38–81.

Umbach, E., Glöckler, K. and Sokolowski, M. (1998). *Surf. Sci.* **402–404**, 20–31.

van der Hoek, B., van der Eerden, J. P., Bennema, P. and Sunagawa, I. (1982). *J. Cryst. Growth* **58**, 365–380.

van Kerckhoven, C., Hony, S., Peeters, E. *et al.* (2000). *Astron. Astrophys.* **357**, 1013–1019.

Vázquez, H., Oszwaldowski, R., Pou, P. *et al.* (2004). *Europhys. Lett.* **65**, 802–808.

Venables, J. A., Spiller, G. D. T. and Hanbücken, M. (1984). *Rep. Prog. Phys.* **47**, 399–459.

Verlaak, S., Steudel, S., Heremans, P., Janssen, D. and Deleuze, M. S. (2003). *Phys. Rev. B* **68**, 195409.

Vescoli, V., Zwick, F., Henderson, W. *et al.* (2000). *Eur. Phys. J. B* **13**, 503–511.

Vollmer, M. S., Effenberger, F., Stecher, R., Gompf, B. and Eisenmenger, W. (1999). *Chem. Eur. J.* **5**, 96–101.

Wagner, V. (2001). *phys. stat. sol. (a)* **188**, 1297–1305.

Wang, H. H., Carlson, K. D., Geiser, U. *et al.* (1990). *Physica C* **166**, 57–61.

Wang, H. H., Stamm, K. L., Parakka, J. P. and Han, C. Y. (2002). *Adv. Mater.* **14**, 1193–1196.

Wang, J., Sun, X., Chen, L. and Chou, S. Y. (1999). *Appl. Phys. Lett.* **75**, 2767–2769.

Wang, Z. Z., Girard, J. C., Pasquier, C., Jérome, D. and Bechgaard, K. (2003). *Phys. Rev. B* **67**, 121401.

Ward, M. D. (2001). *Chem. Mater.* **101**, 1697–1725.

Wickman, H. H., Trozzolo, A. M., Williams, H. J., Hull, G. W. and Merritt, F. R. (1967). *Phys. Rev.* **155**, 563–566.

Wigner, E. and Huntington, H. B. (1935). *J. Chem. Phys.* **3**, 764–770.

Williams, J. M., Wang, H. H., Beno, M. A. *et al.* (1984). *Inorg. Chem.* **23**, 3839–3941.

Williams, J. M., Kini, A. M., Wang, H. H. *et al.* (1990). *Inorg. Chem.* **29**, 3272–3274.

Wirth, K. R. and Zegenhagen, J. (1997). *Phys. Rev. B* **56**, 9864–9870.

Witte, G. and Wöll, C. (2004). *J. Mater. Res.* **19**, 1889–1916.

Wood, E. A. (1964). *J. Appl. Phys.* **35**, 1306–1312.

Wu, Z., Ehrlich, S. N., Matthies, B. *et al.* (2001). *Chem. Phys. Lett.* **348**, 168–174.

Wudl, F., Wobschall, D. and Hufnagel, E. J. (1972). *J. Am. Chem. Soc.* **94**, 670–672.

Yagubskii, E. B., Shchegolev, I. F., Laukhin, V. N. *et al.* (1984). *JETP Lett.* **39**, 12–16.

Yakushi, K., Aratani, S., Kikuchi, K., Tajima, H. and Kuroda, H. (1986). *Bull. Chem. Soc. Jpn.* **59**, 363–366.

Yamada, J. and Sigimoto, T. (2004). In *TTF Chemistry: Fundamentals and Applications of Tetrathiafulvalene*, eds. J. Yamada and T. Sigimoto. New York: Springer-Verlag.

Yamada, J., Watanabe, M., Akutsu, H. *et al.* (2001). *J. Am. Chem. Soc.* **123**, 4174–4180.

Yamada, J., Akutsu, H., Nishikawa, H. and Kikuchi, K. (2004). *Chem. Rev.* **104**, 5057–5083.

Yanagi, H. and Okamoto, S. (1997). *Appl. Phys. Lett.* **71**, 2563–2565.

Yokoyama, T., Yokoyama, S., Kamikado, T., Okuno, Y. and Mashiko, T. (2001). *Nature* **413**, 619–621.

Zangmeister, R. A. P., Smolenyak, P. E., Drager, A. S., O' Brien, D. F. and Armstrong, N. R. (2001). *Langmuir* **17**, 7071–7078.

Zangwill, A. (1988). In *Physics at Surfaces*. Cambridge: Cambridge University Press.

Zeppenfeld, P., George, J., Diercks, V. *et al.* (1997). *Phys. Rev. Lett.* **78**, 1504–1507.

Zhang, Z. and Lagally, M. G. (1997). *Science* **276**, 377–383.

Zhang, J., Wang, J., Wang, H. and Yan, D. (2004). *Appl. Phys. Lett.* **84**, 142–144.

Zwick, F., Jérome, D., Margaritondo, G. *et al.* (1998). *Phys. Rev. Lett.* **81**, 2974–2977.

Index